National Seminar of Physics Education (SNPF 2021)

Journal of Physics: Conference Series Volume 2104

Banjarmasin, Indonesia
11 September 2021

ISBN: 978-1-7138-5075-5
ISSN: 1742-6588

Printed from e-media with permission by:

Curran Associates, Inc.
57 Morehouse Lane
Red Hook, NY 12571

Some format issues inherent in the e-media version may also appear in this print version.

Institute of Physics
Dirac House, Temple Back
Bristol BS1 6BE UK

Phone: 44 1 17 929 7481
Fax: 44 1 17 920 0979

techtracking@iop.org

Additional copies of this publication are available from:

Curran Associates, Inc.
57 Morehouse Lane
Red Hook, NY 12571 USA
Phone: 845-758-0400
Fax: 845-758-2633
Email: curran@proceedings.com
Web: www.proceedings.com

TABLE OF CONTENTS

Preface
National Seminar of Physics Education
Universitas Lambung Mangkurat, Banjarmasin, Indonesia
11 September 2021

The National Seminar of Physics Education/ Seminar Nasional Pendidikan Fisika (SNPF) 2021 was organized by the Physics Education Study Program, Teacher Training and Education Faculty, Universitas Lambung Mangkurat (ULM) in collaboration with the Physical Society of Indonesia (PSI), South Kalimantan branch. The SNPF covers several fields of study, including strategy, models, methods, media, and evaluation in learning physics, chemistry, biology, mathematics, physics research results, artificial intelligence, future science development, and general education.

The theme raised was "Optimizing STEM Learning in Creating a Generation that is Excellent, Character, and Competitive in the Era of Society 5.0". This seminar is a scientific forum that is very appropriate for sharing and disseminating research results and the best contribution in facing the Era of Society 5.0 by strengthening efforts to produce a generation that is Excellent, Character, and Competitive, which are great opportunities for this nation.

On behalf of the committee, we would like to thank the Chancellor of Universitas Lambung Mangkurat, Dean of the Teacher Training and Education Faculty, Head of the Mathematics and Natural Sciences Major, advisory board, steering committee, all committee members, reviewers, presenters, and participants. Special thanks are also given to the Proceedings of the IOP Conference. On behalf of the SNPF 2021 committee, we would like to thank all parties for their participation in supporting this publication. We hope to see everyone at the next conference.

Kind regards,

Dewi Dewantara, M.Pd.
Chair of the SNPF 2021

Committee Members
National Seminar of Physics Education
Universitas Lambung Mangkurat, Banjarmasin, Indonesia
11 September 2021

A. Patron : Abdul Salam M, M.Pd
B. Chair: Dewi Dewantara, M.Pd
C. Secretary: Surya Haryandi, M.Pd
D. Secretarial Team
- Saiyidah Mahtari, M.Pd
- Devi Munawaroh
- Fitri Norkhalisa
- Rizki Maulidia Isnaniah

E. Publications Team Members
- Misbah, M.Pd
- Amalia Rosalina
- Dwi Mugi Lestari
- Raihanah Zulfah
- Wardatul Humairah

F. Equipment Team Members
- Herru Soepriyanto S, SE
- Rahmat Saifuddin Anwar
- Muhammad Choirul Hadi Santoso
- Nur Muhammad Akbar Hamidi
- Muhammad Wildi Firahmi
- Mahrufah Melati Putri
- Sultan Syekh Rifa Rizkya Robby Hidayatullah

G. IT Team Members
- Andy Azhari, S.Pd.
- Napisah
- Octaviani
- Nur'aina
- Fitria Dewi Firdawati
- Dina Selvia
- Isnur Alfiah
- Adindha Maidiana Ariyanti
- Erni Mahrita
- Ayu Hidayati
- Andriana Safitriani
- Hani Sapna
- Lavenia Wulandari

H. Reviewer Team Members:
- Dr. Zulkarnaen, M.Si. (Universitas Mulawarman)
- Dr. Sahrul Saehana, M.Si. (Universitas Tadulako)
- Dr. Mustika Wati, M.Sc. (Universitas Lambung Mangkurat)
- Dr. Suyidno, M.Pd. (Universitas Lambung Mangkurat)
- Dr. Syahmani, M.Si. (Universitas Lambung Mangkurat)
- Dr. Binar Kurnia Prahani, M.Pd. (Universitas Negeri Surabaya)

Journal of Physics: Conference Series **2104** (2021) 011002 doi:10.1088/1742-6596/2104/1/011002

Peer review declaration

All conference organizers/editors are required to declare details about their peer review. Therefore, please provide the following information:

- **Type of peer review: Single-blind / Double-blind / Triple-blind / Open / Other (please describe)**
  The SNPF 2021 article review process was carried out using a single-blind review system. The number of papers submitted was 84 articles. The number of SNPF 2021 reviewers is 14 people. One reviewer reviewed one article. The review process was done via email and or OCS. The article review results were returned to the authors for revision within a certain period of time. The author sent the revised results, and the plagiarism checked results of the article. The editor then rechecked the revision result. If suitable, it would be forwarded to the editor, either for plagiarism, language, or template.

- **Conference submission management system:**
  OCS (http://snpfmotogpe.ulm.ac.id/ocs/index.php/snpf/2021)

- **The number of submissions received:**
  There were 84 articles submitted to SNPF 2021

- **The number of submissions sent for review:**
  There were 84 articles reviewed by reviewers of SNPF 2021.

- **The number of submissions accepted:**
  There were 40 articles accepted for recommendation/publication to JPCS IOP Publishing.

- **Acceptance Rate (Number of Submissions Accepted / Number of Submissions Received X 100):**
  47,61%

- **The average number of reviews per paper:**
  One article was reviewed four times: content review 2x, language review 1x, and template review 1x.

- **The total number of reviewers involved:**
  There were 14 reviewers (content review), 10 editors (content and template review), and 4 people (language review)

- **Any additional info on the review process:**
  The author sent the revised article along with the similarity check (maximum 20%). A similarity check was also carried out using Turnitin (maximum 20%) when an article was declared fit for publication. So, the similarity check was done two times.

- **Contact person for queries:**
  +628975586104 (Misbah)

Journal of Physics: Conference Series **2104** (2021) 011002 doi:10.1088/1742-6596/2104/1/011002

Universitas Lambung Mangkurat
Misbah_pfis@ulm.ac.id

Learning gamma-ray scanning technique through an educational simulation rig

W Wibisono and S Sugiharto*

Center for Isotopes and Radiation Application (CIRA)-National Nuclear Energy Agency of Indonesia (BATAN), Jl. Lebak Bulus Raya No. 49, Kawasan Nuklir Pasar Jumat, Jakarta 12440, Indonesia

* sugi@batan.go.id

Abstract. Gamma-ray scanning is non-destructive testing (NDT) technique to inspect problems in distillation columns of industrial plant. In order to familiarize with the problems' identification and data interpretation, an attempt has been made by constructing an educational rig for simulation of gamma-ray scanning technique in a laboratory scale. The rig is made of cylindrical transparent acrylic material with a diameter and a height of 0.3 m and 2.4 m, respectively. The rig consists of artificial trays and packed bed structures designed in such a way that it represents a duplicate of the distillation unit but in a miniature size. The rig is also equipped with a submerged water pump and water tank which both are located at the bottom part of the rig. Water from the tank are circulated by a water pump through a pvc tube to a distributor at the upper part of the rig. In the rig, water is flowing downward gravitically for passing through packed bed and trays structures until terminated at the tank. The gamma-ray scanning experiment was carried out by moving the radiation detector and Cs-37 gamma-ray source simultaneously from top to bottom for every 5 cm moving step. Artificial problems such as flooding, tray positions, collapsed trays are clearly identified. The lesson learned from this experiment concludes that simulation of gamma ray scanning is very suitable for troubleshooting and diagnosing malfunctions of the internal structure of the distillation column.

1. Introduction

For plant engineers, big five classess of conventional non-destructive testing (NDT) consist of radiography, liquid penetrant, magnetic particles, ultrasonic and eddy current techniques. They are well known as they are frequently used for routine inspection in industrial process plant [1,2]. The inspection is usually carried out when the plant is shut down. The plant components should be inspected thoroughly to find any discontinuities, mostly welding defects, in them. The time required for inspection of one unit of plant using conventional NDT techniques is in the range of 2 – 3 weeks in average.

Although it is not so similar in the inspection way, a nuclear-based technology called gamma-ray scanning technique is a completely non-destructive method [3]. It offers a non-invasive, straightforward measurement, shorter inspection time, and is un-ambiguous in providing essential scanning data [4,5]. It is also cost-effective and safe [5]. The gamma scanning technique uses one gamma radiation source and one radiation detector. The strength of gamma sources used in the gamma scanning is commonly in the range of 0.4 – 40 GBq, which is 100 - 1000 times lower than those usually used in industrial radiography [6]. From a radiological point of view, gamma-ray scanning is safer compared to industrial radiography. The high energy of gamma-ray is capable of penetrating any materials it passes, making that an online inspection mode could be conducted and the operation of the inspected processing plant is not necessarily be stopped. With these capabilities, the gamma scanning technique is superior and shows its state of the art of nuclear techniques for an online application to industrial plants [3,6–8].

The gamma scanning technique is usually used for scanning distillation units in petroleum and petrochemical refineries. The terms that are commonly used in gamma-ray scanning are troubleshooting, diagnosing and malfunctioning of plant internal structures. The term of malfunction may have similar connotation with the terms of defects or indications in conventional NDT techniques. Typical malfunctions that can be identified using the gamma-ray scanning technique are damaged or missing trays in trayed column type, damage or missing supports in packed bed column type, flooding, blockage, and liquid levels. Other problems that are more difficult to be examined are entrainment and weeping due to these problems absorbing small part of radiation intensity [9].

Distillation or fractionating tower is essentially a cylindrical pressurized steel vessel with a typical heigh is about 120 feet [10]. There are two types of distillation unit that is tray type and packed ones [3,5,8,11–14]. The primary function of the distillation unit is to perform the initial separation of crude oil into intermediate products, which will undergo further processing in a downstream unit to become ready-to-use products for consumers [15]. Doing gamma-ray scanning inspection in petroleum or petrochemical plant is not an easy task. Skillful and experience in handling radiation or radioactive material is required. This article aims to introduce a gamma-ray scanning technique as applied to a laboratory simulation rig to study the mechanism of internal process in a distillation column. The rig is made of transparent acrylic material make that artificial malfunctions could be seen during the experiment. This simple experiment is dedicated to educational purposes that may be useful for undergraduate students in physics and engineering.

2. Theory of gamma-ray scanning

Implementation of gamma scanning technique for internal inspection of process materials in distillation column is carried out by moving concurrently in vertical direction a sealed gamma radiation source and scintillation detector, (NaI(Tl)), along the exterior side of the column. The background theory of the gamma scanning techniques is based on the interaction of gamma radiation with the material its passed in the mode of transmission. During the period of investigation, the radiation source is encapsulated and placed permanently in a specially designed housing and makes no contact with a radiation detector or with the process materials in the column. A source holder with an appropriate panoramic collimator is designed to expose the column. A scintillation detector on the other side of the column is employed to record the radiation intensity emitted by the radiation source. Interaction of the gamma radiation with a medium of interest in the column will produce a change in the beam's radiation intensity, which correlates to the properties of the material in the column [3–5].
The theory of gamma scanning technique is based on fundamental relation [7,11,12]

$$I = B.\, I_0 \exp\left(-\sum_i^n \mu_i \rho_i x_i\right) \quad (1)$$

where, I and I_0 are radiation intensity after and before interacting with material or transmitted and incident radiation intensities respectively, in count per second (cps). μ_i is mass absorption coefficient of ith material in m^2/kg, ρ_i is density of ith material in kg/m^3 and x_i is the thickness of the ith material, in cm. B is the buildup factor (dimensionless). In practical application, each medium in the column is assumed to be linear and isotropic. By using collimated beam, it is also assumed that intensity contribution form scattering radiation is considered none, therefore the buildup factor is set equal to one. Based on these assumptions, it is safe to state that Eq. (1) can be simplified as

$$I = I_0 \exp(-\mu_l x) \quad (2)$$

where μ_l is linear constant and apply for each medium. It is worth noting that gamma radiation for column investigation should be capable of penetrating the wall thickness of the column and all materials inside the column.

3. Experiment

Gamma-ray scanning has been demonstrated experimentally to a simulation rig made of cylindrical acrylic material. The rig dimension is 0.3 m in diameter and 2.4 m in height. The rig consists of trays and packed bed structures designed in such a way that it represents a simplified duplicate of the real industrial distillation column, but in a miniature size. The rig is also equipped with a submerged water pump and water tank which both are constructed at the bottom part of the rig. Water in the tank is circulated into the rig by the pump through a PVC tube. At the upper part of the rig water fall gravitically which undergo for further passing through packed bed, tray structures until terminated in the tank. Cesium-137 gamma source with activity around 10 mCi placed in panoramic collimator is used to scan the rig. Panoramically collimated scintillation detector (NaI(Tl)) and 9302 ratemeter (both are produced by Minekin Inc) are used as the data acquisition system. The experiment is carried out for full scan from top to bottom of the rig. Both radiation source and detector are concurrently moved downward for every 5 cm moving step. The moving mechanism is controlled by software applied to the electro-mechanic system. During the experiment, all the radiation intensity is counted by the radiation detector and recorded in a laptop computer for subsequent analysis. Schematic diagram of experiment is shown in Fig.1.

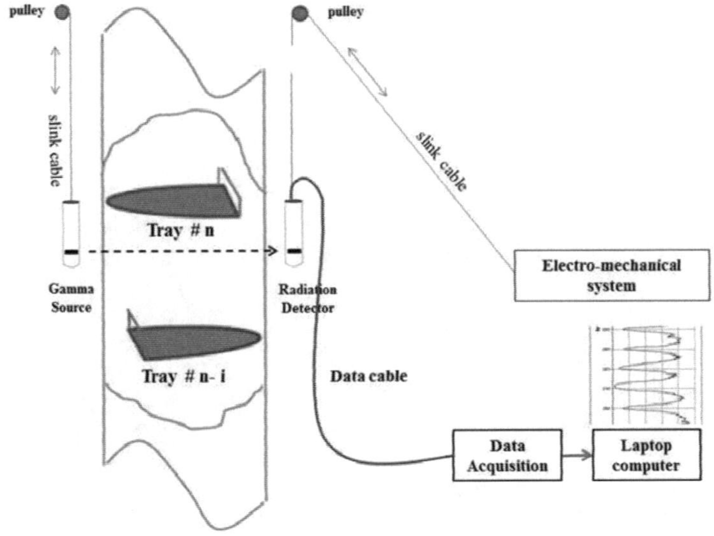

Figure 1. Schemitic diagram of gamma-ray scanning experiment

4. Result and discussion

Before going to further detailed discussion, let us consider the following guidance to have a meaningful conclusion. It is not easy to interpret the scanning data if it solely based on the scan data. One has to learn the corresponding mechanical drawing of the scanned column thoroughly before doing the measurement [3,16,17]. Any useful hydraulic information such as the position of tray, packed bed, inlet-outlet flow and manhole are usually described in the drawing. One has also to consider the position of the mechanical obstacles at the exterior of the column. When a continuous full scan is carried out, the scan data of the obstacles can be a source of false information, leading to misinterpretation [17].

Journal of Physics: Conference Series **2104** (2021) 012001 doi:10.1088/1742-6596/2104/1/012001

Mechanical drawing is essential information for the determination of scan orientation [3,16,17]. Scan orientation is a strategic plan in the placement of radiation source and radiation detector outside wall of the scanned column. It is worth understanding that although the source and detector collimators are designed in panoramic shape, the data obtained from measurement is just the counted radiation intensity where the detector is placed. Therefore, each scan orientation essentially produces information about the hydraulic condition of the material in the column only in scan direction [17]. In most industrial applications, the scan is directed to detect the critical components of the internal structure, such as trays and packed bed. Although it is also critical in intensifying chemical processing, the scan orientation to detect downcomer is sometimes not necessarily be carried out [16].

Scanning data obtained from radiation intensity measurement on the simulation rig is presented in Fig. 2. In principle, interpretation of scanning data is an implementation of Equation (1) for idealized conditions. Equation (1) states that radiation intensity measured by a detector is exponentially proportional to the density of the material that exists between the source and the detector. The dimensionless quantity, $\mu_i \rho_i x_i$, is the most important term because it is related to the formation of scan profile generated from the measurement. μ_i is constant for a given gamma-ray energy and material composition. In this situation, increasing material density and material thickness will reduce the radiation intensity recorded by a detector and vice versa.

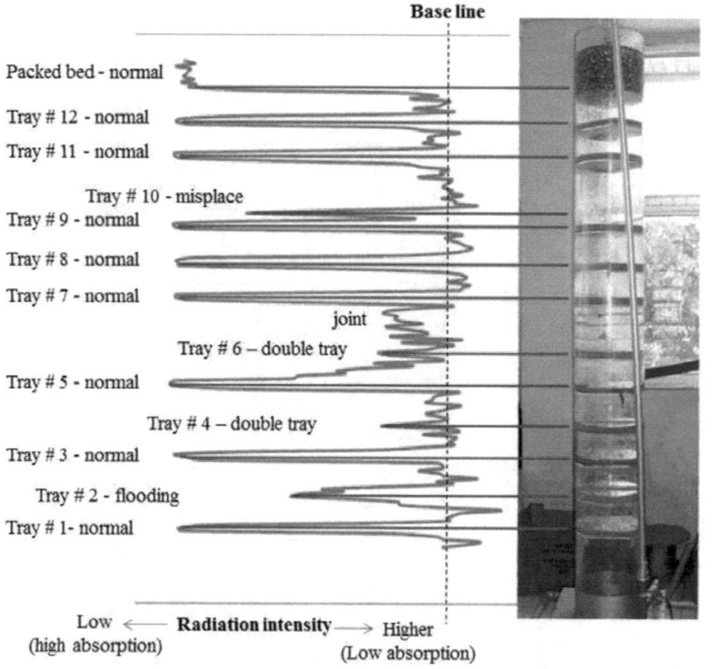

Figure 2. Gamma-ray scanning experiment shows scan profile generated from interaction of gamma radiation with materials in the simulation rig. Scan data is recorded when the experiment is being performed

When the Equation. (1) is applied to interpret the density profiles of materials in the trayed column, there would be three main mechanisms of interaction processes to occur:

- When gamma radiation interacts with space containing air or vapor phase of liquid, a small amount of incident radiation is absorbed and most of the radiation is transmitted. The scan profile produced from this interaction shows a scan profile with the highest (maximum) radiation intensity. In some publications, such scan profile is termed as a baseline [9,14,16].
- When gamma radiation goes through the liquid on the tray, incident radiation is partly absorbed by liquid. The recorded radiation intensity shows an appreciable medium-strength scan profile.

- When gamma radiation interacts with tray or packed bed, the incident radiation is strongly absorbed by these materials, and the transmitted radiation intensity shows weak scan profiles.

As can be seen from Fig. 2, variation of scan profiles are caused by the interaction of gamma-ray with different sizes and orientations of the internal components of the column. Strong absorption of gamma-ray by packing bed results in low transmitted radiation intensity, which produces a low scan profile. A similar result is also indicated by strong absorption of gamma-ray by a material of trays, as indicated by trays 1,3, 5, 7-9, 11, and 12. They show similar scan profiles and also their corresponding adjacent spaced distances. Here, these materials are in their position and are functioning properly.

Figure 2 also shows simulated problems on artificial trays. As can be seen, simulated flooding occured on the tray 2. Trays 4 and 6 show almost similar scan profile, because these trays are doubled like a superimposed tray to another. Scan profile for a joint spot in the space between trays 6 and 7 is also clearly observed as indicated by broadened scan profile along the joint in a vertical direction. Tray 10 collapsed and fell down to the tray 9 is also observed from its corresponding scan profile. The position of the collapsed tray is not perfectly superimposed on tray nine but there is a bit of space gap; therefore the scan profile of the collapsed tray is not the same to the situation on double trays.

To sum up, the gamma-ray scanning method has become a non-destructive method that is routinely applied to industry. The establishment of a simulation rig is an attempt to familiarize the usage of the gamma-ray scanning method on a laboratory scale. We believe that this method is valuable for undergraduate physics and engineering undergraduate students to study the complex structure of petroleum and petrochemical processing units.

5. Conclusion

This experiment demonstrates that gamma-ray scanning is an effective method for studying internal processes and material integrity in a visible simulation rig consisting of artificial packed bed and trays. The process is stimulated mechanically by circulating water from and to the water tank. The tray position and artificial malfunction such as flooding and tray misplace or collapsed trays are clearly observed during the experiment. The experiment also shows that online measurement using high penetrating gamma energy of radioactive material even in low-level radioactivity shows its superiority in providing valuable information about internal structure materials in the column.

Acknowledgment

This experiment was funded by the government of Indonesia under the research program in the field of NDT at the Center of Isotopes and Radiation Application (CIRA).

Authors' statement

All authors have the same contribution and there are no conflict of interest.

References

[1] IAEA 2008 *Training Guideline in Non-destructive TestingTechniques* (IAEA-TECDOC-628/Rev.2 Vienna)
[2] Carroll Q B 2011 *Radiography in the Digital Age Physics Exposure Radiation Biology*
[3] Shahabinejad H, Feghhi S A H and Khorsandi M 2014 *Radiat. Meas.* **61**
[4] Zahran O, Kasban H and Abd El-Samie F E 2010 *J. Electron. Electr. Eng* **2** 279–89
[5] Barati H, Khorsandi M and Feghhi S A H 2019 *Nucl. Instruments Methods Phys. Res. Sect. A Accel. Spectrometers, Detect. Assoc. Equip.* **916** 183–9
[6] IAEA 2005 Technical Data on Nucleonic Gauges IAEA-TECDOC-1459 Vienna
[7] Shahabinejad H and Feghhi S A H 2015 *Appl. Radiat. Isot.* **99** 25-34.

[8] Abdullah J 2005 *Hydrocarb. Asia* **15** 62–5
[9] IAEA 2011 Protocol for Gamma Scanning of Industrial Process Column Preprint RAS/8/111-P2
[10] Cheremisinoff N and Rosenfeld P 2009 *Handbook of Pollution Prevention and Cleaner Production Vol. 1: Best Practices in the Petroleum Industry*
[11] Laraki K, Alami R, Cherkaoui El Moursli R, Bensitel A and El Badri L 2007 *Nucl. Instruments Methods Phys. Res. Sect. A Accel. Spectrometers, Detect. Assoc. Equip.* **578** 340–4
[12] Hammad M E, Dessouky M I, Zahran O, Kasban H, Elaraby S M S and El-Samie F E A 2015 *J. Nondestruct. Eval.* **34** 1–9
[13] Ángeles M A R, Castro F I G, Hernández J G S and Ramírez A R U 2015 *Chem. Eng. Process. Process Intensif.* **97** 55–65
[14] Vasquez P A S, De Mesquita C H, LeRoux G A C and Hamada M M 2010 *Appl. Radiat. Isot.* **68** 658–61
[15] Ibrahim D, Jobson M, Lii J and Gonzalbez G G 2007 *Proc. of the 27th European Symposium on Computer Aided Process Engineering – ESCAPE 27 (Barcelona), October 1st-5th*
[16] Alami R and Bensitel A 2012 *Petrochemicals* **153**
[17] Sugiharto S 2012 *Atom Indones.* **38** 138–46

Determining the current profile along the anode-cathode line in Cu/Ni electrolyte by logger pro

M I Aminudin, M Toifur, D Sulisworo

Physics Education Department, Ahmad Dahlan University, Indonesia

mihya1500007029@webmail.uad.ac.id

Abstract. In electroplating, the broad role of cathode and anode is often neglected. This study aims to simultaneously prove the electric current in the three pairs of probes in the cathode-anode direction. It also describes the effect of variations in the anode-cathode area ratio. This result proves that the greater the ratio, the greater the current density in all pairs of probes. Data retrieval used a current sensor current probe DCP-BTA Vernier displayed on a computer screen with the help of Logger Pro software. The results in the study show that the current at every point in the measurement has a less identical graph. In electroplating, nickel (Ni)-1 has a relatively constant amplitude than nickel (Ni)-2. The variation of the anode area ratio at the first nickel (Ni) is 1: 0.821. The ratio of the anode area to the second nickel (Ni) is 1: 0.343. From the analysis results, the first nickel (Ni) has a relatively constant amplitude compared to the second nickel (Ni).

1. Introduction

In electroplating, the broad role of the cathode and anode is often neglected. At the same time, the amount of electric current used in electroplating is current density (J), which is the ratio between current (I) and surface area (A). Therefore, the difference in cross-sectional area of the anode-cathode determines the current density. If the cathode cross-sectional area is not the same as the cathode cross-sectional area, the current density in each region along the anode-cathode path is not the same. If the anode area is greater than the cathode, the current density near the cathode is greater than the current density near the anode.

Setyowati and Iriani conducted a study on the effect of current density on the thickness and crystal structure of the nickel layer on copper. The results showed that the thickness of the nickel layer increased with increasing current density [1,2].

In this research, an electric current measuring instrument was designed to measure electric current in several locations in line with the anode-cathode direction in the electrolyte solution simultaneously. The probe consisted of 3 pairs made of copper with a distance between the pins of 2.3 mm. This research examined two factors, namely proving: first, the electric current in the three pairs of probes in the cathode-anode direction at the same time, in varying the anode-cathode area ratio to verify that the larger the ratio, the greater the current density in all pairs of probes and, more specifically, the probe closest to the cathode. For data retrieval, a Vernier current sensor was displayed on a computer screen with the help of Logger Pro software.

2. Theoretical Background

2.1. Electroplating

Electroplating is one of the processes to coat metal using the principle of electric current and the help of chemical compounds to produce coating particles that cover the material [3–5].

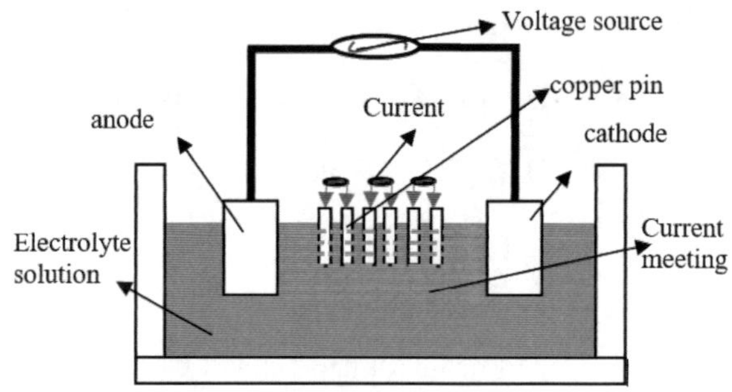

Figure 1. Electroplating process

The electroplating process consists of a positive electrode (anode), a negative electrode (cathode), and an electrolyte solution containing metal ions [1,6,7]. During electroplating, ions move from the anode to the cathode. These ions carry nickel particles towards the copper so that the copper becomes coated.

2.2. Flow Meeting

Electric current density is a quantity that states the amount of electric current flowing per unit area of the electrode surface [9]. Current density is divided into two types, namely, anode current density and cathode current density. In the electroplating process, the current density is the cathode current density, the amount of electric current required to obtain metal atoms in each unit surface area of the coated workpiece. For this electroplating process, the cathode current density plays a critical role. It will affect the efficiency of the coating, the oxidation-reduction reaction, and the diffusion of the coating on the surface of the coated object.

The principle of the electroplating process is based on Faraday's Law of electrolysis. There is a difference between the mass of the sample after and before electroplating. The thickness of the nickel layer formed can be calculated using the Lowenheim equation [7]

$$\delta = \frac{W}{\rho A} \tag{1}$$

with,
δ = thickness of the formed layer (cm)
ρ = coating density (g/cm^3)
W = m_2-m_1 = the mass of the layer formed (g)
A = layer surface area (cm^2)

2.3. Electric current

Electric current is the flow of electrons from an atom to the atom that occurs in a conductor at a certain speed at a particular time. The cause of the emergence of the electric current is the potential difference at both ends of the conductor. It gets the power to push the electrons to move places [8,9].

Current is a characteristic of a conductor [10]. A microscopic quantity associated with current is the current density given the symbol J [11,12]. The current density is a vector and is a feature of a point in the conductor. It is not a characteristic of the conductor as a whole. If the current is distributed uniformly in a conductor whose cross-sectional area is A, then the current density for all points is

$$J = \frac{I}{A} \quad (2)$$

2.4. Logger Pro

Logger Pro is software that has the privilege of presenting quantitative data and graphs simultaneously from natural physical phenomena and providing a bridge between direct observations and abstract representations of various physical phenomena, both conceptually and analytically [13]. The Logger Pro program can be presented together with the occurrence of electric current measurements, namely the relationship between the electric current (A) measured on the Vernier DCP-BTA current probe current sensor as a function of time (t). This software also allows observers to predict and compare the results obtained theoretically with the observed behaviour objectively [14].

3. Method

This study uses the following equipment:

- Laptop windows ten pro 64 bit RAM 4 GB is a suggestion to get research data obtained from Logger Pro.
- An electrolyte solution consists of a mixture of 260.00 grams of NiSO4, 40.00 grams of H_3BO_3, 60.00 grams of $NiCl_2$, and 1 litre of distilled water.
- Copperplate as cathode with a thickness of 0.5 mm with a surface area of 11.5 cm^2.
- Two nickels as anodes with a surface area of 14.0 cm^2 and 33.54 cm^2.
- Power Supply, used as a voltage source for electroplating.
- Container glass is a container for the electrolyte solution.
- Three Current Probe sensors are parts that play an essential role in research that function as current sensors.
- Logger Pro software is a part that plays an essential role in managing the Current Probe's work.
- LabQuest is a liaison between the current sensor and Logger Pro on the Laptop.
- A series of copper pins, with a distance between pins of 2.3 mm, captures electric current, which the current sensor will read.
- A plating rack was used for placing samples of copper, nickel, and a series of copper pins.
- The connecting cable is a liaison between the current sensor with a series of pins and the power supply with the anode and cathode.

4. Experimental Procedure

4.1. Data How to Retrieve Data

The step is as follow:

- Pour the electrolyte solution that has been made into the electrolyte solution container glass.
- Connect the copper plate on the cathode and first nickel on the anode with the power supply
- It is placing copper and nickel with a platting rack that has been installed with a series of pins. A row of copper pins is attached using hot glue to a pallet rack that has been perforated and connected with a connecting cable. A series of rivets are used to capture an electric current in an electrolyte solution. A row of pins is placed along with the anodes that are immersed directly in the electrolyte solution.

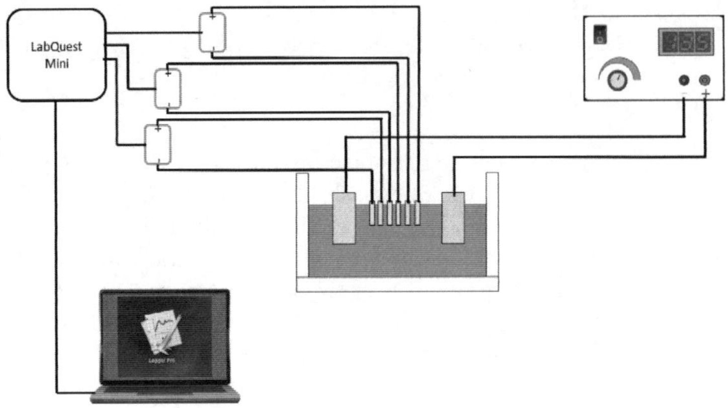

Figure 2. The series of experimental tools

Figure 3. Measuring point on the electrolyte solution

- Three pairs of pins are connected to the Vernier Current Probe sensor using a connecting cable. The Vernier Current Probe is used as an electric current reader in a solution.
- Connecting Vernier Current Probe sensor with LabQuest mini as a transducer and LabQuest mini to Laptop
- Prepare the plating process with a voltage on the 5-volt power supply for 18 seconds.
- Open the Logger Pro software and set up a sampling time of 18 seconds and a sampling rate of 10 samples/s.
- After the current is read on the Logger Pro then press the collect button to collect current data.
- After the plating process is complete, copy the current data obtained from the current sensor and transfer the data to Microsoft Office Excel to be analyzed and graphed between current and time.
- Perform the same steps from a to i by replacing the second nickel anode.

4.2. Data analysis
From the data obtained through the three current sensors, a graph is made between current and time. Furthermore, for one type of anode, the electric currents in the three current probes are compared, and for different anodes, the current density in the three pairs of probes is compared.

5. Result and Discussion
The following is an actual series of the data collection process

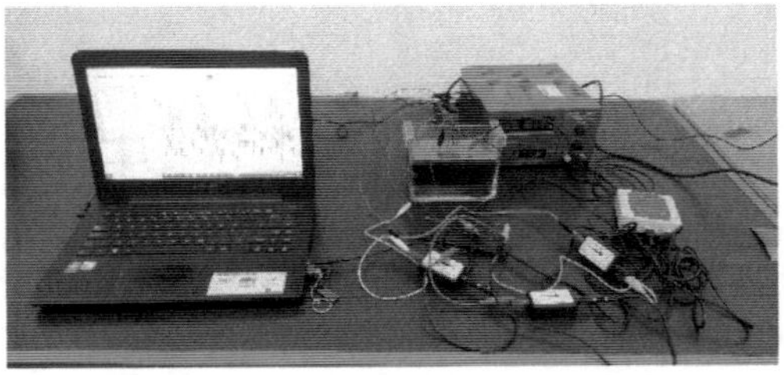

Figure 4. Data collection process

The results of current measurements at three points in the electrolyte solution are shown in Figure 5.

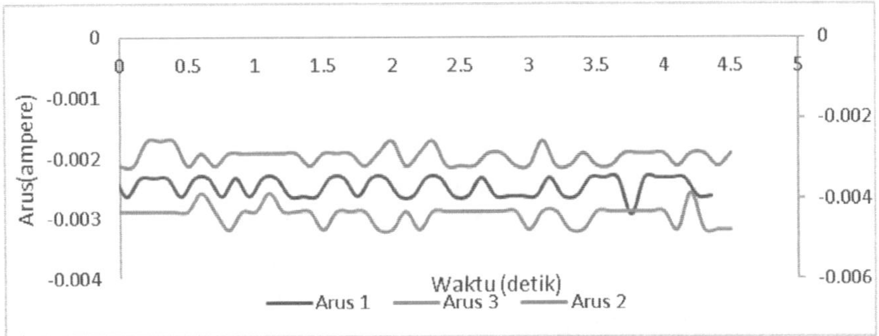

Figure 5. Graph of current versus time on nickel 1

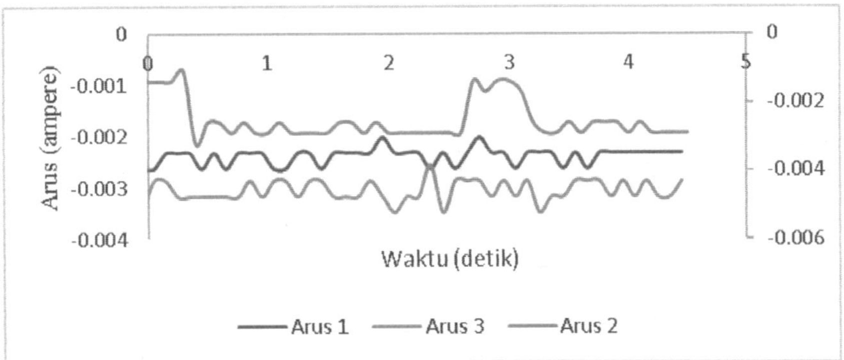

Figure 6. Graph of current versus time on nickel 2

Figure 5 and Figure 6 show that the current at each point in the measurement has a graph that is not always identical. However, as in Figure 5 in the first second, the current in each measure looks identical at some points. Especially in the electroplating of the first nickel, which has a relatively constant amplitude compared to the second nickel. This condition is because the area between the anode and cathode in the first nickel has almost the same surface area compared to the second nickel. The distribution of ions from the anode to the cathode will be more parallel. Whereas in the second nickel, which has a larger surface area than copper (cathode), then the distribution of the ions will be narrowed so that the amplitude on the second nickel will be higher than the first nickel. When compared between the current values at each current point at each point along with the anodes in Figure 5. It shows a graph between the first current, second current, and third current in specific seconds, such as in the first second having identical charts. It can show currents at various points in the electrolyte solution having the same current graph. Figure 6 also shows that the first, second, and third current graphs have fewer identical graphs. This condition is because the distribution of current density is not always regular during the measurement process.

6. Conclusion

In this study, the electric currents in the three pairs of probes in the anode-cathode direction have fewer identical graphs. Still, there are almost identical graphs in specific seconds, such as in the first second in Figure 5 and at the second type in Figure 6. This condition is because the distribution of current density is not always regular during the measurement process. The variation of the anode area ratio in the first nickel is 1:0.821, and the anode area ratio in the second nickel is 1: 0.343. The data obtained show that the first nickel has a relatively constant amplitude compared to the second nickel. So the electric current generated from the first nickel cathode is greater than the current generated from the second nickel cathode.

References

[1] S S, Iriani Y and Ramelan A H 2016 Pengaruh Rapat Arus Terhadap Ketebalan Dan Struktur Kristal Lapisan Nikel pada Tembaga *Indones. J. Appl. Phys.* **2** 1

[2] Noor Setyo V M 2012 Pengaruh Kuat Arus Pada Pelapisan Nickel Dan Nickel-Hard Chromium Plating Terhadap Sifat Fisi Dan Mekanis Permukaan Baja AISI 410 17–22

[3] Toifur M, A. N, Okimustava and Sukarelawan I 2017 Pengaruh Waktu Deposisi Pada Tebal Lapisan, Struktur Mikro, Resistivitas Keping Lapisan Tipis Cu/Ni Hasil Deposisi Dengan Teknik Elektroplating *J. Mater. dan Energi Indones.* **07** 33–43

[4] Buyang Y and Asmaraningrum H P 2015 Pengaruh Voltase Dan Waktu Terhadap Tembaga Menggunakan Metode Elektroplating *Magistra* **2** 226–36

[5] Rasyad A and Budiarto B 2018 Analisis Pengaruh Temperatur, Waktu, dan Kuat Arus Proses Elektroplating terhadap Kekuatan Tarik, Kekuatan Tekuk dan Kekerasan pada Baja Karbon Rendah *J. Rekayasa Mesin* **9** 173–82

[6] Cruz M R A, Zarzoza G O, Castañón G A M and Martínez J R 2012 Thin films from different materials obtained by the Sol-Gel method : study of the morphology through Atomic Force Microscopy (AFM) *Curr. Microsc. Contrib. to Adv. Sci. Technol.* 1370–6

[7] Bayuseno A and Nugroho S Pengaruh konsentrasi larutan dan kuat arus terhadap ketebalan pada proses pelapisan nikel untuk baja karbon rendah

[8] Saolisa R, Aliani D, Syafira I F, Rizki I S, Mardiana E, Maulidia M and Saefullah A 2018 Perbandingan Kekuatan Ikan Lemon (Lubia Caeruleus) Dengan Ikan Lele (Clarias Batrachus) Pada Tegangan 18 Volt **4** 57–66

[9] Safitri I A, Rudiyanto B, Nursalim A and Hariono B 2016 Uji Kinerja Smart Gried Fuel Cell Tipe Proton Exchange Membran (PEM) Dengan Penmbahan Hidrogen *J. Ilm. Inov.* **16**

[10] Toifur M, Dwi P R and Khusnani A 2018 Hubungan suhu anil dengan perubahan nilai resistivitas lapisan tipis Cu / Ni Hasil deposisi menggunakan teknik elektroplating *Semin. Nas. Quantum* **25** 597–604

[11] Winatapura D S, Ariadi W and Sukirman Y P E 2007 Pengaruh Penambahan Ag 2 O Terhadap Rapat Arus Kritis Superkonduktor Yba 2 Cu 3 O 7-X Hasil Proses Pelelehan **8** 144–9

[12] Romadhoni M A R, Agussalim A and Risanti D D 2017 Analisa Pengaruh Perubahan Rapat Arus Terhadap Pembentukan Passive Layer Al2O3 pada Proses Hard Anodizing Material QQA-250/4, AMS 4037 *J. Tek. ITS* **6**

[13] Subhan M 2020 Penentuan Efisiensi Sel Surya Merk Skytec Solar Model Sip-220 dengan Bantuan Software Logger Pro *Gravity Edu (J. Pendidik. Fis.)* **3** 14–7

[14] Wulandari M, Noviandini D and Sudjito D N 2014 Inovasi Pembelajaran Fisika dengan Metode Eyetracking Analysis Based Camera Studi Kasus pada Pembelajaran Hukum Kekekalan Momentum **5**

Briquettes from Biomass Waste

N H Haryanti[1], Suryajaya[1], H Wardhana[2], S Husain[1], R Noor[2], Y Anggraini[1], N Sofi[1] and D Aprilia[2]

[1]Department of Physics, Lambung Mangkurat University, Banjarbaru, Indonesia 70714.
[2]Department of Civil Engineering, Lambung Mangkurat University, Banjarbaru, Indonesia 70714.

ninishadiharyanti@ulm.ac.id

Abstract. This research made three kinds of briquettes from various biomass waste, including alaban wood charcoal and rubber seed shells mixed with coal bottom ash and coal fly ash. The purpose of the study was to obtain the characteristics and quality of briquette combustion. Making briquettes is by drying, grinding, and sifting raw materials then mixed with adhesive, printing and drying. Briquettes were made with variations in composition and pressure and the particle size of the material passing through the 50 and 250 mesh sieves. Briquettes produced from alaban wood charcoal and coal bottom ash, or fly ash, obtained more bottom ash or fly ash composition characteristics. The moisture content and calorific value would be lower while the ash content was higher. While the initial ignition time, the combustion duration is getting longer, but the burning rate would decrease. Briquettes made from rubber seed shells and coal bottom ash obtained variations in composition and pressure that affect the characteristics and quality of combustion. The higher the rubber seed shell composition and pressure, the lower the water and ash content, but the calorific value increased.

1. Introduction

Humans always need energy in carrying out their activities. Non-renewable energy sources such as petroleum, if used continuously, will run out in a short time, so that alternative renewable energy is needed [1]. There are many alternative sources of renewable energy, including biomass or organic waste materials. Biomass includes wood waste, agricultural/plantation/forest waste, organic components from industry and households, which could be reused as the source of alternative energy. The utilization of biomass waste as alternative energy can be produced using simple technology such as briquettes. The use of briquettes, especially briquettes produced from biomass, can replace fossil fuels [2]. Some biomass has considerable potential as a material for making briquettes, including rubber seed shells, coal waste in the form of bottom ash and fly ash and alaban wood charcoal waste.

Several studies of briquettes have been carried out using a mixture of coconut shell and sawdust; oil palm shells with variations in particle size of 7 mesh, 16 mesh, 25 mesh; peat and palm frond charcoal; bamboo; hazelnut skin and tamarind peel with various sizes of 20 mesh, 40 mesh, 60 mesh; rice husk and teak wood powder with size variations of 40 mesh, 60 mesh, 100 mesh [3 - 8]. From several studies, no one has used coal bottom ash, coal fly ash, alaban wood charcoal, and rubber seed shells to make briquettes. In addition, research on briquettes using a particle size of 250 mesh has not been carried out.

Coal bottom ash still has a carbon content value that can be reused by increasing the heat value using biomass mixture [9], [10]. Alaban wood charcoal has several advantages: its quality is equal to briquette charcoal and does not emit smoke [11]. The large number of rubber seeds produced from rubber plants causes the rubber seed shells that have not been utilized optimally to become one of the potential biomass that can produce briquettes as alternative energy. [12]. Several factors need to be considered in the

manufacture of briquettes, including raw materials, adhesive materials, the particle size of the material, the pressure of the printing machine.
The particle size of different briquette materials will determine the value of compressive strength, moisture content, and calorific value. The smaller the particle size of the material, the stronger the adhesion between the particles when an adhesive is added and the higher the calorie [13-15]. In the manufacture of these briquettes, in addition to using alaban wood charcoal, bottom ash and coal fly ash and rubber seed shells, a small particle size of 250 mesh are also used, and this is a novelty of this research.

Some of the benefits of using briquettes are the immense potential of biomass in Indonesia which is a source of raw materials. In addition, briquettes can be increased in density, adjust the shape and size, are not dirty, are easy to transport and are practical as fuel [16]. Briquettes have an economic advantage in manufacturing because they can be produced with simple technology and high calorific value. The price of briquettes is lower and is a renewable alternative energy source. In this study, alaban wood charcoal biomass waste, rubber seed shells, bottom ash, and coal fly ash were used to make briquettes. This study aimed to obtain the characteristics and quality of briquette combustion made from biomass waste. The attributes of briquettes include moisture content, ash content, calorific value, while the rate of briquette combustion has initial ignition time, duration of combustion and combustion speed.

2. Method

This research is experimental research conducted in a laboratory. The raw material used in the manufacture of briquettes is alaban wood charcoal waste obtained from the charcoal company Citra Utama (Ltd) in Banjarbaru, while coal bottom ash and fly ash was obtained from the Asam-Asam power plant located in Jorong District, Tanah Laut Regency, South Kalimantan Province. Meanwhile, rubber seed shell waste was taken from rubber farmers' plantations in Pengaron Village, Simpang Empat District, Banjar Regency, South Kalimantan Province. In this study, three kinds of briquettes made from biomass waste were made with different mixed materials, namely briquettes made from alaban wood charcoal and coal bottom ash, briquettes made from alaban wood charcoal and coal fly ash, briquettes made from rubber seed shells and coal bottom ash. The three kinds of briquettes were made with various compositions, and three samples were made for each composition.

Making briquettes made from alaban wood charcoal and bottom ash and coal fly ash is carried out by drying, grinding and sieving the raw materials. The raw materials are crushed and sieved using a sieve to pass a 250 mesh sieve to obtain a more petite and uniform size. When making powdered raw materials from rubber seed shells, the waste process of rubber seed shell is first cleaned of dirt, then dried for three days in the sun, that it is carbonized at a temperature of 300°C. After the carbonization process is complete, the charcoal is then pulverized and sieved to obtain 50 mesh rubber seed shell waste charcoal powder to manufacture briquettes.

The process of making briquettes includes mixing raw materials with adhesives, molding and drying. Making briquettes is by mixing alaban wood charcoal powder and coal bottom ash until homogeneous with composition variations of 100%:0%, 90%:10%, 80%:20% and 70%:30%. In contrast, the variations in the compositions used in alaban wood charcoal briquettes and coal fly ash are 100%:0%, 90%:10%, 80%:20%. In the process of mixing raw materials for charcoal, rubber seed shell waste and coal bottom ash, with variations in the material composition of 60%: 40%, 70%: 30%.

It was making adhesive by cooking tapioca flour with water to form a gel. Tapioca flour was chosen as an adhesive because it has a higher carbon content than other flour types. The content of carbon content in tapioca flour is 85.20%. Meanwhile, alaban wood charcoal was 76.69%, and coal bottom ash was 41.87%. [17].

The adhesive that has formed a gel is then mixed with alaban wood charcoal powder, bottom ash, coal fly ash, rubber seed shell charcoal as much as 5%. The results of the mixing of the samples were put into a mold with a size of 2 cm x 5 cm and compacted using a hydraulic pump with a pressure of 150 kg/cm^2. Meanwhile, briquettes made from rubber seed shell charcoal and coal bottom ash were molded with pressure variations of 100, 150, and 200 kg/cm^2. The molded briquettes were dried in the oven at 120 ^{0}C for 4 hours and cooled at room temperature for 24 hours. Furthermore, the dried briquettes were tested for characteristics and combustion quality tests.

Characteristic tests include moisture content, the ash content in accordance with SNI 06-3730-1995, while the calorific value is in accordance with ASTM D2015. The calorific value test uses the Bomb Calorimetric method. The quality standard of briquettes is based on SNI 01-6235-2000, namely moisture content (< 8%), ash content (< 8%), calorific value (> 5000 cal/g). The quality test of briquette combustion includes initial ignition time, duration of combustion and combustion speed.

3. Results and Discussions

3.1 Briquettes Made from Alaban Wood Charcoal and Coal Base Ash

The average moisture content, ash content, and calorific value of briquettes made from alaban wood charcoal biomass waste and coal bottom ash from 3 samples for each material composition are as shown in Table 1.

Table 1. The characteristics of briquettes made from alaban wood charcoal and coal ash[18]

	Alaban charcoal + bottom ash				Alaban charcoal + fly ash		
	100%:0%	90%:10%	80%:20%	70%:30%	100%:0%	90%:10%	80%:20%
Moisture (%)	3.91	4.10	2.95	3.86	3.91	4.31	3.96
Ash content (%)	0.38	8.32	18.28	26.67	0.38	8.18	17.39
Calorific value (cal/g)	6.621.07	5,871.53	5,349.03	4,749.60	6,621.07	5,858.73	5,283.93

The highest average water content is 4.10% in 90%:10% briquettes composition, and the lowest average water content is 2.95% in 80%: 20% composition. The water content affects the rate of briquette burning. High moisture content will inhibit the ignition rate so that it will reduce the rate of combustion [19]. In this study, the moisture content of the briquettes obtained did not exceed the SNI because the moisture content test results were 8%. It can be seen that the more composition of coal bottom ash used, the lower the moisture content.

In Table 1, it can be seen that the highest average ash content of 26.67% was obtained in 70%: 30% briquettes and the lowest average ash content was 0.38% in 100%:0% briquettes. The more coal bottom ash composition used, the higher the ash content of the briquettes. The ash content in this study that met the SNI was briquettes with a composition of 100%: 0% and 90%: 10%, with an ash content value of 8%.

The calorific value is the amount of heat released by each kilogram of briquettes if burned completely [19]. Table 1 shows that the highest average calorific value of 6,621.07 cal/g is 100%:0% briquette, and the lowest is 4,749.60 cal/g 70%:30% briquette. In this study, only one briquette did not meet SNI with an average calorific value of 4,749.6 cal/g, SNI for briquettes was 5,000 cal/g. It can be seen that the more composition of alaban wood charcoal waste used, the greater the calorific value of the briquettes.

The briquette combustion quality test includes initial ignition time, duration of combustion and combustion speed, as shown in Table 2.

Table 2. Briquette Combustion Quality Results.

	Alaban charcoal + bottom ash				Alaban charcoal + fly ash		
	100%:0%	90%:10%	80%:20%	70%:30%	100%:0%	90%:10%	80%:20%
Ignition time (mins)	1.32	2.02	2.13	2.22	1.32	2.20	2.52
Burning duration (mins)	74	79	80	84	74	77	84
Burning speed (g/mins)	0.34	0.26	0.24	0.22	0.34	0.33	0.21

From Table 2, the quickest initial ignition time of briquettes, 1.32 minutes, were briquettes with a composition of 100%:0%, briquettes with the longest ignition time of 2.22 minutes were briquettes with a composition of 70%:30%. It is because 70%:30% of briquettes have the lowest calorific value of 4,749.6 cal/g, making them difficult to ignite. At the initial ignition time, it can be seen that the higher the composition of the coal bottom ash used, the longer the initial ignition time of the briquettes.

Based on Table 2, the briquettes that ignite the fastest to become ash are briquettes with a composition of 100%:0%, which were 74 minutes and the longest lit briquettes with a composition of 70%:30%, which was 84 minutes. In the duration of combustion, it can be seen that the higher the composition of the coal bottom ash used, the longer the period of burning the briquettes.

In Table 2, the highest burning rate of briquettes was 0.34 g/mins with a composition of 100%:0%, and the lowest burning rate was 0.22 g/minute of briquettes with a composition of 70%:30%. At the searing speed of briquettes, it can be seen that the higher the composition of the coal bottom ash used, the lower the briquette burning speed. It is because briquettes with a high coal bottom ash composition have tiny pores, and high ash content will make it difficult for outside air to enter because the ash covers the surface of the briquettes, making the combustion speed slow. In research [6] with variations in the composition of peat and palm frond charcoal, it was found that the initial ignition and the fastest briquette burning speed were briquettes with a composition of 90%:10%.

3.2. Briquettes Made from Alaban Wood Charcoal and Coal Fly Ash

The average moisture content, ash content, the calorific value of briquettes made from alaban wood charcoal biomass waste and coal fly ash from 3 samples for each material composition are as shown in Table 1. Table 1 shows the average moisture content of briquettes; the lowest was 3.91% in briquettes with a composition of 100% alaban wood charcoal, while the highest is 4.31% in briquettes with a composition of 90%:10%. The value of the moisture content of all samples of briquettes meets SNI No. 01-6235-2000 ($\leq$ 8%). The moisture content affects whether or not the briquettes are easy to burn [20]. The higher the moisture content, the more complex the briquettes to burn, so the calorific value produced will also be lower.

The results in Table 1 show the highest average ash content of 17.388% in briquettes with 20% coal fly ash composition and the lowest average ash content of 0.383% in 100% alaban wood charcoal briquettes. Briquettes with 100% alaban wood charcoal composition are still below SNI, but briquettes with 10% and 20% coal fly ash composition exceed SNI. The higher the ash content in a briquette, the lower the quality of the briquettes because the high ash content means the lower calorific value of the briquettes.

The calorific value dramatically determines the quality of the briquettes. The higher the calorific value, the quality of the briquettes produced would be better. The results in Table 1 show the highest average calorific value of 6,621.067 cal/g in briquettes with a composition of 100% alaban wood charcoal and the lowest 5,283.933 cal/g found in briquettes with a composition of 20% coal fly ash. The average calorific value of all samples of briquettes according to SNI No. 01-6235-2000, the calorific value of briquettes is 5,000 cal/g. From the study results, the higher the composition of coal fly ash causes the calorific value of the briquettes to decrease.

The initial ignition timing of the briquettes can be seen in Table 2. It can be seen that the briquettes with 100% alaban wood charcoal composition have the fastest initial ignition of 1.32 minutes. In contrast, the briquettes with the longest ignition of 2.52 minutes were briquettes with a 20% coal fly ash composition. The more the addition of coal fly ash, the more difficult it is to ignite the briquettes. [21].

The results in Table 2 show that briquettes with a composition of 100% alaban wood charcoal have the fastest burning duration of 74 minutes. At the same time, the briquettes with the most extended burning duration of 84 minutes were briquettes with a composition of 20% coal fly ash. It was found that the more coal fly ash composition caused the duration of the briquettes to be burned to become longer. The longer the duration of combustion, the more efficient the use of briquettes can be.

The burning rate of briquettes is the speed at which the briquettes run out to ashes. The results in Table 2 show that briquettes with 100% alaban wood charcoal composition have the highest burning rate of 0.34 g/mins. In comparison, the briquettes with the lowest burning rate of 0.21 g/minute were briquettes with a composition of 20% coal fly ash. It seems that the more coal fly ash composition, the lower the burning rate of the briquettes.

3.3. Briquettes made from rubber seed shell and coal bottom ash

The results of the average moisture content, ash content, calorific value of briquettes made from rubber seed shell biomass waste and coal bottom ash from 3 samples for each material composition are as shown in Table 3. A2 and A3 samples with a composition of 60%: 40% and 70%: 30% pressure 100 kg/cm^2, B2 and B3 samples with composition 60%:40% and 70%:30% pressure 150 kg/cm^2, C2 and C3 samples with composition 60%:40% and 70%:30% pressure 200 kg/cm^2.

Table 3. Rubber seed shell biomass waste and coal bottom ash briquette characteristics[22]

Sample	Moisture (%)	Ash content (%)	Calorific value (cal/g)	Ignition time (mins)	Burning Duration (mins)	Burning rate (g/ mins)
A2	10.65	16.78	4,018.66	10.35	80.56	0.62
A3	9.43	12.53	4,536.23	10.55	81.20	0.62
B2	7.57	16.73	4,578.32	12.02	90.21	0.55
B3	6.49	12.94	4,549.88	12.06	91.13	0.55
C2	5.86	17.26	3,909.81	12.19	94.21	0.53
C3	4.35	12.63	3,597.59	12.21	94.22	0.53

In Table 3, the highest moisture content is 10.65% in briquettes with a composition of 60%: 40% pressure of 100 kg/cm^2, while the lowest moisture content is 4.35% in briquettes with a composition of 70%:30% pressure of 200 kg/cm^2. High moisture content in the briquettes will cause the briquettes to be difficult to ignite. The briquetting pressure of 100 kg/cm^2 resulted in high water content for briquettes mixed with rubber shells and coal bottom ash. In contrast to the pressure of 150-200 kg/cm^2, the water content of the briquettes meets the SNI 01-6235-2000 standard. In testing the moisture content, the higher the biomass composition used, the higher the moisture content produced [23]. From the results of the moisture content, it can be seen that the moisture content decreases as the pressure is applied; this is in accordance with other research [24].

In Table 3, the addition of coal bottom ash to the briquette mixture tends to increase the ash content. The ash content of briquettes is much influenced by the chemical composition of the briquette raw materials. One of the constituent elements of ash is silica; the silica content in coal bottom ash is almost 60%. The mineral content of coal bottom ash is SiO_2 61.92%, Al_2O_3 16.00%; Fe_2O_3 6.47%; CaO 6.85%; MgO 7.90% and several other compounds in relatively small amounts [25]. Increasing briquette ash content in the composition variation is due to the silica content in the coal bottom ash. The ash content of the briquettes produced is relatively low compared to mixed biomass briquettes (coffee shells, kapok shells, coconut shells) and coal bottom ash. [23]. In testing the ash content, it can be seen that the higher the pressure, the lower the ash content; this is in line with other research [24].

Based on Table 3, the calorific value is still below SNI 01-6235-2000. The highest calorific value is 4,578.32 cal/g in the composition of 60% rubber seed shell briquettes and 40% coal bottom ash with a pressure variation of 150 kg/cm^2. While the lowest calorific value in briquettes with a composition of 70% rubber shell and 30% coal bottom ash with a pressure variation of 200 kg/cm^2. The greater the pressure applied, the more the calorific value of the briquettes produced [24]. However, the tests' results obtained varied calorific values, an indication of the homogeneity factor at the time of mixing briquettes made the briquette mixture not mixed evenly [26].

Each variation in the composition of briquettes shows an initial ignition time that is not much different but tends to be slow, the initial ignition time is in the range of 10-12 minutes, with other pressures given to each briquette, namely 100 kg/cm^2, 150 kg/cm^2, and 200 kg/cm^2.

In Table 3, the higher the briquetting pressure, the longer the initial ignition time. Besides that, the composition of different materials with the same pressure difference in combustion duration is not too long. Still, it can be seen that the pressure difference gives a significant combustion duration. The greater the pressure applied, the denser the cavities in the briquettes, making it difficult for oxygen to enter the pores of the briquettes so that combustion tends to be slower and lasts longer until the briquettes turn to ash. Good briquettes are briquettes that have a long burning duration. The longer the flame period, it will save the use of briquettes as fuel [28]. The briquettes with the most extended burning duration of 94.22 minutes have a composition variation of 70% rubber shell: 30% coal bottom ash.

The rate of combustion is influenced by the structure of the material, the content of bound carbon, and the level of hardness of the material [29], [30]. The burning rate of briquettes in all compositions is not too far between 0.53-0.62 g/mins. Based on Table 6, it can be seen that the greater the pressure applied to the briquettes, the lower the combustion speed. Because the denser the briquettes, the smaller the pores formed, making less oxygen available so that the combustion process is hampered.

4. Conclusion

The characteristics of the briquettes obtained that the higher the composition of bottom ash and coal fly ash, the lower the moisture content and calorific value. In contrast, the ash content would be higher.

While the quality of briquette combustion, the higher the composition of bottom ash and coal fly ash, the longer the initial ignition time and duration of discharge and the lower the combustion speed. The recommended composition is 90% (alaban wood charcoal): 10% (bottom ash or coal fly ash) because the moisture content, ash content and calorific value are in accordance with the provisions of SNI. Variations in the composition and pressure of briquettes affect the characteristics and quality of combustion of rubber seed shells and coal bottom ash briquettes. The more the composition of the rubber seed shell gave lower moisture and ash content, while the calorific value was higher. Higher pressure on briquetting yielded lower moisture and ash content but higher calorific value, increasing the quality of briquette combustion. Briquettes with good combustion quality are easy to ignite, long burning duration, and have a slow combustion speed. The long burning time of a briquette will affect the quality and efficiency of combustion. A better briquette has a longer burning duration with a constant flame.

Acknowledgements
Thank you to all who have greatly helped in this research, especially to the Institute for Research and Community Service at Lambung Mangkurat University, which has provided funding through Basic Research for Higher Education Excellence.

References

[1] Rajaseenivasan T, Srinivasan V, Qadir G S M and Srithar K 2016 *Engineering Journal* **55** 2833 – 38
[2] Sova D, Porojan M, Bedelean B and Humunic G 2018 *International Journal of Thermal Sciences* **124** 1 - 12
[3] Purwanto D, Utami R and Suryani S D 2015 *Jurnal Riset Industri Hasil Hutan* **7** 1–8
[4] Afriani C D, Yufita E and Nurmalita 2017 *Journal of Aceh Physics Society* **6** 6–9
[5] Setiowati R and Tirono M 2014 *Indonesian Journal of Neutrino* **7** 23–31
[6] Nugraha A, Widodo A S and Wahyudi S 2017 *Jurnal Rekayasa Mesin* **8** 29–36
[7] Aisyah I S, Saifullah A and Satya T 2017 *Prosiding Seminar Teknologi Dan Rekayasa* **4** 1-6
[8] Suryaningsih S, Nurhilal O and Affandi K A 2018 *Jurnal Ilmu Dan Inovasi Fisika* **2** 15–21
[9] Gunawan B and Slamet S 2015 *Jurnal SIMETRIS* **6**
[10] Gunawan *et al* 2015 *Prosiding Snst Ke-6 Tahun 2015 Fakultas Teknik Universitas Wahid Hasyim Semarang*
[11] Kahariayadi A D, Setyawati, Nurhaida, Diba F and Roslinda E 2015 *Jurnal Hutan Lestari* **3** 561 – 68.]
[12] Julian R T 2016 Pemanfaatan Limbah Cangkang Biji Karet Menjadi Briket Sebagai Bahan Bakar Alternatif Dengan Bahan Perekat Amilum (Politeknik Negeri Sriwijaya)
[13] Septian E T, Wijianti E K and Saparin 2017 *Jurnal Teknik Mesin* **3** 22–29
[14] Damayanti R, Lusiana N and Prasetyo J 2017 *Jurnal Teknotan* **11** 51–60
[15] Sudiro and Suroto S 2014 *Indonesian Journal of Sainstech* **1** 1–18
[16] Sudrajat R D, Setiawan and Roliadi H 2004 *Jurnal Penelitian Hasil Hutan* **24** 227 - 240
[17] Haryanti N H, Suryajaya, Wardhana H, Anggraini Y and Andini N S 2018 *Prosiding Nasional Fisika XXXII*
[18] Haryanti N H, Suryajaya, Wardhana H, Husain S, Anggraini Y and Sofi N 2018 *Journal of Physics: Conference Series* **1120**
[19] Kholil A 2017 *Analisis Fisis Briket Arang dari Sampah Berbahan Alami Kulit Buah dan Pelepah Salak* (Universitas Islam Negeri Maulana Malik Ibrahim)
[20] Martynis M, Sundari E and Sari E 2012 *Jurnal Litbang Industri* **2** 35-41
[21] Thoha M Y and Fajrin D E 2010 *Jurnal Teknik Kimia* **1** 34-43
[22] Haryanti N H, Noor R and Aprilia D 2018 *Prosiding Seminar Nasional Pendidikan.* Universitas Lambung Mangkurat 203-9
[23] Slamet S and Gunawan B 2016 *Prosiding SNATIF* **3**

[24] Darvina Y and Asma N 2011 *Upaya Peningkatan Kualitas Briket dari Arang Cangkang dan Tandan Kosong Kelapa Sawit (TKKS) melalui Variasi Tekanan Pengpresan* (Fakultas Matematika dan IPA Jurusan Fisika Universitas Negeri Padang)
[25] Munir M 2008 *Pemanfaatan abu batubara (fly ash) untuk hollow block yang bermutu dan aman bagi lingkungan* (Pascasarjana Universitas Diponegoro)
[26] Risna 2016 *Pengaruh Tekanan Dan Ukuran Partikel Terhadap Kualitas Briket Cangkang Coklat* (Jurusan Fisika Fakultas Matematika dan Ilmu Pengetahuan Alam. Universitas Halu Oleo)
[27] Haryanti N H, Noor R and Aprilia D 2019 *Prosiding Seminar Nasional Pendidikan Universitas Lambung Mangkurat* 211-7
[28] Astini N 2014 *Pemanfaatan Limbah Cangkang Biji Karet (Hevea Brasiliansis Muell.Arg) sebagai Briket Arang* (Fakultas Teknik Universitas Lambung Mangkurat)
[29] Jamilatun S 2008 *Jurnal Rekayasa Proses* **2**
[30] Jamilatun S 2011 *Makalah Pendamping Kimia* **2** 37-8

Digital literacy of indonesian prospective physics teacher: Challenges beyond the pandemic

A Z Rahmawati, Z Haryanto and N F Sulaeman

Physics Education Study Program, Faculty of Teacher Training and Education, Mulawarman University, Indonesia

Azahraarahmawati@gmail.com

Abstract. Pandemic covid-19 reinforces the needs of digital technology in all aspects, including education. Digital literacy skills become essential for all teachers and also prospective teachers. This study aims to find an overview of the digital literacy capabilities of prospective physics teachers. This study was a quantitative descriptive research with data collection using an instrument in a questionnaire that included 29 items from 4 aspects of digital literacy: Accessing content, creating content, transliteracy, and evaluating content. The sample consisted of 171 physics education students from 3 State Universities in Indonesia. The result showed the level of digital literacy in the moderate category. From the four aspects, creating content related to creating or presenting information on various digital platforms showed lower results than accessing content, transliteracy, and evaluating content. In a prospective science teacher curriculum, creating learning content on digital platforms needs to be developed. This research finding suggests that digital literacy needs to be enhanced during the prospective science teachers' program.

1. Introduction

Digital literacy is a life skill that involves using technology, information and communications, social skills, learning ability and attitude, critical thinking, creativity, and inspiration as digital competencies[1]. Literacy culture is a prerequisite for life skills in the 21st century that must be developed through integrated education starting from family, school, and community[2]. In the last few years, information and communication technology development in all education sectors has been notable [3]. Active and contextual learning will be carried out optimally if supported by adequate media, methods, tools, and materials. In the era of this technology development, various forms of media are proliferating. Through virtual learning or online learning, the learning process is not limited by time and space. The interaction between teachers and students takes place anytime and anywhere. Online learning is a distinct advantage for teachers and students that make students more active in constructing knowledge[4].

Meanwhile, as the effect of the COVID-19 pandemic, science teacher educators were faced with deciding how to support prospective teachers learning to teach [5]. To further complicate this rapid transition, educators' autonomy was constrained by a range of institutional, technological, accreditation, and regulatory issues. For many, this means planning for online science teaching methods courses, working closely with accrediting bodies and university clinical placement offices to explore alternatives to face-to-face experiences, and negotiating variances of in-person and online student teaching experiences[6]. In connection with the Covid-19 pandemic in Indonesia, the government has provided regulations regarding distance learning in which prospective teachers and teachers must master technology to support learning. The 21st-century skills are needed as digital literacy is used as a tool for working[7]. This National Literacy Movement was carried out to develop a digital literacy culture. Based on a survey conducted by the Ministry of Information and Information in collaboration with UNICEF on children and adolescents aged 10-19 (as many as 400 respondents) spread throughout Indonesia and representing urban and rural areas, information was obtained that around 79.5% of children and adolescents are internet users and digital media. It has resulted in a transition where school-age children and adolescents like to explore information through the internet and digital media, like exciting lessons that can be used immediately with various available applications, and tend to like the virtual world[2].

Digital literacy in the preparation of prospective teachers is becoming a new concern around the globe. In Israel, the circumstances of the prospective teacher showed that most of the participants had basic research and information search skills overall[8]. Moreover, in Indonesia, this study indicated that not all informants had digital literacy skills because this skill requires continuous and consistent practice to do it well[1]. The teachers perceived that they were ready to manage digital learning due to school closures amidst the pandemic[5]. Therefore, a choice of technology and teaching methodology to adapt and adjust was a solution to the learning problems[9]. On another level, self-perceptions and skills of prospective physics teachers in the Technological Pedagogical and Content Knowledge Knowledge (TPACK) framework to support distance learning. Based on data analysis, self-perceptions of prospective physics teachers within the TPACK framework are insufficient category. There is a significant difference between the self-perception of prospective physics teachers who have taken computer training and those who have never taken computer training. The skills of prospective physics teachers in the TPACK framework are also sufficient [10]. However, the specific measurement of prospective physics teachers' digital literacy in Indonesia is rarely found. Therefore, our research aims to explore the digital literacy level and the aspects of Indonesian prospective physics teachers.

2. Method

2.1 Instrument Preparation

The questionnaires generated our primary data, and then it was made on a measuring scale to produce quantitative data. This research was quantitative descriptive research with an instrument in a questionnaire that included 29 items from 4 aspects of digital literacy. Research subjects were asked to fill out a questionnaire about their basic competencies of digital literacy prospective physics teacher and focus on four contents, as shown in figure 1.

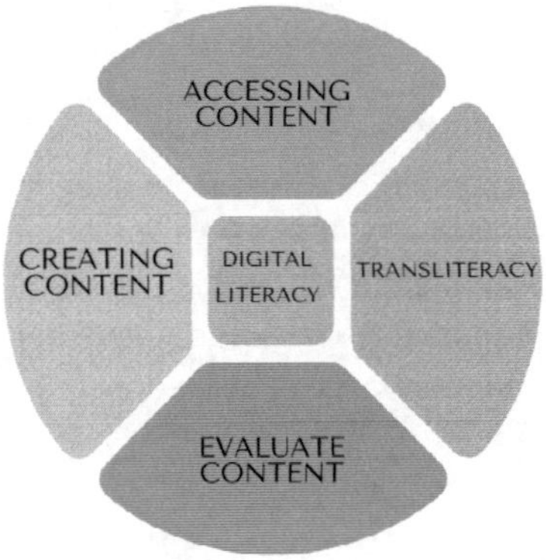

Figure 1. The Aspect of Digital Literacy

In this study, the scale used was the Likert scale. For quantitative data analysis, the answers of the respondents were given a score from 1 to 5. Then, the questionnaire was tested for its validity and reliability. Validity and reliability tests were carried out using the application of *SPSS Statistics 20*. From the test results, 25 questions were valid, and four were revised with provisions for correcting sentence editors. Reliability results were also in the high category[11].

In preparing a questionnaire on digital literacy, a theoretical study from experts is required and analyzed to understand digital literacy. Then, basic competencies are adapted to conditions in the field.

A trial was conducted to measure its validity and reliability. The test was given to 25 Physics education students class of 2018. The calculations used the help of SPSS Statistic 20 for the Windows application. In this calculation, the numbers obtained must be compared with the critical values for the Pearson correlation[11]. In testing reliability, the coefficient value must be at least more than 0.70 for the questionnaire to be ready to use.

2.2 Data Collection

The data collection was conducted by using purposive sampling. The chosen sample which met the requirements amounted to 171 prospective physics teachers from 3 universities. Sample A was located in East Kalimantan, sample B in South Kalimantan, and sample C in Sumatra island. The sampling technique used was purposive sampling, a deliberate choice of research participants because of the quality with specific criteria [12]. The criteria were three study programs that have used curricula by emphasizing the principles of developing science-based physics learning media, contextual technology, especially Information and Communication Technology (ICT).

2.3 Data Analysis

2.3.1 Instrument Response Analysis

Primary data was generated from distributing questionnaires, then made on a measuring scale to produce quantitative data. In this study, the scale used was the Likert scale. For quantitative data analysis, the answers of the respondents were given a score from 1 to 5.

2.3.2 Digital Literacy Level of Students

The results of the research data score were then processed and made on a scale of 100 with the formula:

$$A = \frac{X}{Y} \times 100 \quad (1)$$

A = final score with a scale of 100
X = score obtained by each respondent
Y = Maximum score of 29 questions

Using the Norms Reference Assessment system, the digital literacy level was categorized into very high, high, medium, low, and very low ability levels [13]. Norms Reference Assessment is an assessment that refers to the group norms or the values obtained by students compared to the values of other students in the group. The requirements in determining the category using the Norms Reference Assessment technique are shown in table 1.

Tabel 1. Category of Digital Literacy[13]

Category	*Interval*
Very High	$x \geq \bar{x} + 1.5$ SD
High	$\bar{x} + 0.5$ SD $\leq x < \bar{x} + 1.5$ SD
Moderate	$\bar{x} - 0.5$ SD $\leq x < \bar{x} + 0.5$ SD
Low	$\bar{x} - 1.5$ SD $\leq x < \bar{x} + 0.05$ SD
Very Low	$x < \bar{x} - 1.5$ SD

2.3.3 Aspect of Students Digital Literacy

The results of the research data were then processed based on each aspect. The percentage was obtained with the formula:

$$Z = \frac{X}{Y} \times 100\ \% \quad (2)$$

Z = Percentage score
X = score obtained for each aspect
Y = Maximum score from all aspects

A test was carried out using Pearson product moment with the help of the *SPSS* application to find out the correlation between each asset. After obtaining the results, the data were then analyzed to get a conclusion.

2.3.4 Correlations of Aspect Digital Literacy
In correlational research, researchers seek to determine whether a relationship exists between two (or more) quantitative variables, such as age and weight or reading and writing ability. In this study, correlation test using SPSS 20 for windows application.

3. Result and Discussion

In this study, the results from data collected through the questionnaires were examined. From 171 responses, the analysis clarified the level of digital literacy and the aspects: accessing content, creating content, transliteracy, and evaluating content. Moreover, the correlation of each aspect was also explored.

3.1 Digital Literacy Level
The results of the research data score were then processed, and the average scores were analyzed using Microsoft Excel. The results of students' digital literacy questionnaire responses can be seen in the table of scores. The average score was 73, and the Standard Deviation was 8.84, so the value was in the medium category[13]. The data interpretation of students' digital literacy levels is presented in the following table 2.

Tabel 2. Levels of Digital Literacy Skills. In general, the level of digital literacy skills is in the medium category

Category	*Frequency*	%
Very High	12	**8.22**
High	28	**19.18**
Moderate	63	**43.15**
Low	33	**22.60**
Very Low	10	**6.85**
Total	**146**	**100**

The majority of students were already in the high category at using technology and finding sources of information effectively. However, evaluating an information source and creating work in learning still has to be improved. In the field, students rarely create new and reliable sources of information. The students demonstrated medium to low digital literacy readiness. These results pointed out that prospective teachers with current technological sophistication lack the knowledge, skills, and necessary experience to integrate technology into the classroom to assisting them in teaching and assisting their students in learning. In addition, some students lack basic technology skills, and therefore they need more hands-on practice in the operation of basic technology.

3.2 Aspect of Digital Literacy
The results of data interpretation regarding aspects of digital literacy can be seen in Figure 2.

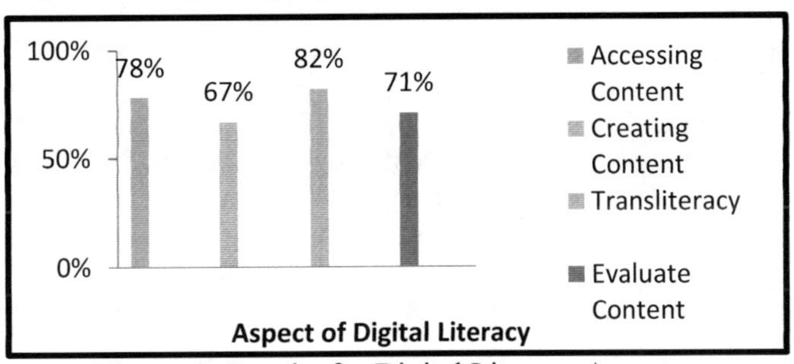

Figure 2. Results for Digital Literacy Aspects.

From the results, the aspect of transliteracy had a more significant percentage, which was 82%, then accessing content was 78%, and evaluating content was 71%. The lowest was the aspect of creating content which was 67%. Then, viewed from the indicators of each aspect, it can be presented in figure 3.

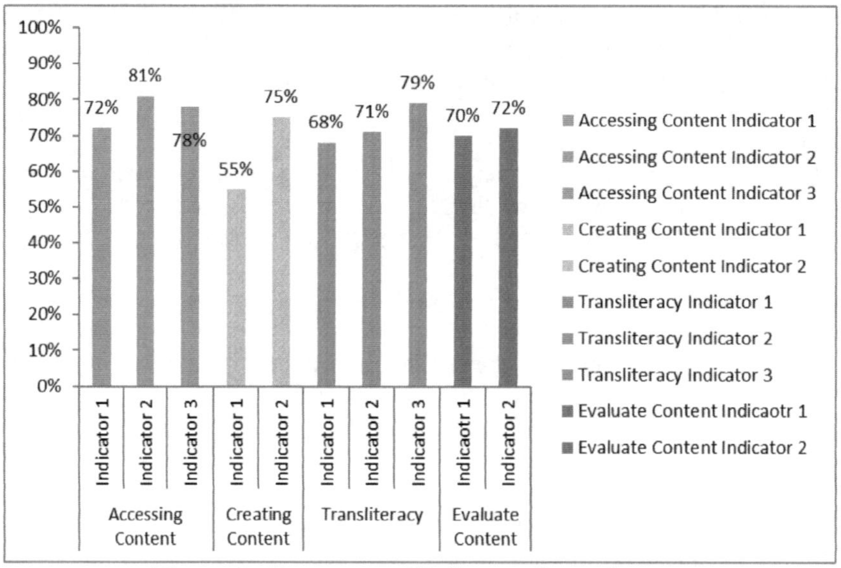

Figure 3. Percentage of indicators for each aspect

Based on the figure above, it can be seen that the abysmal ability is in the aspect of creating content indicator 1. While the highest ability that many students have is in the aspect of accessing content indicator 2.

3.2.1 Accessing Content

Indicators of these aspects are recognized as information needs, access, understanding, and use of information effectively and efficiently, select and sort methods to access the required information and identify the types of information sources in the digital library[8][14]. Referring to how the student can access or find a reliable source of information, most students can select and sort methods to access the information needed and identify the types of sources of information obtained. Meanwhile, the ability to recognize information needs, access, understanding, and use of information effectively and efficiently is still relatively low compared to the other two indicators. Overall, these results indicate that selecting information effectively and efficiently still needs to be improved.

3.2.2 Creating Content

Indicators of these aspects are to create web pages by inserting images, text, and hyperlinks to convey information and to use ICT to create or design information[8][14][15]. This aspect includes questions about student skills in creating or presenting information on various platforms and using various platforms of e-learning. This aspect and indicators can create new sources of information using existing technology. The aspect of creating content can be developed through subjects in the study program. Among them, in one of the Universities on physics education study program, there is a course "Physics Workshop" which aims to facilitate students in doing physics practicum. This lecture teaches about applying physics concepts in everyday activities and designing them into simple technological products. However, this course is an optional course and not required for students.

Meanwhile, there are no similar courses in the other two study programs that can produce products related to physics. However, several replacement courses can develop the aspects of creating content, such as laboratory research subjects and technology and information applications. This aspect needs to be improved for prospective teachers, in line with the increasing technology and information that continues to develop.

3.2.3 transliteracy

Indicators of this aspect may use various learning applications (Quiziz, google classroom, Edmodo, Phet, and others), be able to store information and mark by entering copyright, and understand the ethical implications or consequences of literacy users[16][8][14][15]. It involves the ability to take advantage of all different platforms, especially to create content, collect, share and communicate through various social media, discussion groups, smartphones, and various available online services.

In addition, most students rarely use supporting applications as learning media. For example, the zoom meeting application is still rarely used by students. Some of the reasons are that the teaching place's network is less supportive; some feel less comfortable using face-to-face applications, and it is easier to use group chat. Another example is the virtual lab, in which various virtual lab resources are still unfamiliar to prospective teacher students. This knowledge is reasonably necessary. According to the current situation that makes it impossible to use the usual laboratory, using a virtual lab is beneficial for students' understanding.

3.2.4 Evaluate Content

Indicators of this aspect compare information from various sources and critically evaluate information sources (such as quality, reliability, credibility, and efficiency in use)[16][8][14][15]. It contains the ability to sort the information to be presented and its relationship with classroom learning. Based on the results of the response, students are pretty selective in evaluating the information obtained. However, students also pay less attention to the credibility of the information obtained. It can be seen from the lack of concern for students to improve an inaccurate source of information. Evaluate content needs to be developed to improve valid and reliable information.

3.3 Correlation Levels of Digital Literacy Aspects

The relationship between aspects of digital literacy in this study was carried out by correlation analysis using bivariate with the Pearson correlation test to determine the significance between aspects. Based on the data analysis results, the correlation values shown range from 0.41 to 0.60. These results indicate that a reasonably good relationship exists between aspects. Therefore, the aspects are related to each other.

4. Conclusion

Based on the results of research carried out regarding the ability level of digital literacy aspects of prospective physics teacher students, it can be concluded that the level of digital literacy skills is in the medium category, with an average score of 73%. Students' digital literacy skills are supported by courses related to digital literacy, such as physics learning courses.

The contribution from each aspect of digital literacy dramatically affects the level of digital literacy of students. It can be seen from the results of data analysis that the aspects of creating content still need to be developed to be able to improve students' digital literacy skills. Each aspect of digital literacy and the level of correlation for each aspect shows a mutually constructive relationship between aspects; the higher the contribution of each aspect, the more excellent the digital literacy skills of prospective physics teacher students. This conclusion is supported by the acquisition of calculations that show a sufficient correlation value between aspects of digital literacy.

References

[1] Silvana H and Darmawan C 2018 *Pedagogia* **16** 146
[2] Febliza A and Oktariani 2020 *Jpk Unri* **5** 1–9
[3] Goodfellow R and Lea M 2016 *Literacy and the Digital University* (The SAGE Handbook of e-learning Research)
[4] Syarifudin A S 2020 *J. Pendidik. Bhs. dan Sastra Indones. Met.* **5** 31–4
[5] Qadar R, Mahdiannur M A and Sulaeman N F 2021 *Cypriot J. Educ. Sci.* **16** 1154–74
[6] Campbell T, Melville W, Verma G and Park B Y 2021 *J. Sci. Teacher Educ.* **32** 1–6

[7] Rahayu, Tantri Mayasari F 2018 *Seminar Nasional Pendidikan Fisika* 141–8
[8] Peled Y 2020 *Educ. Inf. Technol.* **26** 2876-96
[9] Aditya D S 2021 *J. Technol. Sci. Educ.* **11** 104–16
[10] Efwinda S and Mannan M N 2021 *J. Phys. Conf. Ser.* **1806**
[11] Fraenkel, J. R., Wallen, N. E., Hyun H H 2012 *How To Design And Evaluate Research In Education* (New York)
[12] Etikan I 2016 *Am. J. Theor. Appl. Stat.* **5** 1
[13] Idris D 2007 *J. Mat. Stat. dan Komputasi* **4** 26–9
[14] Silverblatt A 2018 *Information Literacy in the Digital Age*
[15] Wheeler S 2013 *eLC Researcj Pap. Ser.* 14–25
[16] Belshaw D 2011 *EdD Thesis, Durham Univ. Durham*

Analysis of problem-solving ability of physics education students in STEM-based project based learning

E Susanti, R Maulidah and Y S Makiyah

Physics Education, Siliwangi University, Tasikmalaya, Indonesia
ernita.susanti@unsil.ac.id

Abstract. This study is a quantitative descriptive. This study aimed to describe the problem-solving abilities of physics education students in the 2nd School Physics Laboratory course. The subjects of this study were 65 people consisting of 2 classes with the determination of the sample using the purposive sampling technique. Data collected is done through a problem-solving ability instrument which includes four indicators recognizing problems, planning solutions, solving problems, and evaluating. Based on the results of the data analysis, the conclusion is that students' problem-solving abilities have been at a high level with a percentage average of 79. The caused by learning that applies STEM-based Project-Based Learning (PjBL). The learning helps develop students' abilities in planning, communicating, solving problems, and making the right decisions from the problems given. This learning can train students' skills in planning, organizing, negotiating, and making conventions about the issues of the task to be done, who is responsible for each task, and how the information will be collected and presented.

1. Introduction

Research on the ability to solve physics problems has become the focus of research in recent years. There is because 21st-century education demands various abilities that must be mastered and possessed by prospective 21st-century teacher students, one of which is the ability to solve problems [1]. One of the expected targets of modern education is to educate student teacher candidates who can solve the problems they encounter in everyday life and social life. As prospective teachers, they will have expected to solve the problem they find easily [2]. The ability to solve problems is one of the 21st-century skills that are the focus of physics learning. That needs to will be given to physics education students as professional physics teacher candidates [1].

Problem-solving ability is a fundamental thing that must be owned by students as prospective teachers [3]. Students are said to be successful in the learning process if they can solve problems using basic knowledge. Problem-solving is not only emphasized quantitative aspects such as mathematical equations and procedures. But also emphasizes qualitative analysis aspects in the form of choosing the right concepts and principles in solving problems [4]

Problem-solving ability is a person's ability to find solutions through a process that involves obtaining and organizing information [5]. In the problem-solving process, students need to find solutions based on concepts/laws rationally and reflect on the problem-solving process and solutions [6]. In physics learning, physics problem solving is related to physics concepts [5]. The factors that influence solving physics problems are the knowledge structure possessed by students and the character of the problem faced by students [7].

One of the significant components in the problem-solving of physic is to identify the physics principles that are relevant to the problem [8]. Based on students' ability problem solve that can divide into two, namely those with high abilities and low abilities. Students who have high ability tend to use arguments based on physics concepts. They tend to evaluate solutions and tend to use representational aids. On the contrary, students who have low ability tend to use quantitative arguments such as the use of formulas in solving physics problems [5,9].

Outline of physics problem-solving consists of recognizing problems, implementing strategies, planning strategies, and evaluating solutions [5]. These steps are modified based on the problem-solving steps Young and Freedman and Heller et al. The indicators for each of these steps can be seen in Table 1. These indicators are modified from [5], [10]. This indicator indicates the physics problem-solving ability used in this study [5,10].

Table 1. Stages and Indicators of Physics Problem Solving Ability

Stage	Indicator
Recognizing the problem	Identify problems based on basic concepts
Planning strategy	Determining the right concept for problem-solving
	Create a troubleshooting trial procedure
Implementing strategy	Solve problems according to plan
	Perform data analysis based on experimental results
Evaluating solution	Draw conclusions from the answers obtained and re-check the results obtained

One form of learning that involves a problem-solving process is Science-Technology-Engineering-Mathematics (STEM) Education-based learning. STEM is an approach used by integrating Science, Technology, Engineering, and Mathematics in the learning process. In this case, physics represents the science studied by prospective teacher students. The application of technology is a supporting tool to help students understand the concepts being studied. The application of engineering aims to train students to design, assemble, draw, and of other activities so that students understand the procedures for solving problems. Furthermore, mathematics aimed to simplify the concept of science itself in a more systematic and mathematical form.

STEM learning provides opportunities for students to learn science, mathematics, and engineering by solving problems that have real-world applications [11]. Application STEM in learning can be an alternative to science learning that can encounter 21st-century education [12]. STEM in learning can encourage students to design, develop, utilize technology, sharpen cognitively, and apply their knowledge[13]. STEM can also train students to use their knowledge to create problem-solving designs related to the environment by utilizing technology [12].

STEM learning emphasizes several aspects of the learning process, namely (1) asking questions and defining problems; (2) developing and using models; (3) planning and conducting investigations; (4) analyzing and interpreting the data; (5) using mathematics; information and computer technology; and computational thinking; (6) building explanations (science) and designing solutions (engineering); (7) engage in evidence-based arguments; (8) obtain, evaluate, and communicate information [14]. The application of STEM in learning can be supported by various methods and learning models. That is because STEM is integrative. Many research results explain one of the learning models that can use in STEM learning, namely PjBL. PjBL can make students apply their academic abilities and creativity to solve given problems [15].

PjBL is a learning model based on a problem that involving students in design, problem-solving, decision making, or investigative activities [16]. PjBL can guide students to solve the problems given and more emphasis on the products produced. The products produced by students can be in the form of ideas or devices. Problem-solving and product solving can be done by individuals or groups. Working in groups can encourage students to work together but still be responsible for their work independently. In addition, in groups, students can manage learning independently according to the circumstances of their respective groups [16,17]. In the finishing of products, students can use science and technology. So that students indirectly understand the functions and benefits of science and technology in everyday life [12].

A further characteristic of PjBL is that students can choose the activities and work carried out during the learning process. Students can be communicative and creative in developing practical thinking as they engage in inquiry, active discovery, exploration, and decision making. Knowledge is based on experience and experimentation in real life [17]. In addition, PjBL makes learning meaningful by connecting new learning with the period of experience they have. PjBL is also an inclusive approach, where all students can participate to the best of their ability[18].

Based on the preliminary research in have the shape of learned observations, that known that previously learned had not trained problem-solving skills. The learning is done by explaining the physics phenomena that occurred and concepts used to solve and underlie these phenomena. So that students are less trained to recognize and solve the problems given. Therefore, researchers are interested in applying the STEM-based PjBL model to practice the problem-solving skills of physics education students. This study aimed to describe the problem-solving abilities of physics education students in the School Physics Laboratory 2 course.

The use of PjBL will have expected to be an alternative science learning that can face 21st-century education. A combination of STEM and PjBL will have expected to actualize the problem-solving abilities of physics education students as prospective physics teachers. This thing has must be trained for pre-service teachers because pre-service professional teachers must have skills in problem-solving so that they can design interested and meaningful learning.

2. Methods

This research is quantitative descriptive. Descriptive research aims to describe, interpret or explain a variable or situation. This study aims to describe the problem-solving abilities of physics education students in the School Physics Laboratory 2 course. The subjects of this research were physics education students of 2018, totaling 65 people. The research subjects consisted of 2 classes with the determination of the sample using the purposive sampling technique. The time of the research was have carried in the even semester of 2019/2020.

The data collecting technique was have carried through a problem-solving ability instrument. The instrument includes four indicators, namely recognizing problems, planning solutions, solving problems, and evaluating. Data analysis used a questionnaire was have analyzed with the percentage score generated from the Likert scale calculation. The Likert scale was have used to measure a person's attitudes, opinions, and perceptions about an event by giving a score with a predetermined weight. The average percentage was determined by Equation 1.

$$P = \frac{f}{N} x100\% \qquad (1)$$

P is the percentage average, C is the collected score by students' perception, and M is the maximum score of students' perceptions. The category of students' perception can be confirmed in Table 1 [19]

Table 2. The Category of Students' Satisfaction

Percentage average	Criteria
85 – 100	Very high
70 –84	High
55 – 69	Moderate
40 – 54	Low
0 – 39	Very low

Based on Table 2, the results of the analysis are categorized as good if they have a value of 70.

3. Result and Discussion

The research was have done in the School Physics Laboratory 2 course. Students were given simple problems but contextual with the problem that occurs in everyday life. These problems are in the form of simple physical phenomena that occur in everyday life. Furthermore, students were have asked to solve these problems using the PjBL model. One example of the problems given to students has shown in Figure 1.

Journal of Physics: Conference Series **2104** (2021) 012005 doi:10.1088/1742-6596/2104/1/012005

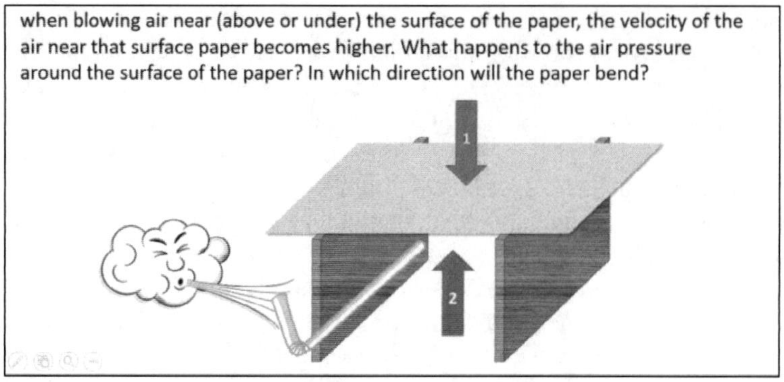

Figure 1. Examples of Problems

In essence, the project-based learning model is designed to solve complex problems. This model is needed in conducting investigations and understanding problems[20]. Implementation of PjBL in the learning by dividing the students into several groups. Each group consists of eight people. Working in groups can encourage students to work together but remain responsible for their work. In addition, students can manage learning independently so that according to the circumstances of each group member [16,17].

After being given a problem, students were have asked to solve the problem. Students are allowed to identify as many problem agendas as possible that are relevant to the given problem. Students were allowed to collect relevant information by reading literature, observing objects, interviewing resource persons, or conducting their trials. Furthermore, students are given independence and flexibility to design project plans that will carry out. Students work in groups to plan how their project will carry out.

In planning the completion of the project, students prepare a schedule so that the project can be completed on time. At this stage, students' problem-solving skills are drill to critical thinking and good at estimating what is needed to be done for preparation, manufacture until the project can be complete on time. The next step after the preparation of the schedule is to monitor the progress of the project. That aims to see whether the project has been running according to plan or not. In completing the project, students can take advantage of science and technology so indirectly that students understand the functions and benefits of science and technology in everyday life [12]. The utilization of science and technology can train students in applying their knowledge to solve problems related to the environment by utilizing technology.

In implementing the PjBL model, process evaluation and evaluation of learning outcomes do out. Evaluations of Processes are carried out through student progress reports regarding the projects that are working. Evaluation of results does out through a problem-solving ability instrument which includes four indicators. That is recognizing problems, planning solutions, solving problems, and evaluating. The results of the analysis of the problem-solving abilities of physics education students are shown in Table 3.

Table 3. Results of Problem-solving Ability Analysis

Stage	Indicator	Percentage	Criteria
Recognizing the problem	Identify problems based on basic concepts	84	High
Planning strategy	Determining the right concept for problem-solving Create a troubleshooting trial procedure	86	Very high
Implementing strategy	Solve problems according to plan Perform data analysis based on experimental results	77	High
Evaluating solution	Draw conclusions from the answers obtained and re-check the results obtained	69	Moderate
	Average	79	High

Based on Table 3, the ability of students to recognize the given problem is at a high level. That is because Students were given simple problems but contextual with the problem that occurs in everyday life. So that students can identify problems based on physics concepts. The ability to plan strategies is at a high level. At this stage, students can determine what concepts will be used to solve the given problem and can make experimental procedures to solve the problem. Students can sequence the work steps and solutions that will be carried out and determine who is responsible for implementing these solutions. Furthermore, the stage of the ability to apply the strategy is at a high level. At this stage, students can solve problems according to the experimental procedures they have compiled. At this stage, students work together in solving problems but remain responsible for their respective work. After conducting the experiment and obtaining the data, the students then analyzed the data. At this stage, the students' ability to analyze data is at a high level.

The stage of the ability to evaluate the solution is at the moderate level. At this stage, the ability students to conclude the answers obtained and re-check the results in moderate levels. That is because students only conclude the results of their experiments, have not connected these results with the physics concepts used, and do not relate them to the purpose of the experiments.

Based on the results of data analysis, it has been concluded that the problem-solving ability of physics education students in STEM-based PjBL learning is at a high level. It is in accordance with previous research [20] and [21] who stated that the PjBL model could improve students' problem-solving abilities. Another study conducted by [22] showed that learning with a STEM approach was able to improve students' problem-solving abilities.

STEM-based PjBL learning provides opportunities for students to solve real-world problems by applying science, math, and engineering. In addition, students actively engage in ill-defined tasks become well-defined outcomes through collaboration in groups [11,23]. STEM-based PjBL learning allows students to solve problems through prototyping and science-based design. When students have trained to solve a problem using the four STEM disciplines, students are encouraged to be more scientifically creative in solving problems [24]. Problem-solving abilities built by students will put students at an advantage in utilizing their potential as well as possible.

This STEM-based PjBL learning ends with a presentation of the results of project assignments that students have done. Based on the results of the project presentation, the researcher assessed that the students' ability to use technology is at a high level. It looks from the presentation of project results using video editing when explaining the problem-solving procedure.

4. Conclusion

Based on the results of the study can be concluded that the problem-solving abilities of physics education students are at a high level. The problem-solving ability carried out includes four indicators, namely a) recognizing problems, b) planning strategies, c) implementing strategies, and d) evaluating solutions. At the stage of recognizing the problem, students' abilities are at a high level. At the planning, the strategy, the student's abilities are at a very high level. At the stage of implementing the strategy, students' abilities are at a high level, and at the evaluating solutions, students' abilities are at a moderate level. STEM-based PjBL learning can train students to apply their knowledge to solve problems related to the environment by utilizing technology. In addition, it encourages students to work together but remains responsible for their respective work. STEM-based PjBL learning is recommended to use in other subjects or courses that have the same characteristics as the School Physics Laboratory 2 course.

References

[1] Jayadi A, Putri D H and Johan H 2020 *J. Kumparan Fis.* **3** 25–32
[2] Sezgin Selçuk G and Çalýskan S 2008 *Latin-American J. Phys. Educ.* **2** 151–66
[3] Rizqa A and Harjono A 2020 *ORBITA: J. Kajian Inov. Aplikasi Pendidik. Fis.* **6** 243–7
[4] Setianingrum L, Perno and Sutopo 2016 Analisis Kemampuan Pemecahan Masalah Fisika Siswa SMK 37–42
[5] Sujarwanto E, Hidayat A and Wartono 2014 *J. Pendidik. IPA Indones.* **3** 65–78

[6] Riantoni C, Yuliati L, Mufti N and Nehru N 2017 *J. Pendidik. IPA Indones.* **6** 55–62
[7] Chi M T H and Glaser R 1985 *Hum. Abil. An Information-Processing Approach* 227–57
[8] Lin S Y and Singh C 2013 *Phys. Rev. Spec. Top. - Phys. Educ. Res.* **9** 11–9
[9] Mason A and Singh C 2011 *Phys. Rev. Spec. Top. - Phys. Educ. Res.* **7** 1–17
[10] Rosid M A and Listyani E 2014 *J. Pendidik. Mat. dan Sains* 1–13
[11] Quinn H, Schweingruber H, Keller T, Framework C, Science N K- and Standards E 2012 *A framework for K-12 science education: Practices, crosscutting concepts, and core ideas* (National Academies Press)
[12] Permanasari A 2016 *Semin. Nas. Pendidik. Sains* 23–34
[13] Kapila V and Iskander M 2014 *J. STEM Educ. Innov. Res.* **15** 46–51
[14] Muyassarah A, Ratu T and Erfan M 2019 *Pros. SNFA (Seminar Nas. Fis. dan Apl.* **4** 1
[15] Holubova R 2008 *Online Submission; US-China Educ. Rev. v5 n12 p27-36 Dec 2008* **5** 27–36
[16] Panasan M, Nuangchalerm P and Muang A 2010 *J. Soc. Sci.* **6** 252–5
[17] Filippatou D and Kaldi S 2010 *Int. J. Spec. Educ.* **25** 17–26
[18] Westwood P 2006 *Teaching and Learning Difficulties: Cross-curricular Perspectives* (Camberwell, Vic: ACER Press)
[19] Riduwan 2008 *Belajar Mudah Penelitian untuk Guru-Karyawan dan Peneliti Pemula* (Bandung: Alfabeta)
[20] Nurfitriyanti M 2016 *Formatif: J. Ilmiah Pendidik. MIPA* **6** 149–60
[21] Makrufi A, Hidayat A and Muhardjito M 2018 *J. Pendidik. Teor. Penelitian, dan Pengemb.* **3** 878–81
[22] Dewi M, Kaniawati I and Suwarma I R 2018 *Quantum Semin. Nas. Fis. dan Pendidik. Fis.* **0** 381–5
[23] Han S, Capraro R and Capraro M M 2015 *Int. J. Sci. Math. Educ.* **13** 1089–113
[24] Siew N M and Ambo N 2018 *J. Balt. Sci. Educ.* **17** 1017–33

Fostering TPACK in pre-service physics teachers during the covid-19 pandemic

S A Putri[1], N F Sulaeman[1], P Damayanti[1] and P D A Putra[2*]

[1]Physics Education Study Program, Faculty of Teacher Training and Education, Mulawarman University, Indonesia
[2]Science Education Study Program, Faculty of Teacher Training and Education, University of Jember, Indonesia

*pramudya.fkip@unej.ac.id

Abstract. The pandemic situation forces the educational system to respond to a variety of issues. During the pandemic, it's essential to learn more about Technological-Pedagogical-Content-Knowledge (TPACK) and how it can be incorporated into pre-service physics teacher preparation. This qualitative research collects data from ten online teaching videos and ten offline teaching videos of 20 pre-service physics teachers. This study attempts to comprehend the appearance of each TPACK component by analyzing the videos with an observation sheet. The observation sheet explored seven components of TPACK, which are Content Knowledge (CK), Pedagogical Knowledge (PK), Technological Knowledge (TK), Pedagogical and Content Knowledge (PCK), Technological and Content Knowledge (TCK), Technological and Pedagogical Knowledge (TPK), and the last is Technological, Pedagogical and Content Knowledge (TPACK) itself. The result showed that pre-service teachers were determined good TPACK skills, but lacked in certain components. From the result, in both online and offline microteaching, participants rarely showed their TPK. In online activities, it is also seen that participants were faced difficulties in applying their PCK. The different nature of online and offline microteaching drives the need for teacher education institutions to plan and implement microteaching in the pedagogical courses to improve pre-service teachers' TPACK in both situations.

1. Introduction

The COVID-19 pandemic has greatly affected higher science education which caused offline learning on campus to be banned [1,2]. Though it is too early to accurately predict the widespread implications for higher education, as the scope and consequences are immense and unclear [1], education systems need to adapt to this new situation in scientific ways. Online learning seems to be the best alternative considering the current health emergency [3]. Before the pandemic, the implementation of online learning was insignificant [2], especially in developing countries such as Indonesia.

Science teacher education programs have now offered online classes for pre-service teachers. The positive result of the pandemic is that with advanced technology, online interaction between lecturers and pre-service teachers has become more frequent than ever [1]. There are still challenges to overcome. One of them is how to teach these pre-service teachers field-based experiences [4]. Microteaching is one of the courses that are best conducted with traditional face-to-face learning. Therefore, the outcome of the microteaching course could be different if done online. This becomes a worrisome issue considering microteaching is a course that helps to develop pre-service teachers' TPACK [5,6] It is essential for teachers to create technical knowledge and incorporate it into content, teaching, and learning in a specific context [7]. Therefore, the ability of pre-service teachers to integrate technology, pedagogy, and content, or TPACK, is strongly needed [6].

TPACK ability is a crucial part of ensuring the proper conduct of online learning during a pandemic [8,9]. Learning activities are based on knowledge of the material to be taught (content knowledge), knowledge of how to teach a material, and knowledge of how to use various technologies, all of which have intersections or intersections to support each other [6] and pre-service teachers need master this knowledge [10]. TPACK is a framework for integrating Technological Knowledge, Pedagogical Knowledge, and Content Knowledge in a learning context. TPACK came from PCK described by Shulman. As science and technology have evolved, so has thought on how teachers' understanding of learning technology is tied to PCK to support successful learning using technology [10]. TPACK aims to define the knowledge teachers need while integrating it in their teaching, which includes seven components [11]. The TPACK diagram includes three core categories of knowledge, including pedagogical knowledge (PK), content knowledge (CK), and technological knowledge (TK), in which these three cores are then combined to create the other components that are Pedagogical Content Knowledge (PCK), Technological Content Knowledge (TCK), Technological Pedagogical Knowledge (TPK), and Technological Pedagogical and Content Knowledge (TPACK) itself [12].

Pre-service teachers with high TPACK skills have an increased chance of achieving their programs' goals to become well-polished teachers [11]. Professional competence is the teacher's ability to master the subject matter broadly and deeply [13]. Many pieces of research have been done to examine TPACK, such as [12] which investigated the role of TPACK in the physics classroom, [14] which analyzed technological pedagogical content knowledge self-confidence of prospective pre-school teachers during the pandemic, and [6] who investigated teaching ability of prospective teacher in terms of TPACK in microteaching courses. However, little has been researched to investigate the fostering of pre-service physics teachers' TPACK during the pandemic. Therefore, this research was carried out to find the TPACK abilities of prospective physics teachers in online and offline microteaching.

2. Method

This qualitative research investigated 20 physics pre-service teachers' TPACK during the COVID-19 pandemic in online and offline learning. The sample selection in this research is purposive sampling. Puposive sampling relies on the researcher's judgment for selecting persons [15] with characteristics that are defined for purpose that is relevant to the study [16]. Purposive sampling procedures differ from random sampling in that they ensure that particular types of instances from those who could potentially be included are included in the final sample in the research study. A relatively small and purposefully selected sample may be used in a qualitative study to increase the depth of understanding [17]. These pre-service teachers were purposively selected among the final year students of Physics Education, Mulwarman University, and made a microteaching video about teaching a specific physics topic. These samples are chosen considering the fact that Physics Education in Mulwarman University conducted the microteaching course in the academic year 2020/2021. The microteaching course was conducted both online and offline so that it is in accordance with the purpose of this research. The microteaching videos were then used to identify the pre-service teachers' skills within the TPACK framework. Preliminary analysis is carried out to classify the videos based on the topics taught, duration and whether it's done online or offline.

This research instrument was investigating seven components of TPACK, which were Content Knowledge (CK), Pedagogical Knowledge (PK), Technology Knowledge (TK), Pedagogical and Content Knowledge (PCK), Technological and Content Knowledge (TCK), Technological and Pedagogical Knowledge (TPK), and Technological, Pedagogical and Content Knowledge (TPACK). The observations were made using observation sheets with two Likert Scales arranged based on the TPACK indicator, which allows observation via video. From the data obtained, data analysis was carried out by calculating the average TPACK value for each component. The observations were processed in the form of scores, and the average scores were analyzed using Microsoft Excel. If the average score is expressed in terms of values, it can be found through the following calculation:

The average value of TPACK skill $= \frac{\text{average score}}{\text{maximum score}} \times 100$ [10]

Then, the comparison of TPACK skills of the pre-service teachers based on online and offline learning is conducted.

3. Result and Discussion

The distribution of 20 microteaching videos is as follows:

Table 1. Microteaching videos distribution

Code name	Learning Type	Lesson Topic	Duration
DR	Online	Moment of Force (Torque)	20 m 30 s
MHS	Online	Work and Energy	23 m 33 s
FFM	Online	Kirchoff's Second Law	19 m 32 s
DS	Online	Archimedes' Principle	20 m 16 s
HN	Online	Moment of Force (Torque)	25 m 43 s
SDTM	Online	Pascal's Law	24 m 25 s
NP	Online	Optical Instrument (Magnifying Glass)	15 m 52 s
Z	Online	Hydrostatics	19 m 28 s
MZOA	Online	Electromagnetic Induction	21 m 19 s
RTA	Online	Mechanical Wave	20 m 50 s
DPA	Offline	Thermal Expansion	27 m 4 s
DO	Offline	Moment of Inertia	23 m 14 s
LJ	Offline	Uniform Circular Motion	28 m 42 s
MY	Offline	Uniform Linear Motion	25 m 47 s
NR	Offline	Heat and Black Principle	26 m 14 s
PAR	Offline	Impulse and Momentum	19 m 29 s
RS	Offline	Hooke's Law	24 m 33 s
ST	Offline	Work and Energy	24 m 36 s
RA	Offline	Uniformly Accelerated Linear Motion	27 m 6 s
FDS	Offline	Surface Tension	19 m 41 s

This study is essential to understand how teachers can apply technological, pedagogical, and content knowledge (TPACK) into online and offline learning during COVID-19. The seven constructs of TPACK are comprehensive because they recognize the interaction between content, pedagogy, and technology. In this study, 20 final-year physics students from Mulawarman University, Samarinda, were observed by analyzing the learning videos they have made. Each component of TPACK data collection was then broken down further, including Content Knowledge (CK), Pedagogical Knowledge (PK), and Technological Knowledge (TK), Pedagogical Content Knowledge (PCK), Technology Content Knowledge (TCK), Technology Pedagogical Knowledge (TPK), and Technology Pedagogical and Content Knowledge (TPACK). The pre-service teacher' TPACK skills is shown in Figure 1.

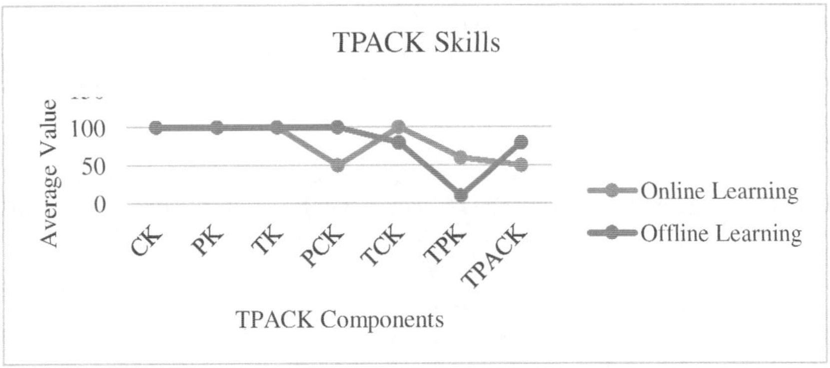

Figure 1. Graph of Physics Pre-service teachers' skills in TPACK components

Out of seven components measured, Content Knowledge (CK), Pedagogical Knowledge (PK), and Technological Knowledge (TK) are the ones that have the highest average value of 100. This result shows that pre-service teachers both in online and offline learning master learning physics content and skills in using technology and creating an effective learning environment. However, pre-service teachers in online learning's average value suddenly dropped in Pedagogical Content Knowledge (PCK) with the value 50, the videos show that the pre-service teachers still struggle to choose learning methods based on the topics taught. They barely had any discussion with the students and corrected students' perceptions on the topics. In contrast, the offline learning side still stands tall with their 100 value. Most of the pre-service teachers chose to conduct physics practice in the laboratory to assist students in gaining a better understanding of the lesson. This allowed them to have a deeper discussion with the students and help them to achieve the purpose of the practice. This result indicates the research in developing exemplary physics lessons is needed. Pre-service teachers need to explore more the implementation of specific teaching model in specific topics, such as in STEM education [18].

In Technological Content Knowledge (TCK) domain, pre-service teachers in online learning show an increase in their value of 100. It is reasonable since in online learning, they have to rely in technology to present the content. All of them have at least used PowerPoint to show the material, and few others have also used virtual laboratory, whereas almost none of the pre-service teachers in offline learning use any technology to present their content. Though the tools used to conduct the physics practice were some of the most recent technologies, explains how the pre-service teachers. The TCK domain is strongly influenced by the CK and TK domain. Pre-service teachers that have a low CK domain will address low conceptualization in their teaching [19]. And the next component which is Technological Pedagogical Knowledge (TPK), the average value of pre-service teachers in online and offline learning has dropped significantly which are 60 and 10 respectively. They failed to integrate technology into their learning environment as shown by taking attendance of the students manually, whereas they could use Google Form, Microsoft Form, or any Learning Management System (LMS). The pre-service teachers also rarely put assignments into a platform so the students can submit their works efficiently.

In the last component which is Technological Pedagogical and Content Knowledge (TPACK) itself, the value of pre-service teachers' in online learning has dropped once again to 50, it shows that the teachers' can not apply the appropriate technology in assessing the learning process and outcomes. In the other hand, the pre-service teachers' in offline learning has redeem themselves with a higher value of 80. This means they have more understanding in applying the appropriate technology in assessing the learning process and outcomes.

The results demonstrate that there is a different pattern of TPACK skills between pre-service teachers that teach in online learning and offline learning. The average value of pre-service teachers in online learning has decreased in PCK (Pedagogical Content Knowledge), where videos demonstrate that pre-service teachers continue to struggle to pick learning methods depending on the topics presented. Whereas almost no pre-service instructors in offline learning utilize any technology to convey their curriculum under the TCK (Technological Content Knowledge) domain. In the TPK (Technological Pedagogical Knowledge) domain, the average value of pre-service teachers in online and offline learning has considerably decreased. They neglected to incorporate technology into their teaching and learning environment. It should be emphasized that pre-service teachers have distinct areas that need to be addressed in both online and offline learning. It is not surprising given that the sessions are held in different settings; thus, the way pre-service teachers perform may also be impacted.

4. Conclusion

From this research results, it can be concluded that the physics pre-service teachers in both online and offline learning have overall good TPACK skills, though the pre-service teachers in offline learning has a slightly higher average value. The pre-service teachers in both conditions of lack the skills in TPK component. This shows that they need more practice to integrate technology in learning. The pre-service teachers that conducted online learning especially lacked the skills in PCK that show that they have yet to learn more about creating the right learning environment for a certain topic. These findings highlight the importance of expanding courses that address components that are deficient in the learning processes of prospective physics teachers.

References

[1] Yang R 2020 *High. Educ. Res. Dev.* **39** 1317–21
[2] Mengistie T A 2021 Context *Res. Glob.* **3**
[3] Erduran S 2020 *Sci. Educ.* **29**
[4] Campbell T, Melville W, Verma G and Park B Y 2021 *J. Sci. Teacher Educ.* **32** 1–6
[5] Mutlu Y, Polat S and Alan S 2019 *Int. J. Technol. Educ. Sci.* **3** 107–18
[6] Turmuzi M and Kurniawan E 2021 *J. Cendekia J. Pendidik. Mat.* **5** 2484–98
[7] Juanda A, Shidiq A S and Nasrudin D 2021 *J. Pendidik. IPA Indones.* **10** 48–59
[8] Sahrir M S, Razif Zaini A, Hamid M A, Hamat Z and Ismail T 2021 *J. Contemp. Issues Bus. Gov.* **27**
[9] Makawawa J C, Mustadi A, Septriwanto J Van, Sampouw F and Najoan R A . 2021 *J. Prima Edukasia* **9** 86–96
[10] Efwinda S and Mannan M N 2021 *J. Phys. Conf. Ser.* **1806**
[11] Koyuncuoglu Ö 2021 *Int. J. Educ. Math. Sci. Technol.* **9** 299–313
[12] Srisawasdi N 2012 *Procedia - Soc. Behav. Sci.* **46** 3235–43
[13] Santika V, Indriayu M and Sangka K B 2020 *IOP Conference Series: Earth and Environmental Science* vol 1808
[14] Gozum A I C and Demir Ö 2021 *Int. J. Curric. Instr.* **13** 712–42
[15] Berndt A E 2020 *J. Hum. Lact.* **36** 224–6
[16] Andrade C 2021 *Indian J. Psychol. Med.* **43** 86–8
[17] Campbell S, Greenwood M, Prior S, Shearer T, Walkem K, Young S, Bywaters D and Walker K 2020 *J. Res. Nurs.* **25** 652–61
[18] Hartini S, Mariani I, Misbah M and Sulaeman N F 2020 *J. Phys. Conf. Ser.* **1567** 42029
[19] Putra P D A and Kumano Y 2018 *New Educ. Rev.* **53** 153–62

The influence of problem based learning model and critical thinking ability on higher order thinking skills (HOTs) of physics prospective teachers students

Nurhayati, Wahyudi*, and L Angraeni
Physics Education, IKIP PGRI Pontianak, Indonesia

wahyudi.kakap@gmail.com

Abstract. This study aims to determine the effect of applying problem-based learning models, critical thinking skills, and their interactions on students' higher-order thinking skills in optical materials. The research method used is a quasi-experimental method with a two-way ANOVA design. This research was conducted at IKIP PGRI Pontianak in the second-semester students of the physics education study program as many as 34 students. Data on students' higher-order thinking skills were obtained using multiple-choice tests with reasons, while data on students' critical thinking skills were measured using multiple-choice tests. Based on the results of data analysis, it can be concluded that: The ability of prospective physics teacher students to complete the HOTS-based physics test is better after the problem-based learning model is applied. Future physics teacher students with high critical thinking skills have better performance in meeting HOTS-based physics tests. There is an interaction between applyingicip2021est the problem-based learning model and the essential skills of thinking of prospective physics teacher students in completing the HOTS-based physics test. Implementation of the problem-based learning model is vital for other physics materials. In addition to improving the understanding of physics concepts, it also enhances higher-order thinking skills for prospective physics teacher students.

1. Introduction

The education sector has a vital role in the development of a nation. An advanced country is supported by education that can develop the potential possessed by students so that students can independently solve the problems they face. Furthermore, the importance of the concept of education that is learned will be felt if one enters social life, especially when one enters the world of work because the ability to apply concepts that have been discovered in school will be used in dealing with the problems of everyday life both useful now and in the future [1]. Therefore, education must be managed professionally by individuals who have superior competence in the field of education.

However, based on the results of observations and interviews conducted with several student teacher candidates, information was obtained that students were less independent and had difficulty solving problems when faced with a problem. One of the causes of these problems is the application of learning that does not focus on students. The paradigm that the lecturer is the only primary source of knowledge who is an expert in his field is still firmly held by students so that students do not get a good process and understanding of knowledge. Knowledge is only obtained from lecture notes. The lack of student involvement in the learning process causes students to become independent. As a result, the thinking process is not carried out properly so that a meaningful learning process is not achieved.

Prospective physics teacher students must master physics content knowledge. Physics is the most fundamental physical science and deals with the basic principles of the Universe. Physical science is the foundation on which other sciences such as astronomy, biology, chemistry, and geology form the basis for most engineering applications [2]. Physics is also a science-based on experimental observations and measurements of physical quantities [3]. Studying physics will be challenging, sometimes frustrating, sometimes painful, and often very rewarding [4]. As a fundamental science, physics is helpful for understanding applied science and as a basis for technological development. In studying physics, students are given knowledge and experience of physics concepts. In addition, students are trained to be skilled in scientific work to solve various problems in everyday life [5]. In other words, the purpose of studying physics is to provide knowledge and understanding of physics concepts, apply them in solving various problems of daily life, and communicate them well in society and the world of work. Then, the implementation of learning physics is also expected to increase the development of science and technology, environmental conservation, and Indonesia's natural wealth.

Introductory Physics Lecture is one of the subjects that prospective physics teacher students must take. This is intended to master the basic concepts of physics to equip students to learn more complex physics material or concepts. However, the first-year students still have difficulty understanding these basic concepts. Physics concepts that are considered difficult for students include visual concepts. Based on research conducted by Ref [6,7], shows that students experience misconceptions about visual concepts. The misconceptions found include the concepts of (1) image formation on concave mirrors, convex mirrors, plane mirrors, and concave lenses; (2) identifying the properties of light; and (3) determining the image distance, image location, and image properties on a convex lens. Stead also carried out the same research, and Osborne [8], whose results showed that many students thought that light was stationary (not propagated) or only propagated if it was in a bright place, the speed of light propagation did not depend on the light source but only on the light propagation medium. The difficulties experienced by students certainly result in low learning outcomes [5]. This is indicated by the GPA results obtained by prospective teacher students, which are less than three on average.

Based on the results of initial interviews with lecturers at a teacher teaching institution, four factors are suspected to be the cause of the difficulty of prospective physics teachers in understanding physics concepts, including (1) common understanding of abstract physics concepts; (2) the primary physics learning method applied seems informative and passive so that students are not trained to improve high-level critical thinking skills according to their competencies; (3) quizzes with structured feedback have never been carried out in introductory physics lectures; (4) students who take fundamental physics lectures are very heterogeneous in terms of initial abilities, learning speed, interest, and motivation to learn. In addition, the results of interviews with several prospective physics teacher students who have taken fundamental physics courses show that most students are challenged to solve problems, especially optical materials. They said that they were able or able to understand the material presented by the lecturer. Still, when they were given more complex visual material issues, they had difficulty determining the solution to the problem. This is undoubtedly an indication that the learning being carried out is still limited to the transfer of knowledge from lecturers to students, not how students learn to find concepts. This causes the concepts transferred to students to be short-lived and sustainable. As a result, students will have difficulty relating the concepts they have acquired to understand new physics concepts.

The level of the intellectual development of students in higher education is already at the level of high-level formal operations. According to Bloom's taxonomy (revised version), cognitive processes are divided into lower-order thinking and higher-order thinking. Students' ability at the low-level thinking level is the ability at the cognitive level to remember, understand and apply. In contrast, students' ability at the higher-order thinking level is the ability at the cognitive level to analyze, evaluate and create [9]. So, it can be said that higher-order thinking skills are also called Higher Order Thinking, namely the ability to analyze, evaluate, and create. According to Ref [10], higher-order thinking is; (1) thinking skills that are at the top of Bloom's cognitive taxonomy, (2) abilities whose teaching objectives behind the cognitive taxonomy can equip students to transfer knowledge, (3) part of thinking skills to apply the knowledge and skills developed during learning in a new context. The new context in question is the application of concepts that students did not previously think about. Still, the concepts have been taught, so it is not necessarily something new universal. For students, higher-order thinking can also be interpreted as a student's ability to connect concepts that have been learned with other things that have never been taught.

The relevant learning model to assist and facilitate and facilitate students in mastering the concepts of physics is the problem-based learning model. Through this learning model, students can develop thinking process skills [11]. According to Ref [12], the Problem Based Learning (PBL) model is a student-centered learning model that emphasizes collaboration in solving problems contextually. The problem-based learning model is considered to be able to help students in developing their ability to understand physics concepts because this model has the following characteristics: (1) The learning process starts from a problem related to student's daily lives, (2) Organize the concepts (materials) that are studied based on problems, not around scientific disciplines, (3) Train students to be responsible for shaping and carrying out the learning process directly and independently, (4) Presentation of learning in small groups, and (5) Guiding students to create products or performances from the concepts they have learned.

In other words, learning through the problem-based learning model begins with the disclosure of problems found in student life, where issues can be raised from students or lecturers. Then, students seek and determine methods to solve problems by deepening the knowledge and what needs to be known in solving problems. Problems so that they are motivated and active in learning. Issues that are used as the focus of learning can provide diverse learning experiences for students, such as collaboration and interaction in groups, in addition to learning experiences related to problem-solving such as making hypotheses, designing experiments, conducting investigations, collecting data, interpreting data, making conclusions—presenting, discussing, and making reports. In physics learning, especially optical material, problem-based learning models are expected to increase students' understanding of the relationship between knowledge about optics and the real world. Meanwhile, familiarize students with critical thinking and problem-solving skills, acquire essential knowledge and concepts from the subject matter, create a democratic classroom environment, and effectively deal with student diversity.

Based on the above background, the focus of the problems in this study are (1) Is there any effect of the application of the problem-based learning model on the ability of prospective physics teacher students to solve HOTS-based physics tests? (2) Is there an effect of critical thinking skills (high and low) on the ability of prospective physics teachers to solve HOTS-based physics tests? Moreover, (3) Is there an interaction between the application of problem-based learning models with critical thinking skills on the ability of prospective physics teachers to solve problems HOTS-based physics tests?

2. Method

The quasi-experimental design was used in this study. The participants of this study were 34 prospective physics teacher students at the 4th-semester level and attended fundamental physics courses. Participants were grouped into one experimental group of 19 people and one control group of 15 people. The experimental group was given treatment through Problem Based Learning, and the control group was assigned conventional learning. The two groups were also distinguished based on high and low critical thinking skills. The initial ability equivalence test of the experimental and control groups using the independent sample t-test showed that the initial abilities of the two groups were equivalent [13].

The technique used to collect data in this study is the test technique. The test technique measures students' ability to solve HOTS-based physics exam questions and students' critical thinking skills. The data collection tools used are (1) optical material HOTS-based exam questions in the form of two-level multiple choice (answer choices and reasons for choosing answers) with a total of 8 questions consisting of analyzing, evaluating, and creating questions; and (2) critical thinking ability test questions in the form of multiple-choice with a total of 30 questions. The test questions used in the research have gone through expert review and trials to determine the quality of the items. As a result, the criteria for the question instrument have met the valid criteria, high reliability, good discriminating power, and varying levels of difficulty (difficult and moderate) [14]. Data on higher-order thinking skills and students' critical thinking skills were analyzed by inferential statistical tests, starting with the data normality test (Kolmogorov-Smirnov test) and the data homogeneity test (Levene Test) [15].

The different effects know two conditions of the implementation of problem-based learning and the conventional model and critical thinking skills (high and low). First, the parametric statistical test (t-independent) is used when the two-group data are distributed and homogeneous. However, if one of the conditions is not met, the data is analyzed using a non-parametric test (Mann Whitney U test) [16]. Second, based on the results of the homogeneity and normality of the data (Table 1 and Table 2), the comparative hypothesis test of higher-order thinking skills based on the treatment group was analyzed using parametric statistics (sample independent T-test). Meanwhile, the comparative hypothesis test based on critical thinking ability was analyzed using non-parametric statistics (Mann Whitney U Test). Analysis of the overall statistical test using IBM SPSS Statistics v.23 with an analysis scheme as shown in Figure 1.

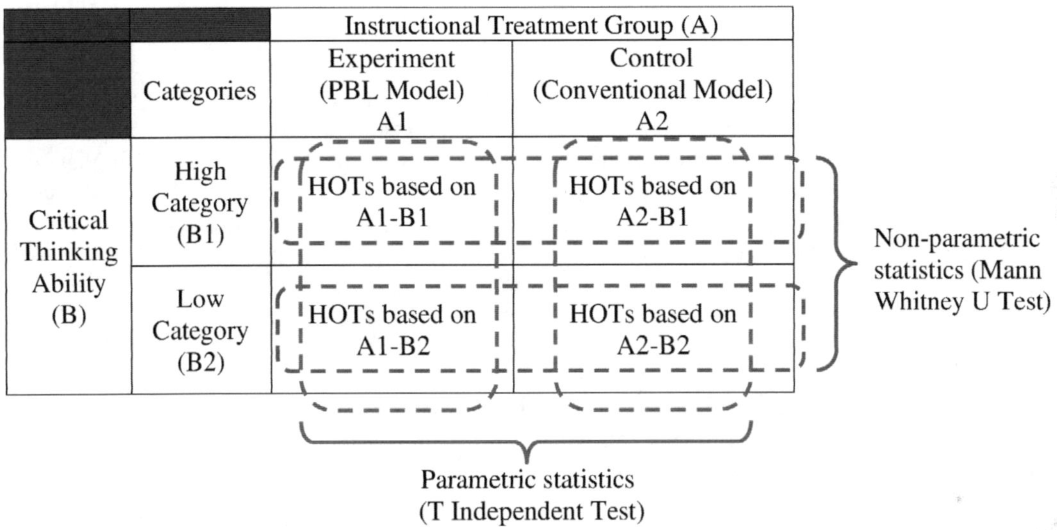

Figure 1. Schematic data analysis techniques for comparative and interaction tests of higher-order thinking skills based on instructional treatment groups and critical thinking ability

3. Results and Discussion

3.1 Result Data Analysis

The data obtained in the study are data on higher-order thinking skills and data on students' critical thinking skills. Data on higher-order thinking skills were measured using a two-tier multiple-choice test given to students after applying the problem-based learning model in the Experimental Group (EG) and applying the conventional model in the Control Group (CG). In contrast, data on critical thinking skills were measured by a multiple-choice test given before treatment. The data for higher-order thinking skills is presented in Figure 2. Based on Figure 2, it can be seen that the average ability of students in solving HOTS questions given learning using the problem-based learning model is higher than students who are taught using the conventional model.

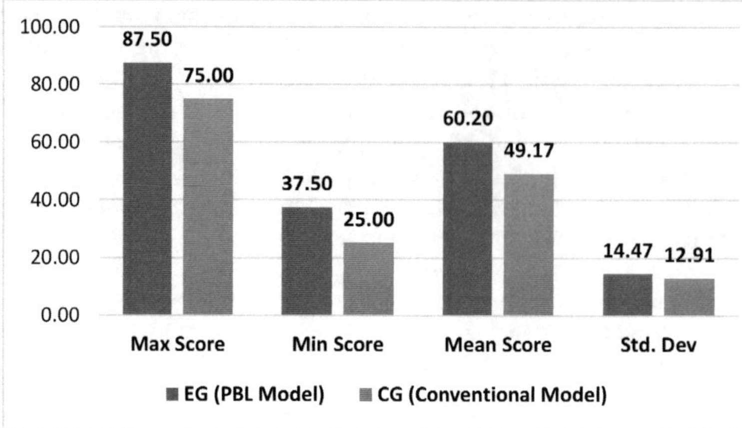

Figure 2. Descriptive Data of Students' Higher Order Thinking Skills

Before the comparative hypothesis test, data on higher-order thinking skills were analyzed using the normality test (Kolmogorov-Smirnov test) and homogeneity test (Levene test) as prerequisites for determining the type of hypothesis statistics used. The results of the homogeneity test of higher-order thinking skills data by class (experiment-control) obtained a significance value of 0.542 (> = 0.05) which indicates that the data is homogeneous. Furthermore, data on higher-order thinking skills based on critical thinking ability (high-low) is also homogeneous (sig-0.356 > =0.05). The results of the data normality test are presented in Table 1.

Table 1. Results of data homogeneity test higher-order thinking skills (HOTS) based on the mean of treatment groups and critical thinking skills

Data of HOTS	Levene Statistic	df_1	df_2	Sig.
based on treatment group (experiment-control)	0.380	1	32	0.542
based on critical thinking ability (high-low)	0.877	1	32	0.356

Table 2. Normality Test Results of Higher Order Thinking Skills Data for each group and category

Data of HOTS	Categories	Kolmogorov-Smirnov		
		Statistic	df	Sig.
based on treatment group	Experiment	0.186	19	0.081
	Control	0.175	15	0.200
based on critical thinking ability	High	0.226	19	0.012
	Low	0.156	15	0.200

Table 2 shows that the data on higher-order thinking skills for both the experimental and control classes came from a population with a normal distribution (sig > 0.05). However, the critical thinking ability data in the high category was not normally distributed (sig < 0.05). Based on the results of the homogeneity and normality of the data (Table 1 and Table 2), the comparative hypothesis test of higher-order thinking skills based on the treatment group was analyzed using parametric statistics (sample independent T-test). Meanwhile, the comparative hypothesis test based on critical thinking skills was analyzed using non-parametric statistics (Mann Whitney U Test).

The hypotheses (Ho) in this study include: (a) there is no effect of the application of the problem-based learning model on the ability of prospective physics teacher students to complete the HOTS-based physics tests, (b) there is no effect of critical thinking skills (high and low) on the ability to student physics teacher candidates in solving HOTS-based physics tests, (c) there is no interaction between the application of problem-based learning models and critical thinking skills on the ability of physics teacher candidates to complete HOTS-based physics tests. The summary of the results of the research hypothesis testing can be seen in Table 3 below.

Table 3. Comparative and Interaction Hypothesis Test Results

Source Data	Statistical Test	Sig.
Group Treatment (CT)	T-independent sample test	0.011
Critical Thinking Ability (CTA)	Mann Whitney U Test	0.049
Interaction CT*CTA	F-Test	0.035

Based on Table 3, the significance value is less than 0.05 (α=5%), so it can be concluded that Ho is rejected (Ha is accepted). This means (a) there is an effect of applying the problem-based learning model on students' abilities in completing HOTS-based physics tests, (b) there is an influence of high critical thinking skills and low critical thinking skills on students' abilities in completing HOTS-based physics tests and (c) there is an interaction between the application of the problem-based learning model with critical thinking skills on the students' ability to complete the HOTS-based physics test.

3.2 Discussion

A problem-based learning model has a significant effect on students' higher-order thinking skills, in this case, the ability to solve HOTS-based questions, indicating that the application of problem-based learning models can improve students' higher-order thinking skills. This is in line with the research conducted by Ref [17]. His study concluded that the higher-order thinking ability of students taught by problem-based learning strategies was significantly higher than traditional learning. In addition, Students' ability to solve higher-order thinking questions after being taught using a problem-based learning model had good criteria for analyzing and evaluating and cannot create [18]. Students who experience a problem-based learning process have higher-order thinking skills. They can easily answer difficult physics questions in contrast to conventional learning models, where students are not less trained in constructing knowledge.

Students' average higher-order thinking ability is higher than the average higher-order thinking ability in classes taught with conventional models. The problem-based learning model requires students to carry out authentic investigations to find real solutions to real problems. Ref [19] revealed that students' high-level thinking patterns in problem-oriented situations, especially learning, can be supported through problem-based learning models. In the problem-based learning model, students are taught in five stages, starting from the student orientation stage on the problem. First, students are given problems that students often encounter in their real lives [20]. Issues are prepared as new learning contexts. Third, analysis and resolution of the problem result in the acquisition of knowledge and problem-solving skills. Issues are faced before all relevant knowledge is acquired, not only after reading a text or listening to a lecture on the subject matter behind the problem.

The second stage is to organize students to learn. The lecturer forms study groups of 3 to 4 people with different initial ability qualifications at this stage. This is so that students with high abilities can help their groupmates with moderate and low skills to form good teamwork. At this stage, group members have their respective roles in the group to have responsibilities within the group. The division of tasks in groups will increase student confidence. Students will be more flexible in expressing ideas or opinions when they feel needed in their groups [21]. Then each group discusses the solutions to the problems presented at the initial stage and plans the problem-solving procedures offered. When planning problem solving, students are trained to analyze the issues and find solutions to the problems given.

The third stage is guiding individual and group investigations. At this stage, each group investigates experiments for problem-solving. In conducting experimental activities, students are assisted with worksheets. At this stage, the lecturer encourages students to collect appropriate information through literature review, then conduct experiments to solve problems and get explanations from problem-solving.

The fourth stage is developing and presenting the work. Each group prepares the results of the experiments that have been carried out in the form of reports and products. The work is created collaboratively and independently by each group, which is then presented in front of the class and responded to by other groups. At this stage, students are trained to create a work by connecting the concepts they have with the concepts they found during learning so that students will be able to develop thinking skills at the level of creating.

The fifth stage is analyzing and evaluating the problem-solving process. Lecturers help students to reflect or assess their investigations and the methods they use. This stage trains students in evaluating all learning that has been done. This is evident from the results of student completion in solving HOTS questions, and it appears that students can plan and carry out problem-solving plans appropriately, understand the intent of the question correctly, and provide the right reasons/evidence so that the written answer will answer the question in question [22].

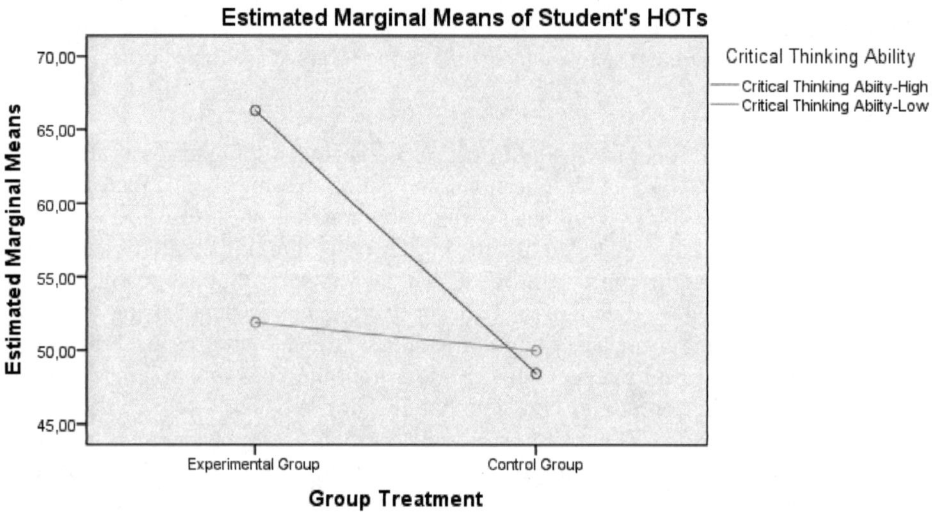

Figure 3. Graph of Interaction between Instructional Treatment (EG/PBL Model and CG/Conventional Model) with Critical Thinking Ability (High and Low)

The results of the hypothesis test also show that critical thinking skills affect students' higher-order thinking skills. This means that students who have high critical thinking skills will be able to solve HOTS questions better than students who have low critical thinking skills after being taught with problem-based learning models. These results indicate an interaction between critical thinking skills and learning models applied in the classroom (Figure 3). According to Ref [19], critical thinking is a systematic and organized process that allows students to formulate and evaluate their own opinions and the evidence, assumptions, logic, and language that underlie other people's statements that they can ultimately solve. Problems encountered in everyday life. This means that students who have critical thinking skills tend to have systematic thinking patterns to go through the stages of analyzing, evaluating to the highest stage of creating in solving problems. The problem-based learning model is constructivist [23], so that students can learn to construct knowledge and apply the principles of physics when learning in class [24], and also students can improve their science process skills [25,26] and problem-solving skills [27,28] through this model.

4. Conclusions

The impact of the application of the problem-based learning model has been investigated for its effect on the ability of prospective physics teacher students to complete the HOTS-based physics test. The ability of prospective physics teacher students to complete the HOTS-based physics test is better after the problem-based learning model is applied. Prospective physics teacher students with high critical thinking skills have better performance in completing HOTS-based physics tests. There is an interaction between the application of the problem-based learning model and the critical thinking skills of prospective physics teacher students in completing the HOTS-based physics test. Implementation of the problem-based learning model is essential for other physics materials. In addition to improving the understanding of physics concepts, it also enhances higher-order thinking skills for prospective physics teacher students.

References

[1] Abidin Z 2005 Strategi pembelajaran di perguruan tinggi (Optimalisasi kinerja dosen dalam pembelajaran di Fakultas Agama Islam Universitas Muhammadiyah Surakarta) *SUHUF* **XVII** 75–8
[2] Serway R A and Jewett J W 2014 *Physics for scientists and engineers with Modern Physics 9th Edition* (USA: Cengage learning)
[3] Halliday D, Resnick R and Walker J 2018 *Fundamentals of physics 10th Edition* (USA: John Wiley & Sons)
[4] Young H D, Freedman R A and Ford A L 2020 *University Physics with Modern Physics Modern Physics 15th Edition* (USA: Pearson Education)
[5] Barus E L and Sani R A 2018 *INPAFI (Inovasi Pembelajaran Fis.* **5**
[6] Handayani S L 2018 *Prosiding Seminar Nasional Pendidikan* **1**
[7] Sheftyawan W B, Prihandono T and Lesmono A D 2018 *J. Pembelajaran Fis.* **7** 147–53
[8] Berg E Van Den 1991 *Miskonsepsi Fisika dan Remediasi* (Salatiga: Universitas Kristen Satya Wacana)
[9] Krathwohl D R 2002 A Revisions of Bloom's Taxonomy *Theory Pract.* **41** 212–9
[10] Brookhart S M 2010 *How to assess higher-order thinking skills in your classroom* (Alexandria: ASCD)
[11] Lin K M and Leng L W 2008 *in 10th Asia-Pacific Conference on Giftedness, Singapore* 1–15
[12] Wijnen M, Loyens S, Smeets G, Kroeze M and van der Molen H 2017 *Interdiscip. J. Probl. Learn.* **11**
[13] Budiyono 2009 *Statistik untuk Penelitian Edisi Ke-2* (Surakarta: Universitas Sebelas Maret Press)
[14] Sumarna S 2009 Analisis, validitas, reliabilitas, dan interpretasi hasil tes implementasi kurikulum 2004 *Bandung: Remaja Rosdakarya*
[15] Wahid S 2005 *Statistik Non-Parametrik contoh kasus dan pemecahannya menggunakan SPSS* (Yogjakarta: Yogjakarta: Penerbit ANDI)
[16] Ananda R and Fadhli M 2018 *Statistik Pendidikan (Teori dan Praktik dalam Pendidikan)* vol 3322018 (Cv. Pusdikra Mitra Jaya/Cv. Widya Puspita)
[17] Ulfa S W 2014 Pengaruh Strategi Pembelajaran Berbasis Masalah terhadap Kemampuan Berpikir Tingkat Tinggi (Manganalisis, Mengevaluasi, Mencipta) dan Keterampilan Proses Sains Mahasiswa STIPAP LPP Medan
[18] Prasetyani E, Hartono Y and Susanti E 2016 *J. Gantang* **1** 34–44
[19] Rahayu S and Hidayat A 2017 Penerapan Model Pembelajaran Problem Based Learning (PBL) Terhadap Kemampuan Berpikir Kritis Siswa Kelas X IPA SMAN 1 Sukawangi Pada Materi Pencemaran Lingkungan *ilib. uinsgd. ac. id* 1–8
[20] Sofyan H and Komariah K 2016 *J. Pendidik. Vokasi* **6** 260–71
[21] Lestari N N S 2012 *J. Teknol. Pembelajaran Indones.* **1**
[22] Nurhayati N and Angraeni L 2017 *J. Penelit. Pengemb. Pendidik. Fis.* **3** 119–26
[23] Kuvac M and Koc I 2019 *Educ. Stud.* **45** 72–94
[24] Shishigu A, Hailu A and Anibo Z 2017 *Eurasia J. Math. Sci. Technol. Educ.* **14** 145–54
[25] Saputri D F, Trisianawati E and others 2021 *J. Phys. Conf. Ser.* **1760** 12017
[26] Prameswari A and Wahyudi W 2019 *J. Pendidik. Sains dan Apl.* **2** 25–30

[27] Kertinus R, Darma Y and Wahyudi W 2019 *Edukasi J. Pendidik.* **17** 135–44
[28] Amalia N, Darma Y and Wahyudi W 2020 *Prosiding Seminar Nasional Pendidikan MIPA dan Teknologi II* **1** 389–98

Physics e-module: A review and bibliometric analysis

D Dewantara[1], E W N Sofianto[2], Misbah[1] and D Munawaroh[1]

[1]Physics Education Study Program, Universitas Lambung Mangkurat
[2]Tadris Fisika, UIN Antasari Banjarmasin

Dewantara_pfis@ulm.ac.id

Abstract. The Covid-19 pandemic has brought significant changes in education and the tendency of face-to-face learning to be online. Therefore, several studies have developed e-modules to support learning, one of which is physics learning. This research is the result of a review based on a bibliometric analysis of the physics e-module. The data was obtained based on articles published in journals and proceedings on Google Scholar in 2020-2021, namely when the Covid-19 pandemic happened. These were selected via Publish or Perish (PoP) software. The data has been sorted until found 74 articles from 142 articles. The data were processed in depth using VOSviewer software. The results showed how the network, overlay, and density were between pieces. It also showed other information regarding the distribution of the authors and their affiliations. Based on this analysis, it was suggested that there is a need to strengthen collaboration between writers and research focused on students' abilities and skills.

1. Introduction

During the COVID-19 pandemic, many schools are conducting distance learning or online. The problem that occurred during the Covid-19 pandemic in the world of education in Indonesia was the lack of student interest in carrying out the learning process [1]. In the current pandemic conditions, information technology to support the learning process is massively used [2].

Physics is one of the branches of Natural Sciences which has a vital role in everyday life [3]. Physics learning in the future is expected to be more meaningful and adapted to the development of science and technology [4,5]. Effects in the era of globalization itself are characterized by the production and use of information technology quickly and modernly to realize the learning process [6].

Technological developments in the educational aspect have given rise to a new paradigm: students are not only recipients of knowledge information. Still, they can act as constructors in managing the knowledge information they acquire [2]. Developments in learning can be seen from several things, including curriculum, learning models, learning methods, teaching materials used and so on [1].

As for the smooth teaching and learning process, teachers can use appropriate and appropriate teaching materials. One of them is through the development of teaching materials and utilizing the right learning media [7–10]. Educators are expected to design, develop, and use media and teaching materials as a form of effort to increase students' motivation, interest, and attention [1,11,12]. Therefore, the development of learning media is needed to make it more attractive to students. One media that can be developed is a learning module in the form of an electronic module (e-module)[13,14].

The use of e-modules can foster creativity and be active in learning [3]. E-Modul is a digital module that will assist students in understanding a concept [15,16]. Learning using e-modules aims to learn independently and measure their abilities through learning outcomes so that learning is more effective in achieving learning objectives [12,17,18]. The electronic modules used in education must be designed by the teacher very interestingly, such as the teacher displaying several pictures and videos in the module so that students do not feel bored in the learning process and the learning objectives can be achieved easily [19]. Electronic modules as teaching materials that utilize technology can change a boring learning atmosphere into fun [11]. Interactive e-module teaching materials in electronic form can help students develop their knowledge in learning material and improve independent learning, both in class and distance learning [20]. This study aims to analyze and describe trends in the research and development of electronic modules in physics learning.

2. Metode

The research method used in the bibliometric analysis adopts the five-step method [21], as shown in Figure 1.

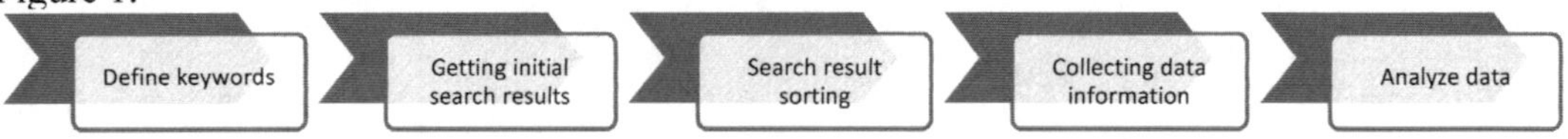

Figure 1 Five-Step Bibliometric Analysis Method

2.1 Find keywords

The literature search uses the keyword "e-module" based on Google Scholar in 2020-2021 (during the Covid 19 Pandemic).

2.2 Get search results

Based on the search results using the keyword, obtained 142 articles. The results of this search are stored in a *Research Information Systems* (RIS) format on a *PoP* .

2.3 Search result sorting

The articles obtained are 142 and then sorted because some are not in the scope of learning physics. In addition, it is re-sorted that only analyzes which journal articles and proceedings articles. Thus there are a total of 74 pieces. The results of this sorting are again saved in RIS format. The RIS-formatted data is then imported into the *Mendeley Desktop* software for further data analysis.

2.4 Collecting data information

On the *Mendeley Desktop,* the RIS-formatted data is reviewed by checking the metadata one by one. If there is incomplete metadata, then editing is done so that the metadata is in accordance with the published article. Edits are also saved in RIS format.

2.5 Analyze data

The RIS format data edited by *Mendeley Desktop* underwent a further analysis process using *VOSviewer* software.

3. Results and Discussion

The output of the PoP Viewer and VOSviewer will display any keywords used by the author. For example, here are the data for the Top 10 articles about e physics modules, which can be seen in Table 1.

Table 1 Top 10 PoP. Sorting Articles

Rank	Writer's name	Year	Title	Journal
1	Fadieny, N, Fauzi, A	2021	Validitas E-Modul Fisika Terintegrasi Materi Bencana Petir Berbasis Experiential Learning [22]	*Jurnal Penelitian Pembelajaran Fisika*
2	Susilawati, S, Pramusinta, P, & Saptaningrum, E	2020	Penguasaan konsep siswa melalui sumber belajar e-modul gerak lurus dengan software flipbook maker [15]	*UPEJ Unnes Physics Education Journal*

3	Agustia, FS, & Fauzi, A	2020	Efektivitas E-Modul Fisika SMA Terintegrasi Materi Kebakaran Berbasis Model Problem Based Learning [23]	*Jurnal Penelitian Pembelajaran Fisika*
4	Ramadayanty, M, Sutarno, S, & Risdianto, E.	2021	Pengembangan E-Modul Fisika Berbasis Multiple Reprsentation Untuk Melatihkan Keterampilan Pemecahan Masalah Siswa [24]	*Jurnal Kumparan Fisika*
5	Haspen, CDT, Syafriani, S, & Ramli, R	2021	Validitas E-Modul Fisika SMA Berbasis Inkuiri Terbimbing Terintegrasi Etnosains untuk Meningkatkan Kemampuan Berpikir Kreatif Peserta Didik [25]	Jurnal Eksakta Pendidikan
6	Nisa, WL, Ismet, I, & Andriani, N	2020	Development of e-modules based on multi-representations in solid-state physics introductory subject [26]	*Berkala Ilmiah Pendidikan Fisika*
7	Ramadhan, MA, Handoyo, SS, & Alfarisi, MM	2020	Pengembangan E-Modul Fisika Dasar untuk Mahasiswa Calon Guru SMK Teknik Konstruksi dan Properti [17]	Jurnal Pendidikan Fisika dan Teknologi
8	Kurnia, R	2020	Validitas E-Modul Fisika Terintegrasi Bencana Gunung Meletus Berbasis Model Inquiry Based Learning untuk Meningkatkan Sikap Kesiapsiagaan Peserta Didik [27]	*Jurnal Penelitian Pembelajaran Fisika*
9	Latifah, N, Ashari, A, & Kurniawan, ES	2020	Pengembangan e-Modul Fisika Untuk Meningkatkan Kemampuan Berpikir Kritis Peserta Didik [28]	*Jurnal Inovasi Pendidikan Fisika*
10	Ningsih, AT, Ruhiat, Y, & Saefullah, A	2020	*EMOSETS: Pengembangan E-Modul Berbasis Science, Environment, Technology, and Society (SETS) Materi Fluida Dinamis* [29]	Prosiding Sendikfi

VOSviewer will display a network visualization between keywords used by the author. Figure 2 shows the visualization of the overlay.

In Figure 2, it can be seen that there are five clusters marked with five different colours, namely purple, green, blue, red, and yellow. In the red cluster, there are 10 items, namely abilities; assessment; critical thinking skills; electronic modules; experts; material experts; media experts; need; problems; and skills. In the green cluster there are 9 items, namely: development; ethnoconstructivism; evaluation; forms; implementation; modules; motivation; perception; r&d. In the blue cluster, there are 8 items, namely inquiry; instruments; learning models; model; products; teaching materials; use; validity. In the yellow-coloured cluster, there are 7 items, namely: approach; developed e-module; lacks; medium; science, stem; technology. The last cluster is purple with 7 items, namely: analysis; e-module; e-modules; learning; physics; physics education; integrated Islam.

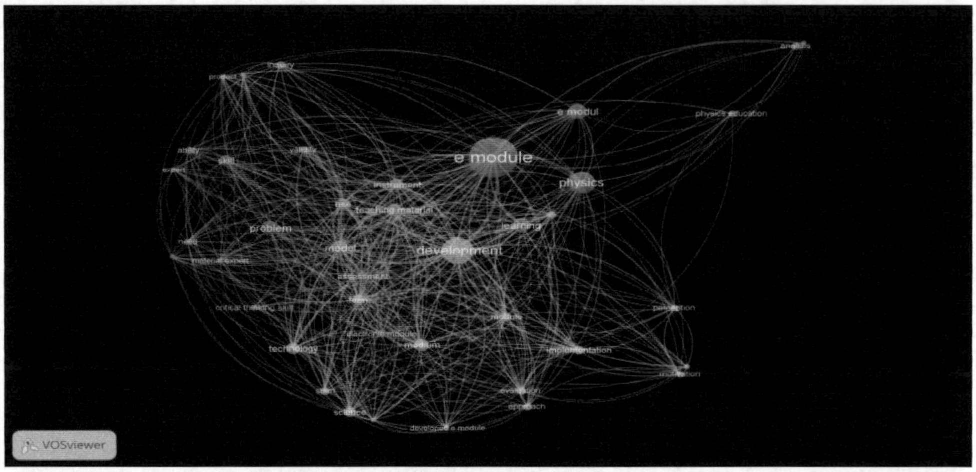

Figure 2 Network Visualization on Google Scholar Database

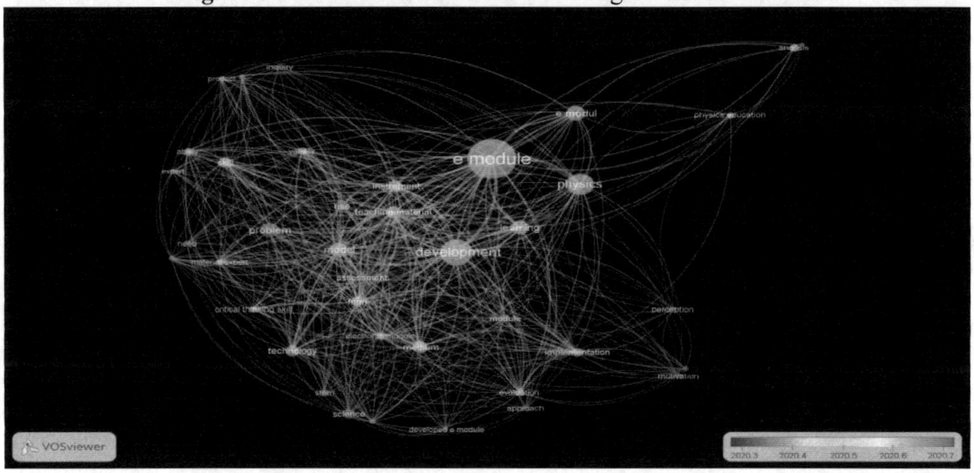

Figure 3 Overlay Visualization on Google Scholar Database

Each cluster contained in Figure 3 will analyse how to spread the article's focus at any time. This way, you can see what the latest trends are. Figure 4 shows a visualization of the density of the keywords used by the author. Density will be indicated by the intensity of the colour in the image. The lighter the colour, the more often the keyword is used by the author.

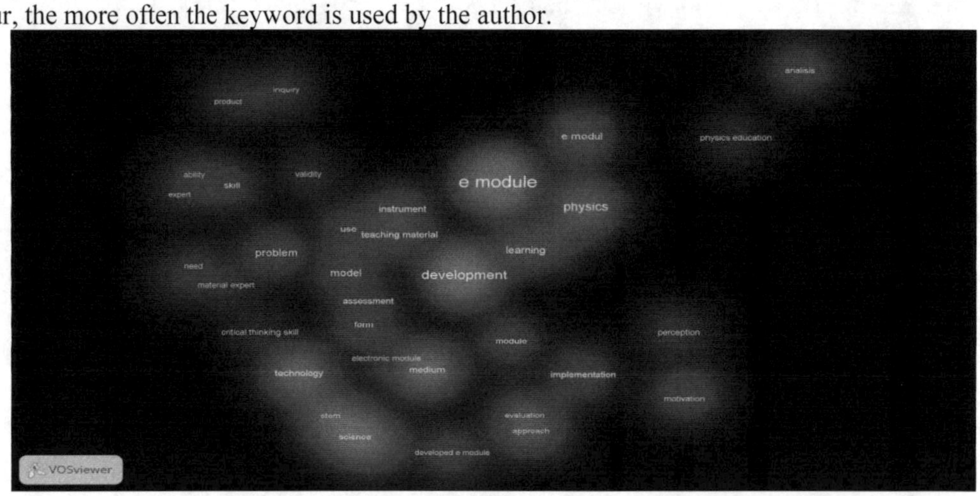

Figure 4 Density Visualization in Google Scholar Database

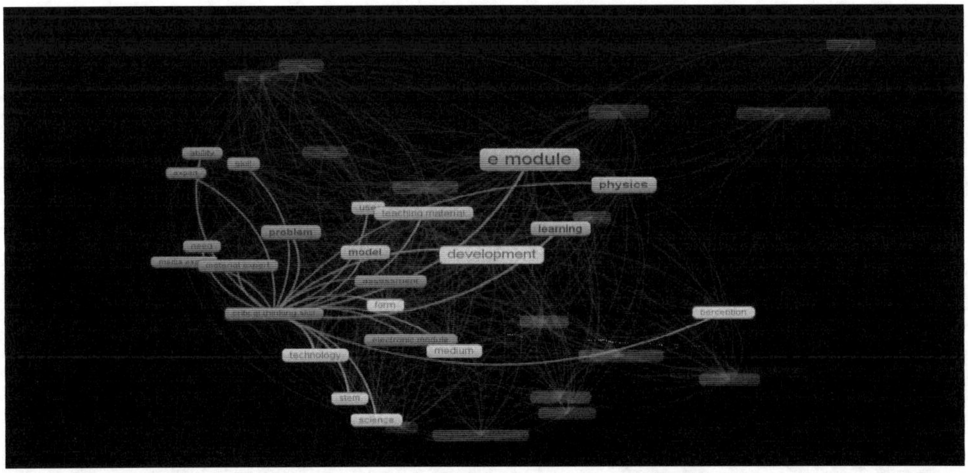

Figure 5 Network Visualization on Google Scholar Database with Keyword Critical Thinking Skill

In this section, the author focuses on observing specifically the keyword “critical thinking skills”. In this case, it can be seen that some research and development have focused on e physics modules to strengthen students' critical thinking skills [20,28,30–33]. There is also a physics module developed, which is part of the teaching material [33]. Other studies also show how e-modules to strengthen students' critical thinking skills are packaged in STEM learning [30,34,35].

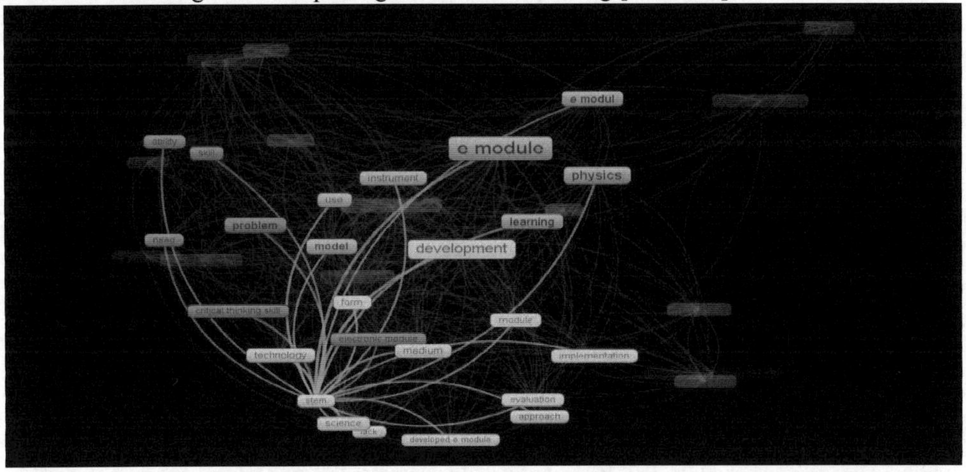

Figure 6 Network Visualization on Google Scholar Database with STEM Keywords

Next, we will focus on STEM keywords. STEM-oriented physics modules have been widely developed [1,14,36]. Existing research does not only focus on development [1,35,37]. However, some examine how it is used in learning, such as through project-based learning [35], problem-based learning [34], and self-directed learning [38]. Also, how to approach the target of student skills which is the primary goal of this development, includes critical thinking skills [30,34,35].

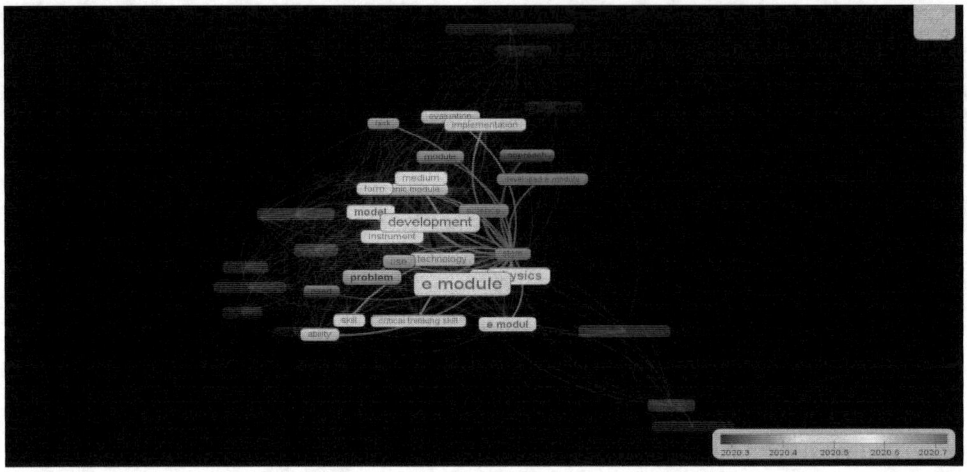

Figure 7 Overlay Visualization on Google Scholar Database with Keyword Critical Thinking Skill

Figure 7 shows how the visualization of overlay is focused on the keyword "critical thinking skills" from the image. The picture shows that the e-module research that focuses on critical thinking skills will be carried out around **2020.5**. And the most recent is how to package e-modules in stems to strengthen students' necessary thinking skills.

Based on this bibliometrics, the latest research related to physics modules is about STEM e-modules that improve students' critical thinking skills. This is good for further development. So it is recommended to innovate e STEM modules so that they have a good update value. Therefore, it is suggested to develop ethno-STEM in the physics module, which aims to improve students' critical thinking skills.

4. Conclusion

Based on the results of bibliometric analysis, it can be concluded that articles with the keyword e-module in the period 2020-2021 can train various aspects of student skills, such as critical thinking skills. In further research, e-modules can be developed by combining them with ethno-STEM to improve students' critical thinking skills.

References

[1] Syahiddah D S, A P P D and Supriadi B 2021 *J. Literasi Pendidik. Fis.* **2** 1–8
[2] Rani L and Maarif S 2021 *J. Medives J. Math. Educ. IKIP Veteran Semarang* **5** 377–89
[3] Sidik F D M and Kartika I 2020 *J. Penelit. Pembelajaran Fis.* **11** 185–201
[4] Prihatiningtyas S and Sholihah F N 2020 *J. Pendidik. Fis.* **8** 223–34
[5] Dewantara D, Misbah M, Mahtari S, Azhari A, Sasmita F D, Melisa, Rusmawati I, Kusuma L W, Ridho M H and Lutfi M 2021 *J. Phys. Conf. Ser.* **1760**
[6] Isnia H W A, Wahyuningtyas D T and Yulianti Y 2020 *Pros. Semin. Nas. PGSD UNIKAMA* **4** 311–9
[7] Aryanti E D and Arief M 2021 *J. Ekon. Bisnis dan Pendidik.* **1** 24–33
[8] Hartini S, Misbah M, Dewantara D, Oktovian R A and Aisyah N 2017 *J. Pendidik. IPA Indones.* **6** 313–7
[9] Kairmo G A, Ratu T, Hermansyah, Yahya F and Fitriyanto S 2021 *J. Sci. Instr. Technol.* **1** 29–33
[10] Malina I, Yuliani H and Syar N I 2021 U *Silampari J. Pendidik. Ilmu Fis.* **3** 70–80
[11] Piyana S O 2020 *J. JPSD (Jurnal Pendidik. Sekol. Dasar)* **7** 21–8
[12] Kurniawan A R, Budiono H, Hariandi A, Marlina M, Kurniawati E F, Meidiawati R and Piyana S O 2020 *Profesi Pendidik. Dasar* **1** 93–104
[13] Zaka A M and Supraptono S 2020 *Automot. Sci. Educ. J.* **9** 12–9
[14] Aulia D M, Parno P and Kusairi S 2021 *J. Ris. Pendidik. Fis.* **6** 7–12
[15] Susilawati S, Pramusinta P and Saptaningrum E 2020 *UPEJ Unnes Phys. Educ. J.* **9** 36–43
[16] Dewi M S A and Lestari N A P 2020 *J. Imiah Pendidik. dan Pembelajaran* **4** 433–41
[17] Ramadhan M A, Handoyo S S and Alfarisi M M 2020 *J. Pendidik. Tek. Sipil* **2** 117–23

[18] Kurniawati E F 2020 *J. Penelit. Ilmu Pendidik.* **13** 10–21
[19] Asrial A, Syahrial S, Maison M, Kurniawan D A and Piyana S O 2020 *JPI (Jurnal Pendidik. Indones.* **9** 30
[20] Sintawati N P and Margunayasa I G 2021 *Int. J. Elem. Educ.* **5**
[21] Haryandi S, Suyidno S, Misbah M, Dewantara D, Mahtari S and Ibrahim M A 2021 *Momentum Phys. Educ. J.* **5** 10–20
[22] Fadieny N and Fauzi A 2021 *J. Penelit. Pembelajaran Fis.* **7**
[23] Agustia F S and Fauzi A 2020 *J. Penelit. Dan Pembelajaran Fis.* **6** 1–8
[24] Ramadayanty M, Sutarno S and Risdianto E 2021 *J. Kumparan Fis.* **4** 17–24
[25] Haspen C D T, Syafriani S and Ramli R 2021 *J. Eksakta Pendidik.* **5** 95–101
[26] Nisa L W, Ismet and N A 2020 *Berk. Ilm. Pendidik. Fis.* **8** 73–81
[27] Kurnia R, Ayu F and Fauzi A 2020 *J. Penelit. Pembelajaran Fis.* **6** 73–80
[28] Latifah N, Ashari A and Kurniawan E S 2020 *J. Inov. Pendidik. Sains* **1**
[29] Ningsih A T, Ruhiat Y and Saefullah A 2020 *Pros. Semin. Nas. Pendidik. Fis. Untirta* **3**
[30] Handayani E S, Saregar A and Wildaniati Y 2021 *J. Phys. Conf. Ser.* **1796**
[31] Suryaningtyas A, Kimianti F and Prasetyo Z K 2020 *International Conference on Educational Research and Innovation (ICERI 2019)* **401** 65–70
[32] Sujanem R, Suswandi I and Yasa P 2020 *Proceeding Senadimas Undiksha* 793–801
[33] Misbah M, Sasmita F D, Dinata P A C, Deta U A and Muhammad N 2021 *J. Phys. Conf. Ser.* **1796**
[34] Kurniati R D, Andra D and Wayan Distrik I 2021 *Conf. Ser. Earth Environ. Sci.* **1796**
[35] Widiyanto R, Herlina K and Andra D 2021 *J. Phys. Conf. Ser.* **1796**
[36] Diani R, Yanti Y, Hartati N S, Fujiani D, Hasanah I F and Alamsyah 2021 *Conf. Ser. Earth Environ. Sci.* **1796**
[37] Kurniati R D, Andra D and Distrik I W 2021 *J. Phys. Conf. Ser.* **1796**
[38] Ramadhan A, Jalinus N, Ta'ali T and Mulianti M 2021 *JINOTEP (Jurnal Inov. dan Teknol. Pembelajaran) Kaji. dan Ris. Dalam Teknol. Pembelajaran* **8** 91–100

Development of student worksheets based on STEM approach to improve students' critical thinking skills

Sutarto, J Prihatin, S Hariyadi, and I Wicaksono*

Faculty of Teacher Training and Education, University of Jember, Jember, Indonesia

iwanwicaksono.fkip@unej.ac.id

Abstract. The development of Worksheets based on the STEM approach pays attention to the elements of writing, the hierarchy of materials, and the selection of questions as an efficient and effective stimulus. This research aims to produce Worksheets based on the STEM approach that meets the feasibility of improving students' critical thinking skills. The development model in this research refers to the 4D (four-D) research and development model by taking sample data from high school science students in the former Besuki Residency, namely Jember, Situbondo, Banyuwangi, and Bondowoso Regencies. The research instrument uses tests, observations, and documentation. The research results show that Worksheets based on the STEM approach meet the valid, practical, and effective criteria for improving students' critical thinking skills. This research implies that the availability of Worksheets is expected for students to research teaching materials independently. In Worksheet, students will get materials, summaries, and assignments related to the material.

1. Introduction

Science is the science of nature that studies objects, phenomena, and processes that occur in nature, and science is concerned with systematically finding out about nature. In essence, science is built based on scientific methods, attitudes, and products. The importance of researching science is to discuss, examine, and prove the existence of facts and natural phenomena related to science learning. Science learning is related to how to find out about nature systematically. Science learning is memorizing and understanding concepts and facts that occur in nature and a process of discovery. The research results indicate that learning science that allows students to find an idea independently will provide direct experience to understand the natural surroundings scientifically [1].

Science learning based on the STEM Approach consists of four elements: integrated science, technology, engineering, and Mathematics [2]. The STEM approach is a multidisciplinary approach to learning, where accurate scientific concepts are combined with the natural world through the practical application of students' science, technology, engineering, and mathematics. Through the STEM approach, students are not just memorizing concepts but also how they understand science concepts and their relation to everyday life. Learning physics in class will be more meaningful for students [4]. Students who learn through the STEM approach can also be independent and develop themselves to gain confidence and work within a specific time [3].

The results of the Research Group research based on road maps in 2018, 2019, 2020 provide an overview of follow-up that is relevant to needs, including: (1) there are no standardized Worksheets so that they are not optimal in supporting the achievement of learning objectives; (2) has not adopted a

STEM approach that is following the development of the globalization era in the learning process; (3) the need for Worksheets that can improve critical thinking skills through the STEM approach. The STEM approach develops when it is connected to the environment and everyday life to present real-world learning experienced by students. The availability of Worksheets based on the STEM approach will assist teachers in preparing lesson plans, as a guide for teachers and students to add information about the concepts learned through systematic learning activities, help students to add information about the concepts learned through organized learning activities, and train students to find and develop critical thinking skills [4].

Seeing the current conditions, the progress of science, science, and technology continues to develop, it is critical to think that the Worksheet, which is also included as teaching materials that are very much needed, must be designed in such a way as to adapt to the times, to prevent learning problems that arise due to insufficient knowledge and skills. Student skills due to the undeveloped teaching materials used [5]. The existence of the Worksheet has a considerable influence on the teaching and learning process. Compiling Worksheet must meet specific requirements to become a good quality Worksheet so that it can support increasing student activity in the learning process and optimize learning outcomes [6].

Students need to have critical thinking skills to face various problems, solve problems, and make the right decisions critical thinking skills will train students to think logically and not accept things easily [7]. Critical thinking skills have several indicators, including (1) giving simple explanations; (2) building basic skills; (3) concluding; (4) providing a further explanation; and (5) setting strategy and tactics. Developing students' critical thinking skills in learning is an effort to improve student learning outcomes critical thinking skills are thoughtful and reasoned ways of thinking focused on making decisions to solve problems. Critical thinking allows students to analyze their minds in making choices and drawing conclusions correctly.

Therefore, the Research Group needs to produce appropriate STEM Approach-based Worksheets based on valid, practical, and effective criteria to improve students' critical thinking skills.

2. Method

The type of research used is Research and Development research. This type of research is used to develop or validate products used in education and learning [8]. The sampling technique used is purposive sampling based on specific considerations with criteria so that the data obtained will be more representative. The sample in this research were high school students at the Ex-Residency of Besuki. The research implementation time is in the Odd Semester of 2021/2022. The development model in this research refers to the 4D (four-D) research and development model [9]. The 4D research and development model consists of 4 main stages: define, design, develop, and disseminate [10].

Data collection techniques include: (1) tests conducted to determine student mastery of learning materials and critical thinking skills; (2) interviews were conducted with teachers and students conducted before and after learning; (3) a questionnaire in the form of a validation sheet used to obtain assessment data on the quality of Worksheet; (4) photo documentation of learning activities and other documents that support the implementation of this research.

Data analysis techniques Validation through quantitative data is obtained from the results of the checklist (√) of each validation sheet as follows [11].

$$P = \frac{\sum X}{\sum Xi} x\ 100\%$$

Explanation : P = Percentage
$\sum X$ = Respondent's score in one item
$\sum Xi$ = The ideal number of scores in one item
100% = Constanta

Further analysis for the criteria for the level of validity is as follows [12].

Table 1. Level of product validity and revision

Percentage	Explanation	Score
81% - 100%	Valid	4
61% - 80%	Quite valid	3
41% -60%	Less Valid	2
<40%	Invalid	1

Practical data analysis technique through the analysis of the implementation of learning is done by describing the observation of the implementation of learning as follows [4].

$$Percentage = \frac{Score\ obtained}{Total\ Score}\ x\ 100\%$$

Table 2. Criteria for learning implementation observation scores

Criteria	Interval
Very good	81,25 < 100
Good	71,5 < 81,25
Medium	62,5 < 71,5
Low	43,75 < 62,5
Very Lows	0 < 43,75

Effectiveness data analysis techniques are through indicators of students' critical thinking skills, namely interpretation, analysis, evaluation, inference, explanation, and self-regulation. The N-gain test to determine the improvement of students' critical thinking skills is as follows [13].

$$N - Gain = \frac{(Post\ test\ Score)-(Pre\ test\ Score)}{(Maximum\ Score)-(Pre\ test\ Score)}$$

Table 3. *N – Gain* criteria

N – Gain Value	Interpretation
$g \geq 0{,}70$	High
$0{,}30 \leq g \leq 0{,}70$	Medium
$g \leq 0{,}30$	Low

3. Result and Discussion

3.1. Validity

Data on the results of the validity of the developed product were obtained using a validity sheet which was assessed by three validators, including material substance, learning design, and use of Worksheets. Validity was also carried out on the instruments used to support this research, including the questions of pre-test and post-test.

Table 4. Validation assessment results

No	Assessment Aspect	Interval Score			Percentage (%)	Criteria
		Validator 1	Validator 2	Validator 3		
1	Material Substance	0,94	0,98	0,78	90%	Valid
2	Learning Design	0,95	0,90	0,80	87%	Valid
3	Worksheet Utilization	1,00	1,00	0,97	99%	Valid
	Average Score	0,96	0,96	0,85	92%	Valid

Based on Table 4 above shows that most of them are in the valid criteria. This is because the validity indicators used are 90% material substance, 87% learning design, and 99% use of Worksheets that have been developed are feasible to test students. The results showed that the feasibility of the product produced in the form of Worksheets according to the effects of reviews from experts was very good with several aspects, namely the content of the material, elements of the use of learning media, and aspects of Worksheet design. In addition, the research results stated that the feasibility of the developed Worksheet product was good with general characteristics, aspects of material substance, and aspects of learning design. The validity of the media obtained from the material aspect and the aspect of media utilization have met the valid criteria and are suitable for use in learning activities [12].

3.2. *Practicality*

Worksheets based on the valid STEM Approach are continued at the trial stage. The data on the results of the learning development test was obtained by conducting a development test in high school. The development test was carried out in 3 meetings. The development test data obtained in this development research is data on the implementation of learning using Worksheet based on the STEM Approach to improve students' critical thinking skills. The practicality of learning using Worksheets based on the STEM approach was obtained through observers on the implementation of learning during the learning process.

Table 5. The results of the *observation* of the implementation of learning

Activity	Meeting-			Average Score	Criteria
	1	2	3		
Preliminary	0,87	0,94	0,87	0,89	Very good
Core	0,96	0,96	0,93	0,95	Very good
Closing	0,95	0,90	0,90	0,92	Very good
Average	0,93	0,93	0,90	0,92	Very good

Table 5 above shows that the implementation of learning using Worksheet in three meetings based on aspects of the preliminary, core, and closing activities got an average score of 0.92. So that on average each activity has been carried out very well. This follows the research results that the implementation of learning with preliminary, core, and closing activities can be carried out well. The implementation of learning is applied because STEM-based Worksheets are easy for students to use and easy to use by teachers to implement learning and minimize time [14]. Seeing the current conditions, the progress of science, science, and technology continues to develop, it is necessary to think that the Worksheet, which is also included in the teaching materials that are very much needed, must be designed in such a way as to be adapted to the times, to prevent learning problems that arise due to insufficient knowledge and skills. Student skills due to the undeveloped teaching materials used [6].

3.3. *Effectiveness*

The effectiveness in this trial is the effectiveness of the product developed in the form of Worksheets based on the STEM approach by looking at student test results to see the level of development of students' critical thinking skills. The data used to determine the effectiveness after using Worksheet based on the STEM Approach is necessary to do a test. The tests carried out were in the form of a pre-test which was carried out before being given the Worksheet based on the STEM approach that was developed, and the post-test was carried out after.

Table 6. Results of critical thinking skills

No	Indicators of Students' Critical Thinking Skills	Score		N-gain	Criteria
		Pre-test	*Post-test*		
1	Interpretation	54%	71%	0,37	Medium
2	analysis	52%	75%	0,48	Medium
3	Evaluation	62%	86%	0,63	Medium
4	Inference	35%	61%	0,40	Medium

5	Explanation	62%	95%	0,87	High
6	Self Regulation	42%	51%	0,15	Low
	Average	51%	73%	0,45	Medium

The critical thinking skills test is used as many as six questions, and each question contains one indicator of critical thinking skills. Based on Table 6 above shows that the explanation indicator is the largest; this is because students practice their critical thinking skills by explaining or making arguments that are supported with the right reasons. In the pre-test and post-test questions, students were asked to explain the material. After observing the picture, students can explain or make an argument correctly. In addition, this explanation indicator obtained the highest N-Gain score. The average N-Gain for students' critical thinking skills on the explanation indicator is 0.87 with high criteria. This is following the research that the explanatory indicator with a score range of 81-100 has very high criteria. In his research, he carried out activities to train students in stating results and providing appropriate arguments [15]. These activities are always trained to students so that students get used to giving opinions appropriately, and these activities can prepare students to develop critical thinking skills [9].

The lowest increase is in the self-regulation indicator, and this is because the questions given to this self-regulation indicator are students control themselves to solve a problem regarding important things from the occurrence [16]. From the students' answers, the average N-Gain score for students' critical thinking skills activities on this self-regulation indicator is 0.15. This is in accordance with the research that self-regulation indicators carry out self-control activities in associating problems with each other's concepts. However, the low number of students in this indicator is caused by several factors, one of which is not optimal learning to utilize self-regulation critical thinking skills [11]. In addition, this is following the results of the research showing that the problem that students must solve is in their ability to control themselves in dealing with a problem by applying their ability to analyze and evaluate what is obtained by themselves so that their ability to think critically increases [17]. Even though students know the concept, they may not know how to control themselves to apply it, so they have difficulty relating concepts to one another.

4. Conclusion

Based on the results of data analysis and discussion, it can be concluded that the Worksheets are based on the STEM approach that meets the valid, practical, and effective criteria for improving students' critical thinking skills. This research implies that the availability of Worksheets is expected for students to research teaching materials independently. In Worksheet, students will get materials, summaries, and assignments related to the material.

Acknowledgments

The successful implementation of this research cannot be separated from the help of various parties, for that the research team of Inovasi Media Pembelajaran Sains Non-Elektronik Research Group would like to thank LP2M Jember University through the Research Group Grants and Research Partner Schools with Assignment Agreement Number 2702/UN25.3.1/ LT/2021.

References

[1] Sutarto, Indrawati, Prihatin J and Dwi P A 2018 *J. Pendidik. IPA Indones.* **7** 376–82
[2] Corlu M S, Capraro M M and Capraro R M 2014 *Eğitim ve Bilim* **39** 74–85
[3] Erlina N, Susantini E, Wasis, Wicaksono I and Pandiangan P 2018 *J. Balt. Sci. Educ.* **17**
[4] Porter A L, Roessner J D, Oliver S and Johnson D 2006 *J. Eng. Educ.* **95** 13–24
[5] Connor D and Holmes P 2002 *Teach. Stat.* **24** 55–8
[6] Wicaksono I, Wasis W and Madlazim M 2017 *J. Balt. Sci. Educ.* **16** 549–61
[7] Facione P A 1990 *Critical Thinking : A Statement of Expert Consensus for Purposes of Educational Assessment and Instruction Executive Summary " The Delphi Report* **423**

[8] Borg W . and Gall M . 1983 *Educational Research an Introduction fourth edition* **1**
[9] Anderson G and Arsenault N 2005 *Fundamentals of educational research* (Routledge)
[10] Ketut Budiastra A A, Erlina N and Wicaksono I 2019 *New Educ. Rev.* **57** 187–99
[11] Ekantini A and Wilujeng I 2018 *Univers. J. Educ. Res.* **6** 1339–1347
[12] Wicaksono I, Supeno and Budiarso A S 2020 *Int. J. Instr.* **13** 157–70
[13] Hake R R 1998 *J. Physics. Am. Assoc. Phys. Teach.* **66** 64–74
[14] Zulhelmi Z, Adlim A and Mahidin M 2017 *J. Pendidik. Sains Indones. (Indonesian J. Sci. Educ.* **5** 72–80
[15] Kennedy T J and Odell M R L 2014 *Sci. Educ. Int.* **25** 246–58
[16] Budiastra A A K, Wicaksono I and Erlina N 2020 *J. Educ. Gift. Young Sci.* **8** 1291–304
[17] Bybee R 2010 *Technol. Eng. Teach.* **70** 30–8

Science literacy-based sound wave e-worksheet: Validity aspects

P C Lestari, S Miriam, and M Misbah

Physics Education Study Program, Faculty of Teacher Training and Education, Universitas Lambung Mangkurat, Indonesia

sarah_pfis@ulm.ac.id

Abstract. This study aims to describe the validity of e-worksheets based on science literacy in sound-bearing materials. This research is research and development with the ADDIE model. This science literacy-based e-worksheet validator numbers three people, consisting of two academics and one practitioner. The validation results using an e-worksheet validation sheet based on science literacy show that the validity of e-worksheets in sound wave material has good criteria. Therefore, e-worksheets based on science literacy can be used to learn physics in sound wave material.

1. Introduction

Science literacy is a crucial ability to be mastered by every individual. Science literacy is closely related to understanding the environment and the problems facing modern society with the ongoing development of knowledge and technology [1]. These problems include the environment, such as energy sources, pollution, and climate change [2]. Awareness of the importance of science literacy will also accelerate scientific and technological knowledge [3,4]. Science literacy skills can be trained in learners through learning in school as a means of formal education [5–8]. One of the subjects in the school related to the rapid development of the world of technology lately is physics. In training science, literacy skills in school learners, teachers, learning environments, and teaching materials that support science literacy are needed [9–11].

However, the current pandemic conditions require schools to conduct distance learning with learners through online learning. This situation encourages teachers to innovate in using teaching materials, one of which is the student worksheet. The student worksheet was chosen because it is a teaching material that can encourage learners to be actively involved in learning [12,13]. The student worksheet is an activity sheet containing instructions and information about the tasks that learners must do to achieve learning goals [14]. Tasks in the student worksheet can be practicums or problems that learners must solve. Current technological advances allow the use of worksheets presented in electronic form (e-worksheets). So that its use remains efficient because it can be used anytime and anywhere as long as the internet network is available [15,16]. In this case, e-worksheet as a tool used in learning plays an essential role in the success of the learning process [14,17]. The application used in this development research is Flip PDF Professional. This application converts the shape of a PDF into a flipbook. Flipbook is a learning medium with voice and page transition effects such as turning the pages of a book.[18]. In addition, there are also various features such as images and videos from Youtube with pop-ups, hyperlinks, quizzes, and more [19]. Products produced from this application can be used on various devices without installing because it is website-based.

Based on previous research, the use of the student worksheet that applies discovery learning is considered to improve learners' science literacy skills. [20]. In addition, the student worksheet, which contains a link between science and everyday life problems, can increase learners' literacy [21]. Learning that uses the student worksheet based on science literacy is considered to improve critical thinking skills, obtained if you have good science literacy skills [22]. Other research that develops e-worksheets based on science literacy using interactive media gets a positive response from teachers and learners. It is classified as very feasible to be used in learning [23]. The study draws on three cognitive domains of science literacy. It can be concluded that overall the use of e-worksheets has a positive effect and can improve the science literacy of learners. However, no research has developed e-worksheets by concocting on aspects of science literacy, namely the role of science, scientific thinking and doing, science and society, and mathematics and science [24]. These aspects are then used as the basis in the preparation of activities and components in this e-worksheet. Sound waves were chosen because they have many applications in everyday life, making it easy to connect concepts with appropriate problems. Based on the description above, developing an e-worksheet based on sound wave science literacy is feasible to be used in physics learning during pandemics.

2. Method

This research is research and development with the ADDIE model (Analyze, Design, Development, Implementation, and Evaluation). The stage of e-worksheet development is shown in Figure 1.

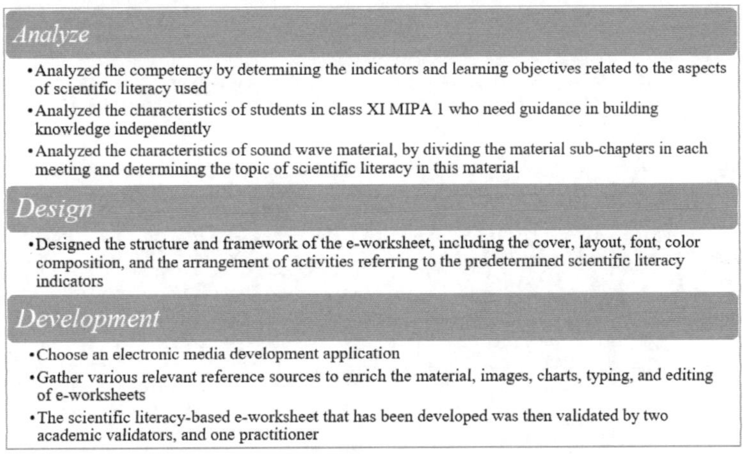

Figure 1. ADDIE model development stage

This science literacy-based e-worksheet validator numbers three people, consisting of two academics and one practitioner. The research instrument used is an e-worksheet validation sheet based on science literacy. The data analysis technique used is a validity test with assessment criteria in Table 1 below [25].

Table 1. Assessment criteria validity

Interval	Criteria
$X > 3.4$	Excellent
$2.8 < X \leq 3.4$	Good
$2.2 < X \leq 2.8$	Pretty good
$1.6 < X \leq 2.2$	Less good
$X \leq 1.6$	Bad

The results of the assessment of the three validators are then carried out a reliability test, which is calculated using the Formula Alfa Cronbach. The reliability coefficient obtained is categorized based on the criteria Widoyoko [26].

3. Result and Discussion

This research produces a product in the form of an e-worksheet based on science literacy on sound wave material as a guide for learners in conducting science literacy activities. The components contained in this science literacy-based e-worksheet are the front cover, instructions for use, experimental tools and materials, and bibliography. In addition, the element of science literacy that is loaded on aspects of the role of science is the text of scientific phenomena and the formulation of problems. Aspects of thinking and working scientifically contain components in hypothesis formulation, variable identification, experimental procedures, experimental results data, data analysis, predictions, test predictions, and conclusions. Finally, the details of questions and stabilization are contained in aspects of science and society and mathematics in science. Each element of this science literacy is written as a subheading in the e-worksheet.

E-worksheets based on science literacy are divided into two parts. The first part contains the characteristic subject matter of sound waves and the degree &intensity of sound waves. The second part of the e-worksheet contains a sub-material rapid propagation of sound waves, Doppler effects, as well as sound sources. Each section consists of literacy activities, namely reading, practicum, and working on problems to solidify the concept obtained through previous activities. An example of a developed e-worksheet display is shown in Figure 2.

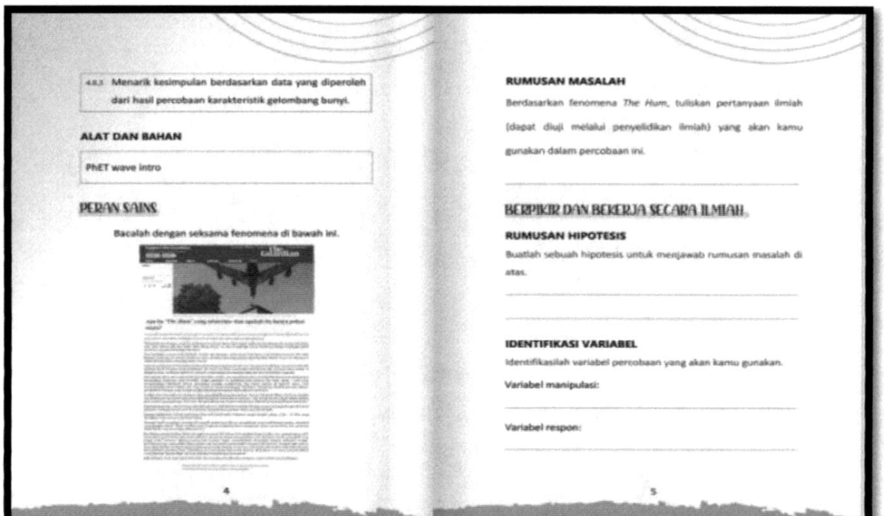

Figure 2. Scientific literacy-based e-worksheet display

Reading activities are complemented by a discourse on scientific phenomena presented in the form of pop up pictures. Practicum activities use the lab's virtual simulation service, PhET Colorado so that learners get a meaningful learning experience and can still be done in groups even remotely. Virtual lab simulations are selected to get the same results even though each learner does them in a different place. This becomes one of the alternatives to overcome distance learning obstacles during pandemics [27]. Virtual lab simulations can also improve science literacy skills, which require applying knowledge and scientific performance skills [8]. Then, learners can work on and discuss with group colleagues through *Whatsapp* groups and send the answers through Google Form, which is connected to an e-worksheet using a hyperlink and barcode features. The advice given by validators in improving the quality of e-worksheets is presented in Table 2.

Table 2. Validator Suggestion

Suggestion	Revision
1. The cover image must match the material studied, and the identity on the cover must contain complete information (e.g., class, semester, etc.).	1. The image on the cover is changed, and the cover's identity is equipped according to the validator's advice.
2. Improve language usage in *the e-worksheet* instructions to make it more effective.	2. Improve sentence arrangement and word usage in *e-worksheet* instructions so that they are clearer and easier to understand.
3. Allow learners to identify some research questions of the phenomena presented and then choose one of them to investigate.	3. They were switching questions about problem formulation to a form that allows learners to identify scientific questions independently.
4. Complete the identity of *the e-worksheet.*	4. Add a day and date column to an *e-worksheet* identity.
5. Provide a bibliography in each section of the *e-worksheet.*	5. Add the bibliography to each section of an *e-worksheet.*

These suggestions are then used as a reference to improve the e-worksheet based on science literacy developed. The difference between the e-worksheet before it is fixed and after it is fixed is shown in Table 3.

Table *3*. The Difference Between The E-Worksheet Before and After is Fixed

Before		After
	Front Cover	
	E-worksheet Identity and Instructions	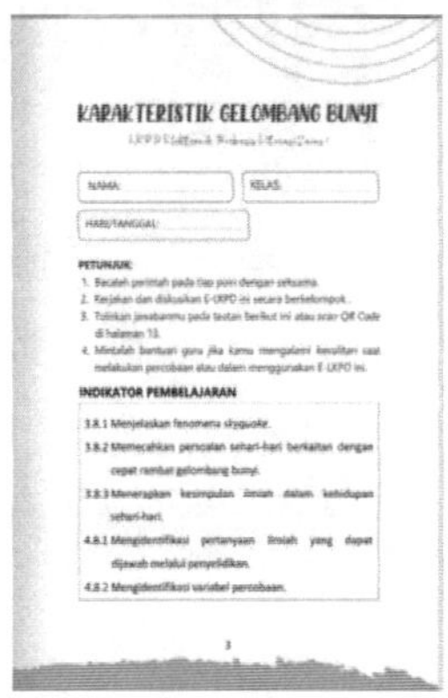

Problem Formula

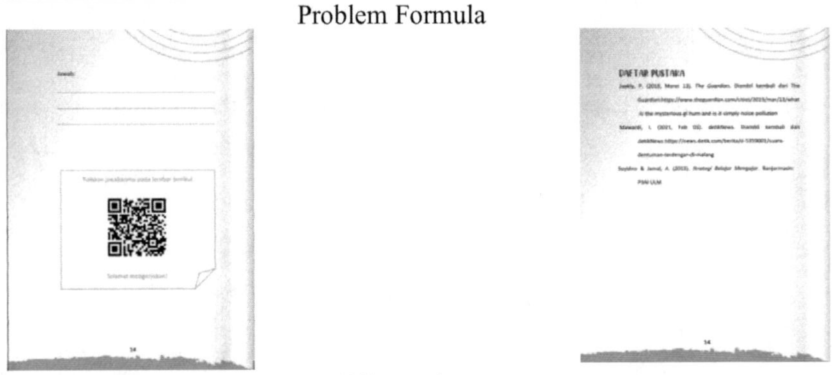

Bibliography

The results of the e-worksheet validation assessment by the validator are presented in Figure 3.

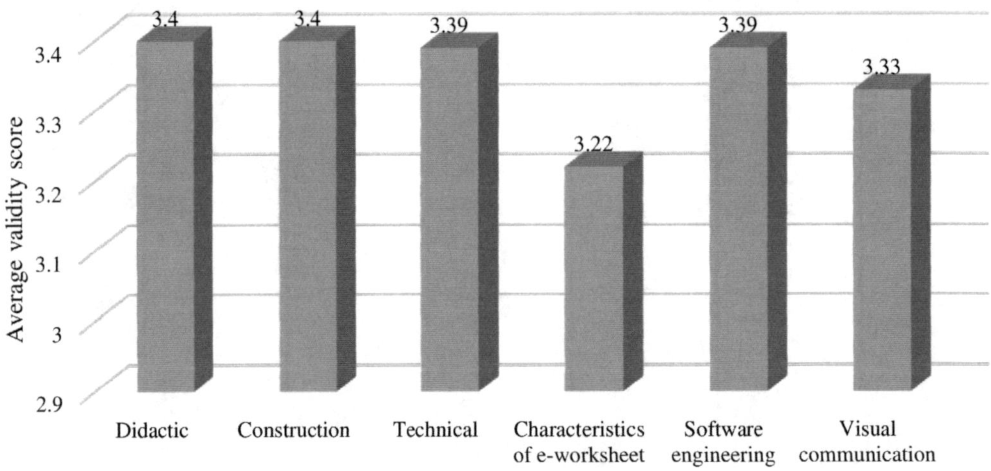

Figure 3. E-worksheet Validity Results

Based on Figure 3 above shows that the average results of the validation aspect of the e-worksheet developed include valid with very high reliability.

E-worksheet can be good if it meets the active, construction, and technical aspects [28]. These aspects are detailed into several assessment criteria, namely can be used by learners as a whole, emphasis on the process of concept discovery, variety of stimulus through various activities, practicing skills, providing a learning experience based on goals, appropriate language, clarity of sentences, simple construction towards complex, simplicity of crawl formulation, conformity of material, there is sufficient space, effective illustration, the existence of clear goals and identity, suitability of capital letters, size, distance, the separator between clues and answers, effective images, interesting colors [28–31]. The color used in this e-worksheet design is white as a primary color with black, red, orange, and yellow elements. This selection is because white is a color with a bright and clean impression so that the writing and image are visible. At the same time, the color red, orange and yellow is a unique, warm, active, cheerful, spirited color that tends to bring closer to the user and suitable combined with the white background color [32].

E-worksheets based on science literacy also meet aspects of science literacy, detailed into several criteria [22]. The criteria are the conformity of science literacy materials and indicators, the existence of science literacy stages, information/images/science literacy activities, and available descriptions/texts that can train learners to identify scientific questions, identify experimental variables, analyze information, solve mathematical problems, make conclusions based on evidence and use of application problems in everyday life. This criterion follows the indicators of science literacy set in learning. This aspect of science literacy is also the basis for preparing components in the e-worksheet [33]. The text of discourse and the problems contained in the e-worksheet apply material learned in everyday life. The everyday phenomena presented concerning the material in this e-worksheet increase the learning motivation of learners [34].

In addition, as an electronic medium, e-worksheet must meet the aspects of software engineering and visual communication.[35]. Details of these aspects are, the media used is effective and efficient, reliable, easy to manage, simple in operation, the accuracy of application selection of media development, communicative, attractive and straightforward display, and navigation icons function correctly. This interactive electronic media used makes e-worksheet an exciting and effective learning medium and is considered to influence the interest and motivation of learners towards the learning presented [15], [13]. E-worksheet can also be used as an alternative learning medium to support students' digital literacy skills [34]. Thus, e-worksheets as teaching materials can help learners learn at home during pandemic times and conduct experimental activities to improve their science literacy [13,36,37]. Therefore, e-worksheets can be developed on different physical materials or other branches of science for further research.

4. Conclusion

Based on the study results, it can be concluded that e-worksheet sound waves based on science literacy are considered valid with very high reliability. Thus, this e-worksheet can be used in learning the physics of sound wave material during the period, especially to improve the science literacy skills of learners.

References

[1] Rahayuni G 2016 *J. Penelit. dan Pembelajaran IPA* **2** 131
[2] Şentürk C 2017 *IOSR J. Res. Method Educ.* **07** 51–62
[3] Vandegrift E V H, Beghetto R A, Eisen J S, O'Day P M, Raymer M G and Barber N C 2020 *J. Scholarsh. Teach. Learn.* **20** 15–30
[4] Rautela G S and Chowdhury K 2016 *Indian J. Hist. Sci.* **51** 494–510
[5] Cordon J M and Polong J D B 2020 *Integr. Sci. Educ. J.* **1** 72–8
[6] Arifuddin M, Aslamiah M, Misbah M and Dewantara D 2020 *Journal of Physics: Conference Series (Vol. 1422, No. 1, p. 012001). IOP Publishing*
[7] Misbah M, Dewantara D, Hasan S M and Annur S 2018 *Unnes Sci. Educ. J.* **7** 19–26
[8] Wen C T, Liu C C, Chang H Y, Chang C J, Chang M H, Fan Chiang S H, Yang C W and Hwang F K 2020 *Comput. Educ.* **149**
[9] Kurnia F, . Z and Fathurohman A 2014 *J. Inov. Dan Pembelajaran Fis.* **1** 43–7

[10] Ratini, Muchtar H, Suparman M A, Tamuri A H and Susanto E 2018 *J. Pendidik. IPA Indones.* **7** 458–66
[11] Parno, Yuliati L, Hermanto F M and Ali M 2020 *J. Pendidik. IPA Indones.* **9** 159–68
[12] Yildirim N, Kurt S and Ayas A 2011 *J. Turkish Sci. Educ.* **8** 44–58
[13] Febriansyah F, Herlina K, Nyeneng I D P and Abdurrahman A 2021 *Integr. Sci. Educ. Teach. Act. J.* **2** 59–73
[14] Utami R and Pujiriyanto 2020 *Proc. Int. Conf. Online Blended Learn. 2019 (ICOBL 2019)* **440** 169–73
[15] Arifin H R 2014 *Lang. Circ. J. Lang. Lit.* **9** 73–84
[16] Haryanto, Asrial, Dwi Wiwik Ernawati M, Syahri W and Sanova A 2019 *Int. J. Sci. Technol. Res.* **8** 1073–9
[17] Riyadi B, Ertikanto C and Suyatna A 2018 *Int. J. Res. -GRANTHAALAYAH* **6** 223–33
[18] Searmadi B P H 2016 *It-Edu* **1** 42–8
[19] Seruni R, Munawaroh S, Kurniadewi F and Nurjayadi M 2019 *J. Tadris Kim.* **4** 48–56
[20] Izzatunnisa I, Andayani Y and Hakim A 2019 *J. Pijar MIPA* **14** 49–54
[21] Prasetya C, Gani A and Sulastri S 2019 *J. Pendidik. Sains Indones.* **7** 34–41
[22] Sholihah N and Indana S 2018 *BioEdu Berk. Ilm. Pendidik. Biol.* **7** 177–86
[23] Koderi K, Latifah S, Fakhri J, Fauzan A and Sari Y P 2020 *J. Phys. Conf. Ser.* **1467**
[24] Fives H, Huebner W, Birnbaum A S and Nicolich M 2014 *Sci. Educ.* **98** 549–80
[25] Arikunto S 2013 *Prosedur Penelitian Suatu Pendekatan Praktik* (Jakarta: Rineka Cipta)
[26] Widoyoko E P S 2019 *Evaluasi Program Pembelajaran: Panduan Praktis Bagi Pendidik dan Calon Pendidik* (Yogyakarta: Pustaka Pelajar)
[27] Ismail I, Permanasari A and Setiawan W 2016 *J. Inov. Pendidik. IPA* **2** 190
[28] Widjajanti E 2008 Kualitas Lembar Kerja Siswa *Makal. disampaikan dalam Kegiat. Pengabdi. pada Masy. dengan judul "Pelatihan Penyusunan LKS Mata Pelajaran Kim. Berdasarkan Kurikulum Tingkat Satuan Pendidik. bagi Guru SMK/MAK" di Ruang Sidang Kim. FMIPA UNY pada tanggal 22 Agustus 2008*
[29] Silvianti R 2017 *Pengembangan Lembar Kerja Peserta Didik Berbasis Pendekatan Contextual Teaching and Learning (CTL) untuk Meningkatkan Kemampuan Komunikasi Matematis dan Self-Efficacy Siswa* (Universitas Lampung)
[30] Lestari L, Alberida H and Rahmi Y L 2018 *J. Eksakta Pendidik.* **2** 170
[31] Depdiknas 2008 *Panduan pengembangan bahan ajar* (Jakarta: Depdiknas)
[32] Purnama S 2010 Elemen Warna Dalam Pengembangan *J. Pendidik. Dasar Islam* 113–30
[33] Susiani S, Indana S and Indah N K 2017 *BioEdu Berk. Ilm. Pendidik. Biol.* **6** 60–7
[34] Sari D S, Widiyawati Y, Nurwahidah I, Masykuri M and Budiyanto C W 2021 T *Proc. 7th Int. Conf. Res. Implementation, Educ. Math. Sci. (ICRIEMS 2020)* **528** 647–54
[35] Supriadi R 2012 *J. Chem. Inf. Model.* **53** 1689–99
[36] Subekti M A S and Suparman 2019 *Proc. Steeem 2019* **1** 185–92
[37] Altun A and Kalkan Ö K 2021 *Educ. Stud.* **47** 403–21

The development of electronic teaching materials on linear impulse and linear momentum to improve students' scientific literacy

R Hidayat, A Salam M and D Dewantara

Physics Education Study Program, Faculty of Teacher Training and Education, Lambung Mangkurat University, Banjarmasin, Indonesia

rizkihidayatnm07@gmail.com

Abstract. Students' low scientific literacy competence and the unavailability of electronic teaching materials that match scientific literacy indicators that can be used in online learning encourage researchers to develop electronic teaching materials to improve students' scientific literacy. This research aims to produce and describe the appropriateness of electronic teaching materials used in the learning process to enhance scientific literacy. The electronic teaching materials developed are based on scientific literacy indicators, namely explaining scientific phenomena, interpreting data and providing scientific evidence, designing and evaluating scientific investigations. The type of research used is research and development of the ADDIE model. The appropriateness of electronic teaching materials in terms of validity, practicality, and effectiveness. Research data was obtained from validation sheets given to three validators, student response questionnaires and learning outcomes tests. The data were analyzed by validity test, practicality test, and effectiveness test using N-gain score. The results showed that: 1) the validity of teaching materials obtained a score of 3.52 in the very high category, 2) the practicality of electronic teaching materials based on student response questionnaires obtained a percentage of 80.81% in the practical category, 3) the effectiveness of electronic teaching materials obtained a score of 0.44 in the medium category and the students' scientific literacy ability 44.11% in the medium category. Based on the results of the study, impulse and linear momentum electronic teaching materials have been produced to improve students' scientific literacy that is appropriate for use in the learning process. This research has implications, namely increasing students' scientific literacy, especially on impulse and linear momentum material.

1. Introduction

The 21st century is marked by the development of science and technology in the field of life in society, especially information and communication technology which is very rapid, where everything can be regulated by technology [1]. The development of science and technology will bring education to face increasingly severe challenges, one of which is that education is expected to produce human resources who have good abilities in facing various challenges in life in the era of globalization [2]. In the 21st century, the development of the world is becoming increasingly rapid and complex. A paradigm shift in the education system is needed to provide the 21st-century skills that learners need to face the era of globalization [3].

One way to face the challenges of the 21st century is by "science literacy". Individuals who are scientifically literate can solve problems in everyday life and produce useful scientific products by using scientific information. Science education plays a vital role for individuals to live their world of life [4]. Developed society believes that literacy is an essential aspect for the world community during the rapid development of the times. Literacy is one of the rights for every citizen in developed countries that the state must facilitate. Indonesia, a developing country, also strives for literacy as a sustainable agenda [5].

Human development and quality can be improved with literacy education which plays a vital role in improving the quality of students. It must be familiarized to students from an early age starting from the closest environment, namely family, and the educational environment, namely schools to the community environment [6]. Based on this, students must master literacy to achieve and live a better quality of life for the present to the future [7]

Literacy skills have an essential role in the success of the younger generation. The younger generation with good literacy skills will easily understand information verbally and in writing. The competencies possessed in life can be supported by the ability of the younger generation to master literacy. The competencies possessed can become more potent if the younger generation masters literacy, namely being able to sort out information that can support the success of their lives [8].

Scientific literacy can be defined as scientific skills and knowledge to identify questions, acquire new knowledge, explain scientific phenomena, draw conclusions based on facts, understand the characteristics of science, have an awareness of how science and technology can shape a natural, intellectual, and cultural environment, and desire to participate in and care about issues related to science [9]. Scientific literacy is a 21st-century life skill that can make quality human resources and improve living standards a determining factor for the progress of a nation. Every individual must understand scientific literacy because scientific literacy can make individuals learn to understand the surrounding environment. In a modern era such as the 21st century, the problems faced by individuals and society are often related to science and technology [10].

Globalization makes the world seem borderless, resulting in comparisons between schools in the form of curricula, assessment methods, and student achievements on an international scale. It makes schools urged to be excellent and be competitive [11]. International programs to compare student achievement, such as Trends in International Mathematical and Science Study (TIMMS), started in 1995 and is carried out every four years. Programmes for International Student Assessment (PISA), which began in 2000, are carried out regularly every three years. The year is an example of an international student comparison program, especially in science and mathematics [12].

Based on a survey conducted by PISA in 2018, released on December 9, 2019, Indonesia was ranked 70th out of 78 with a score of 396 [13]. The causes of the low level of scientific literacy of students in Indonesia related to the ongoing education process include the applied education system, the selection of models, approaches, methods and learning strategies, selection of learning resources, student learning styles, and learning infrastructure [14]. Based on the results of pre-research in class X of SMAN 1 Gambut which was distributed using Google Form, it showed that students' scientific literacy skills were still low; 92.31% of students were still unable to explain scientific phenomena, 86.23% of students were still unable to evaluate and design scientific investigations, and 96.15% of students were not able to interpret data and provide scientific evidence.

Based on an interview with one of the teachers at SMAN 1 Gambut, the learning used during the Covid-19 pandemic uses Google Classroom which is intended not to spend a lot of students' internet quota. The teaching materials are in the form of files in .docx or .pdf formats downloaded via the internet, and assignments in the form of questions in the students' worksheets are then distributed to students. It is due to teachers who do not master the technology that supports the creation of online learning materials. The learning carried out does not refer to scientific literacy indicators but only follows teaching materials on the internet and students worksheets. Based on the results of the needs analysis conducted by researchers, it is necessary to use teaching materials to support online learning. The teaching materials are expected to be able to train students' scientific literacy.

One solution to train scientific literacy at SMAN 1 Gambut is with learning media. Learning media can convey messages from the teacher to students in learning so that it can stimulate the thoughts, feelings, and attention of students to achieve learning objectives [15]. One solution to train scientific literacy at SMAN 1 Gambut is with learning media. Learning media can convey messages from the teacher to students in learning so that it can stimulate the thoughts, feelings, and attention of students to achieve learning objectives [16].

Learning media that can be developed to train literacy are digital teaching materials. Digital teaching materials are all materials that teachers must prepare to support the learning process in non-printed form, namely digital [17]. Digital teaching materials can be created using flipbook software. Flipbook has the advantage that it can convert pdf files in an interactive form in which images, audio, video, animation and hyperlinks can be inserted [18]. A similar study conducted by Firdaus, Pursitasari, Permana, & Suhardi shows that scientific literacy electronic teaching materials are suitable for use as teaching materials [19] The electronic teaching materials used can improve students' scientific literacy competence. Likewise, Sumantri & Khaliq concluded that the 3D Page Flip-based science literacy E-Book is suitable for use as a medium to improve students' scientific literacy [20]. The electronic teaching materials developed are different from previous research, namely all local cultures and virtual practicums integrated with learning materials. Teach electronically without having to install additional applications, and use the hyperlink feature to go to practice questions and move pages quickly.

Based on the background of the study, the researchers developed impulse and linear momentum electronic teaching materials to improve students' scientific literacy. This research aims to produce and describe the appropriateness of electronic teaching materials used in the learning process to improve students' scientific literacy.

2. Method

The development model used for this research is the ADDIE (Analysis, Design, Development, Implementation and Evaluation) model [21]. ADDIE has the following steps:

Table 1. ADDIE Development Model Steps

Step	Explanation
Analysis	Curriculum analysis, needs analysis, student characteristics analysis and teaching material analysis.
Design	Designing electronic teaching materials starts from documents, converting them into pdf form and saving them on the Flip PDF Professional.
Development	Develop electronic teaching materials and upload them so that they can be accessed online by students.
Implementation	The application of teaching materials that have been made in class X MIPA 1 SMAN 1 Gambut was carried out for three meetings where each meeting discusses one sub-chapter.
Evaluation	A formative evaluation was carried out to collect data at each stage used to improve the developed teaching materials, and a summative evaluation was carried out at the end of the study to determine the effectiveness of electronic teaching materials to improve students' scientific literacy.

The research subject is linear momentum impulse electronics teaching materials. The research object is the appropriateness of impulse and linear momentum electronic teaching materials to improve students' scientific literacy in terms of validity, practicality, and effectiveness. Data collection techniques were carried out with validation sheets, student response questionnaires and learning outcomes tests. Before the research was conducted, the researcher conducted a preliminary study by interviewing and giving questions with scientific literacy indicators to find out the problems in SMAN 1 Gambut and the scientific literacy competence of students.

Student response questionnaires were given through a google form containing 18 statements. The response questionnaire given to students had nine positive statements and nine negative statements with options, namely strongly agree, agree, disagree, and strongly disagree. The response questionnaire given is the basis of the practicality assessment of the developed electronic teaching materials. Learning outcomes test is provided through Google Form to students. The questions in the learning outcomes test refer to the scientific literacy competency indicators by the OECD (Organization for Economics Co-operation and Development) in the form of essays. Eligibility of electronic teaching materials is stated from the results of the validation sheet assessment with a minimum category of good [22], for the practicality of electronic teaching materials based on a response questionnaire with a minimal practical category [23], and the minimal effectiveness of the moderate category based on the N-Gain obtained from the results of the pre-test and posttest [24].

3. Result and Conclusion

The developed electronic teaching materials are to improve students' scientific literacy on impulse and linear momentum materials. Electronic teaching materials were developed using Flip PDF Professional software on linear impulse and momentum materials that can be accessed online via Android iOS-based smartphone devices, laptops, and computers. The following is an example of the display of electronic teaching materials accessed on computers in figure 1.

Figure 1. Cover Display of Electronic Teaching Materials Accessed Through Computers

Display of electronic teaching materials accessed on smartphone Android and iOS shown in Figure 2.

Figure 2. Initial Display of Electronic Teaching Materials Accessed Through Android Smartphones (left) and iOS Smartphones (right)

Electronic teaching materials were validated by three validators consisting of two academics and one practitioner. The results of the validation of electronic teaching materials by validators are shown in Table 2.

Table 2. The results of the validity of electronic teaching materials

No	Indicator	Assessment Aspect	Average	Category
1	Content Aspect	1. Software Engineering 2. Content Quality 3. Organization 4. Language 5. Evaluation	3.49	Very Good
2	Display Aspect	1. Visual Communication 2. Format 3. Attractiveness 4. Font Shape and Size	3.56	Very Good
Validity			3.52	Very Good
Reliability			0.97	Very High

Based on the table above, the results of the validator's assessment for the validity of electronic teaching materials are categorized as very good with a score of 3.52 and very high reliability, namely 0.97. It shows that the developed electronic teaching materials can be tested on students of class X MIPA 1 SMAN 1 Gambut.

The validation of electronic teaching materials on the content aspect is in the very good category. Assessment on the content aspect includes aspects of software engineering, content quality, language, organization, and evaluation. The electronic teaching materials developed have clear instructions for use. The use of RAM based on the Device Info application is 38 MB on the Google Chrome browser; this means that there is not too much RAM usage because the devices owned by students have RAM capacities ranging from 2 GB to 8 GB. The software used in the manufacture of electronic teaching materials, namely Flip PDF Professional, is also considered appropriate in manufacturing electronic teaching materials. The output of this application is in the form of HTML, which can be appropriately run using a browser without using additional applications. Electronic teaching materials developed with the Flip PDF Professional application also do not experience errors when accessed.

The electronic teaching materials developed are adjusted to the learning indicators to be achieved as well as scientific literacy indicators by the OECD. Electronic teaching materials are also equipped with pictures, videos and links that aim to attract students' interest. The material in electronic teaching materials is systematically arranged which is divided into three subchapters. The language used in these electronic teaching materials is systematic, simple and easy to understand by students. It is in line with Prastowo, which states that electronic teaching materials must be arranged systematically using language that students easily understand according to students' level of knowledge [25].

Another part of this electronic teaching material is evaluation. Evaluation aims to find out whether students have met the standards or not. The standards are determined based on the learning objectives to be achieved. The evaluations made are also adjusted to the OECD scientific literacy indicators [26].

The appearance of electronic teaching materials depends on visual communication, format, attractiveness, and the shape and size of the letters. The layout and color of electronic teaching materials are consistent in each sub-chapter. Further, the layout of electronic teaching materials is dominated by blue. Blue is often associated with nature, such as the sky and the sea, which gives a relaxed and calm impression. Hideako Chijiwa, in his book entitled "Color Harmony", classifies blue as a cool color. Cool colors tend to make it more comfortable to look at [27]. It makes it possible to make access to electronic teaching materials feel comfortable viewing the display of them being accessed.

The column format used in electronic teaching materials is a double-column that divides the page into two columns. Daryanto suggests that to produce good and attractive electronic teaching materials, it is necessary to design and develop them by considering the appropriate format, attractiveness, and font size. The column format used must be proportional to both the single-column format and the double-column format. The use of multiple columns must have a proportional ratio of the distance between columns [28].

One of the attractiveness of electronic teaching materials is on the cover page. The front cover combines various colors, images and appropriate font sizes. The colors used on the cover page are black and blue. The images used are Newton's Swings and billiards balls, which definitely, are closely related to the material of impulse and linear momentum. The white color is used as the title of the electronic teaching material because the writing background is black. The experimental results of Fonda stated that white writing on a black background and white writing on a red background has a high level of clarity with almost the same value [29].

The contents of the electronic teaching materials that are made contain pictures and illustrations that are presented attractively and the right bold and italics as an emphasis on certain words. The fonts used in this electronic teaching material are Sans Serif fonts, namely Comic Sans MS and Pineapple Grass. Sans Serif fonts were chosen because they are easier to read than Serif fonts. It is supported by Dogusoy, Cicek and Cagiltay, who state that reading electronic text in Serif and Sans Serif typography when compared is generally found that the reading of Sans Serif typefaces is faster and more accurate than the Serif type [30]. The use of the Sans Serif typeface is also supported by Fitrialny & Ardoni. Sans Serif letters have the characteristics of not having hooks, only stems and stems, simple and easy to read, suitable for websites, e-books, databases, and display monitor base. Sans Serif fonts look more straightforward and easier to read in a variety of sizes [31]

The practicality of electronic teaching materials developed based on student response questionnaires distributed through a google form, as shown in Table 3.

Table 3. The Results of the Questionnaire of the Practicality of Electronic Teaching Materials

No.	Indicator	Percentage	Category
1	Benefit Aspect	84.31%	Very Practical
2	Efiiciency Aspect	71.16%	Very Practical
3	Convenience Aspect	85.95%	Very Practical
Average		80.81%	Very Practical

On the aspect of benefits, 94.4% of students stated that electronic teaching materials helped students learn physics. 72.2% of students stated that the electronic teaching materials made the learning atmosphere fun and not boring. A pleasant learning atmosphere can make students have interest and motivation in learning [32]. It is in accordance with the results of the response questionnaire which stated that 88.9% of students were motivated to study the material of impulse and linear momentum. Electronic teaching materials can help students learn independently because they have a communicative and two-way language, making it easier to learn. 83.3% of students stated that electronic teaching materials helped students in learning independently. 83.3% of students stated that electronic teaching materials made it easier to understand learning materials. The benefits aspect obtained a percentage of 84.31%, which is included in the very practical category [33].

The practicality of electronic teaching materials is also assessed based on the aspect of efficiency. Learning using electronic teaching materials is assessed on time by 72.2% of students. This time efficiency is because students can use hyperlinks to move to the desired subchapter or page quickly. The hyperlink feature also allows students to access other content on electronic teaching materials simultaneously. Video features that can be accessed without having to leave electronic teaching materials also make time more efficient. Based on the website speed test via tools.pingdom.com, the first page loading time is 2.24 s, and the time it takes to switch pages is 1.2 s. During the use of electronic teaching materials, 94.4% of students stated that electronic teaching materials were not wasteful in the use of power, and 61.1% of students stated that electronic teaching materials did not consume a lot of internet quota. In the aspect of efficiency, the percentage obtained is 71.16% in the practical category.

The last aspect of the practicality assessment is the aspect of convenience. The language used in electronic teaching materials is considered easy to understand by 94.4% of students. 100% of students stated that the size of the letters in electronic teaching materials did not complicate the learning process. It is due to the use of Sans Serif fonts which are quite easy to read, and the size is 14 so that the size is appropriate for use in the developed electronic teaching materials. Electronic teaching materials that are made can be accessed easily on various networks according to 94.4% of students.

Based on the discussion of the three practical aspects of electronic teaching materials, the overall practicality of electronic teaching materials through student response questionnaires obtained an average of 80.81%, so it can be concluded that electronic teaching materials are practically used in learning. The practical electronic teaching materials can help students understand learning materials [34].

The effectiveness of electronic teaching materials is obtained by calculating the N-gain score obtained based on the pretest and posttest to students of class X MIPA 1 SMAN 1 Gambut. The following are the results of the calculation of the N-gain score.

Table 4. The Result N-Gain Score

Pre-test Average	Post-test Average	h	<g>	Category
0.00	44.11	0	0.44	Medium

The average of scientific literacy competence is shown in table 5.

Table 5. The Result N-Gain Score

Pre-test Average	Category	h	Post-test Average	Category
0.00	Low	0	44,11	Medium

The developed electronic teaching materials train students to be able to solve problems on the Learning Outcomes Test. The Learning Outcome Test is given in the form of 7 essay questions. The level of questions on The Learning Outcome Test ranges from C2 to C5. Table 4 shows the average score of the students' pre-test for scientific literacy competence is 0.00%. During the pre-test, none of the students met the Minimum Completeness of Criteria at school, which was 75. The average posttest score for scientific literacy competence showed an increase to 44.11%. Based on the N-gain score, the effectiveness of electronic teaching materials is found in the medium category.

The indicator explaining scientific phenomena gets 52.94%, receiving the highest percentage compared to other indicators. The scientific phenomenon indicator gets the most elevated rate because the questions used are at the lowest level compared to other scientific literacy indicators. It is in accordance with Nabilah & Sahala findings on the material of linear impulse and momentum stating that the lower the level of the question, the higher the percentage of students' cognitive abilities [35]. Indicators interpreting data and scientific evidence get a percentage of 50%. The lowest percentage was obtained on the indicators of designing and evaluating scientific investigations, namely the percentage was only 29.41%. It is similar to Setiawan's research which shows that the ability to design and evaluate scientific investigations has the lowest increase compared to other scientific literacy competencies. It happens because students are doing practical activities for the first time, so they are not used to designing scientific investigations [36].

The results obtained are in accordance with the behavioristic learning theory. Behavioristic learning theory holds that learning is a change in student behavior due to the learning process itself. The learning process is a stimulus that the teacher can control to obtain the expected student response. Based on the three indicators of scientific literacy, it has been found that students' behaviour who initially had low scientific literacy skills turned into a medium category of scientific literacy skills after the teaching materials were applied.

Overall, the developed electronic teaching materials can improve student learning outcomes. It is in line with the results of Lestari's research which states that using electronic teaching materials is proven to be significantly effective in improving student learning outcomes [37]. It shows that electronic teaching materials have been developed to be an excellent online learning tool during the current Covid-19 pandemic.

The previous description shows that the electronic teaching materials developed can be said to be effective because they are in accordance with the expected goals, namely to increase students' scientific literacy. It is in line with Tegeh opinion that the effectiveness of a product is related to the extent to which the product can achieve certain goals [38].

The developed electronic teaching materials are declared appropriate to be used in the learning process to improve students' scientific literacy. It is based on the validity of the very good category, the practicality of the very practical category, and the effectiveness and competence of scientific literacy in the medium category. The drawbacks contained in the developed electronic teaching materials are not knowing whether students access the electronic teaching materials provided or not. This research has implications, namely increasing students' scientific literacy, especially on impulse and linear momentum material. The limitation in this research is that it has not included all the sub-indicators of scientific literacy in the learning outcome test; further researchers are expected to include all the sub-indicators of scientific literacy in the learning outcome test.

4. Conclusion

Based on the results of the study, electronic teaching materials are suitable for use in the learning process to improve students' scientific literacy. It is based on a) the validity of electronic teaching materials in the very high category, b) the practicality of electronic teaching materials as seen from the questionnaire responses of students included in the very practical category, c) the effectiveness of electronic teaching materials seen from the N-gain score is in the medium category, and the scientific literacy competence of students is in the medium category so that it can be used in the learning process to improve students' scientific literacy. This research has implications, namely increasing students' scientific literacy, especially on impulse and linear momentum material. Based on the shortcomings in the developed electronic teaching materials, it is hoped that similar research can find out whether students have access to electronic teaching materials.

References

[1] Ojimba 2013 *Educational Research International* **2**
[2] Yuliati Y 2017 *Jurnal Cakrawala Pendas* **3** 21-18
[3] Soh T, Arsad N, and Osman K 2010 *Procedia Social and Behavioral Sciences*, 546-554
[4] Nofiana M and Julianto T 2017 *Jurnal Sains Nasional dan Humaniora* **1** 77-84
[5] Masitoh S *Proceedings of The ICECRS* **1** 13-34
[6] Ningsih A T, Ruhiyat Y, and Saefullah A 2020 *Jurnal Untirta* **3** 341-347
[7] Kemendikbud 2017 *Peta Jalan Gerakan Literasi Nasional* (Jakarta : Kemendikbud)
[8] Irianto P O and Febrianti L Y 2017 *The 1st Education and Language International Conference Proceedings Center for International Language Development of Unissula* 640-647
[9] OECD 2016 *PISA 2015 Assesment and Analytical Framework : Science, Readning, Mathematic and Financial Literacy* (Paris : OECD Publishing)
[10] Astutik S Lesmono A D and Adani D A 2019 *Saintifika* **2** 9-22
[11] Amin M 2017 *Prosiding Seminar Nasional L III Tahun 2017* 9-20
[12] Taroreh B and Noviyanti H 2020 *Prosiding Seminar Nasional Pendidikan Fisika Universitas Pekalongan,* **1** 271-280
[13] Schleicher A 2019 *PISA 2018 Insight and Interpretations* (Paris: OECD)
[14] Wahyu E, Fathurrohman A and Markos S 2016 *Jurnal Inovasi dan Pembelajaran Fisika.*
[15] Duludu U 2017 *Buku Ajar Kurikulum Bahan dan Media Pembelajaran PLS* (Yogyakarta: Deepublish)
[16] Jauhari M I 2018 *Jurnal Piwulang* **1** 54-67
[17] Nana N and Surahman E 2019 *Prosiding Seminar Nasional Fisika dan Aplikasinya* **4**

[18] Maf'ula A, Hastuti U S and Rohman F 2017 *Jurnal Pendidikan* **2** 1450-1455
[19] Firdaus Z F, Pursitasari I D, Permana I and Suhardi E 2020. *Pancasakti Science Education Journal* **5** 30-37
[20] Sumantri F N 2020 *Jurnal Inovasi Pendidikan Fisika* **9** 479-483.
[21] Januszewski A and Molenda M 2008 *Technology: A Definition With Commentary* (New York: Lawrence Erlbaum Associate)
[22] Widoyoko E 2016 *Evaluasi Program Pembelajaran* (Yogyakarta: Pustaka Pelajar)
[23] Sudijono A 2009 *Pengantar Statistik Pendidikan* (Jakarta: Rajawali Press)
[24] Hake R. R 1998 *American Journal Of Physics* **66** 64-74
[25] Prastowo A 2012 *Panduan Kretaif Membuat Bahan Ajar Inovatif* (Yogyakarta : Diva Press)
[26] Riadi A 2017 *Jurnal Kopertis Wilayah XI Kalimantan* **15** 2-9
[27] Wantoro S 2013 *Jurnal Ilmiah Unikom* **11** 188-193
[28] Daryanto 2013 *Menyusun Modul : Bahan Ajar untuk Persiapan Guru dalam Mengajar* (Yogyakarta: Gava Media)
[29] Fonda D 2013 *Jurnal Titra* **1** 1-8
[30] Dogusoy B Cicek F and Cagiltay K 2016 *International Conference of Design* 578-586
[31] Fitrialny R and Ardoni 2013 *Ilmu Informasi Perpustakaan dan Kearsipan* **2** 259-266
[32] Hartanti D 2019 *Prosiding Seminar Nasional PEP* **1** 78-84
[33] Laili I 2019 *Jurnal Ilmiah Pendidikan dan Pembelajaran* **3** 306-315
[34] Farenta A S 2016 *Jurnal Pendidikan* **1** 1159-1168
[35] Nabilah M and Sahala H S 2020 *Jurnal Inovasi Penelitian dan Pembelajaran Fisika* **1** 1-7.
[36] Setiawan A R 2019 *Journal of Natural Science* **2** 83-94.
[37] Lestari H D 2020 *Journal of Education Technology* **4** 73-79
[38] Tegeh I, Jampel I and & Pudjawan K 2014 *Model penelitian pengembangan* (Yogyakarta: Graha Ilmu)

Publications related to virtual laboratories during the covid 19 pandemic: A bibliometric review and analysis

S Saparini[1]*, N Andriani[1], and M Misbah[2]

[1]Sriwijaya University, Palembang, Indonesia
[2]Lambung Mangkurat University, Banjarmasin, Indonesia

saparini@fkip.unsri.ac.id

Abstract. During the COVID-19 pandemic, virtual laboratories are instrumental in learning physics concepts. A virtual laboratory is one media that can be used as a solution to carry out concept discovery activities through simulation using computer software. This article was created to explain the trend of publications related to virtual laboratories during the COVID-19 pandemic for 2020-2021 through bibliometric analysis of the literature. The papers analyzed were obtained using the Publish or Perish (PoP) software on the Google Scholar database. The following publication trend expresses in the form of graphic visualization, which is analyzed using VOSviewer software. Overall, articles with the keyword virtual laboratory published in the late 2020-2021 period focus on its relationship with Covid 19 and evaluation. This focus of research shows that during the past year, the focus of research is how to evaluate learning through virtual laboratories.

1. Introduction

Students' experience in carrying out experiments in the laboratory is a factor in engineering and science education [1]. Science learning accompanied by the implementation of experiments in the laboratory is expected to be able to provide experience for students in the form of experience in proving scientific concepts that have been studied or discovering new ideas through experimental activities, as well as being able to remember and understand science concepts better [2]. In addition, laboratory activities can teach performance skills to find an idea and communicate it [3]. Students can carry out laboratory activities face-to-face in the laboratory or virtual laboratories.

An experimental activity carried out using computer software through experimental simulations that resemble practical exercises directly is referred to as a virtual laboratory [4][5][6]. There have been many studies related to the need for virtual laboratories in science education. A literature study conducted in 2010 showed that a virtual laboratory based on a constructivist learning approach at the high school and university level showed several advantages of a virtual laboratory [7]. The study results showed that students better reported experimental results through virtual laboratories, requiring shorter practical times than face-to-face laboratories. Students became more focused on carrying out experiments, all students could participate in experiments, and the implementation of experiments became more efficient. Interesting because students can repeat the investigation if there is an error in data collection without coming to the laboratory [7].

Along with the development of science and technology, especially information and communication technology, the description of the integration of the implementation of virtual laboratories continues to grow. There is a need for a new set of systems to build a modern virtual laboratory system that can create exciting experiences through a virtual laboratory environment [8]. Before the COVID-19 pandemic at the end of 2019, the focus of research related to virtual laboratories was the need for virtual laboratories to complement conducting experiments in laboratories. However, this opinion shifted after the COVID-19 pandemic occurred.

Currently, schools are struggling to change the education system from face-to-face learning to distance learning. The focus of implementing science learning during the COVID-19 pandemic is how to develop appropriate science teaching strategies and techniques related to the implementation of distance learning [9]. However, the performance of a virtual laboratory has not been able to provide the same experience as a face-to-face laboratory [2]. With the passage of time and the emergence of several research publications related to virtual laboratories, concerns about the implementation of virtual laboratories began to decrease. This concern is accompanied by the emergence of recognition that social and interpersonal interactions can also be developed and occur through distance learning [10]. During the COVID-19 pandemic, several research results reported solutions related to the implementation of science learning, especially in the performance of experimental activities. One way to find out the trend of these publications is through bibliometric analysis.

Bibliometric analysis is a quantitative study of bibliographic materials that can provide an overview of studies grouped by articles, authors, and journals [11]. This study discusses bibliometric analysis with the keywords virtual laboratory in the Google Scholar database. The purpose of this study is to determine the trend of publications related to virtual laboratories during the covid 19 pandemics, seen from the keywords, and the number of citations in articles published during the 2020-2021 timeframe.

2. Method

This research is a literature study to determine the trend of publications related to virtual laboratories during the covid 19 pandemic. This research was carried out in 4 stages, namely: (1) selection of the research criteria, (2) selection of a period and type of documents, (3) bibliometric analysis, (4) co-occurrences analysis [12].

2.1. Selection of the research criteria

The scientific articles analyzed came from the publication of the Google Scholar database with the keyword "virtual laboratory." Article search is carried out using the Publish or Perish (PoP) software.

2.2. Selection of a period and type documents

The covid 19 pandemics have been going on since 2019, so scientific articles related to the analyzed virtual laboratory keywords were selected for the 2020-2021 period. The publication in 2021 is limited to articles published until August 2021.

2.3. Bibliometric analysis

The bibliometric analysis stage is the stage to analyze scientific articles related to virtual laboratories, including the number of publications related to these keywords in the 2020-2021 period, the most relevant publications related to virtual laboratories during the covid 19 pandemic, and analysis of publication trends related to virtual laboratories during the covid 19 pandemic.

2.4. Co-occurrences analysis

The last step in this research is to visualize the data obtained using VOSviewer software. The data visualization in question is a visualization of the relationship between publication keywords, publication trends related to virtual laboratories during 2020-2021, and the density of relationships between keywords and virtual laboratories. Data analysis was carried out by paying attention to the virtual laboratory keywords in the titles, keywords, and contents of publications related to virtual laboratories for the 2020-2021 period.

3. Result and Discussion

The research results obtained are expected to provide an overview of publication trends related to virtual laboratory research during the covid 19 pandemics. Table 1 shows articles obtained using Publish and Pheris (PoP) software based on the Google Scholar database.

Table 1. PoP metric data output

Metric Data	Output
Keyword	"virtual laboratory"
Year of publication	2020-2021
Number of articles	500
Citation	1009
Citation/year	1009
Citation/article	2,02
Author/article	2,82
h_index	14
g_index	24
hl_normal	9
hl_yearly	9

Based on the data obtained through the PoP software, the ten articles most relevant to the research objectives were then selected, presented in table 2. The relevance of these articles was determined based on the existence of a virtual laboratory relationship with covid 19. Table 2 shows that the pieces with the highest level of relevance have been cited nine times, although the article was published in 2021. The number of citations shows that the implementation of virtual experiments through virtual laboratories is essential as a reference for similar research.

Table 2. Top 10 articles with the highest relevance by PoP

No	Year of Publication	Author	Title	Journal	Citation	Publisher
1	2021	Kapilan, N Vidhya, P Gao, X Z	Virtual laboratory: A boon to the mechanical engineering education during a covid-19 pandemic[13]	Higher Education for the Future	9	Sage Journals
2	2021	Radhamani, R Kumar, D Nizar, N Achuthan, K	What virtual laboratory usage tells us about laboratory skill education pre-and post-COVID-19: Focus on use, behavior, intention, and adoption[14]	Education and Information Technologies	1	Springer Link
3	2020	M. García-Vela1, J.L. Zambrano, D.A. Falquez, W. Pincay-Musso, K.B. Duque, N.V. Zumba, M.B. Barcia, J.I. Méndez, P.E. Valverde,	Management of virtual laboratory experiments in the geosciences field in the time of COVID-19 pandemic[15]	Proceedings of ICERI2020 Conference	2	Researchgate.net

		P.L. Romero-Crespo, L. Jordá-Bordehore				
4	2021	Yap, W H Teoh, M L Tang, Y Q Goh, B H	Exploring the use of virtual laboratory simulations before, during, and post COVID-19 recovery phase: An Animal Biotechnology case study[16]	Biochemistry and Molecular Biology Education	0	Wiley Online Library
5	2020	Sherrer, S M	A virtual laboratory module exploring photosynthesis during COVID-19[17]	Biochemistry and Molecular Biology Education	0	Wiley Online Library
6	2020	Vasiliadou, R	Virtual laboratories during coronavirus (COVID-19) pandemic[18]	Biochemistry and Molecular Biology Education	24	Wiley Online Library
7	2021	Bunyamin E Siregar E Kusumawardani D	Analysis of Development Needs Virtual Laboratory for Learning Vocational School Practices in the Pandemic Time Covid-19[19]	Journal of Education Research and Evaluation	0	Ejournal. undhiksa.ac.id
8	2021	Mafudi, I Handhika, J	Virtual Laboratory: Using Electronic Workbench as Alternative Learning Physics in Covid-19 Mass Pandemic[20]	Impulse: Journal of Research and Innovation in Physics Education	0	e-journal UIN Sunan Kalijaga Yogyakarta
9	2020	Lohrasbi, M	Chemistry Training with the help of Virtual Laboratory in the Corona Era[21]	Research in Chemistry Education	0	Scientific Information Database (SID.Ir)
10	2021	Tamura, N Taira, H Umemura, K	Challenges of a "Virtual" Biophysics Laboratory during the "Stay home" period caused by COVID-19[22]	Journal of Physics: Conference Series	0	IOP Publishing

VOSviewer is a computer program that can create, visualize, and explore data mapping in bibliometric science and can be used to analyze large amounts of data [23]. Data visualization using VOSviewer in this study are network visualization, overlay visualization, and density visualization.

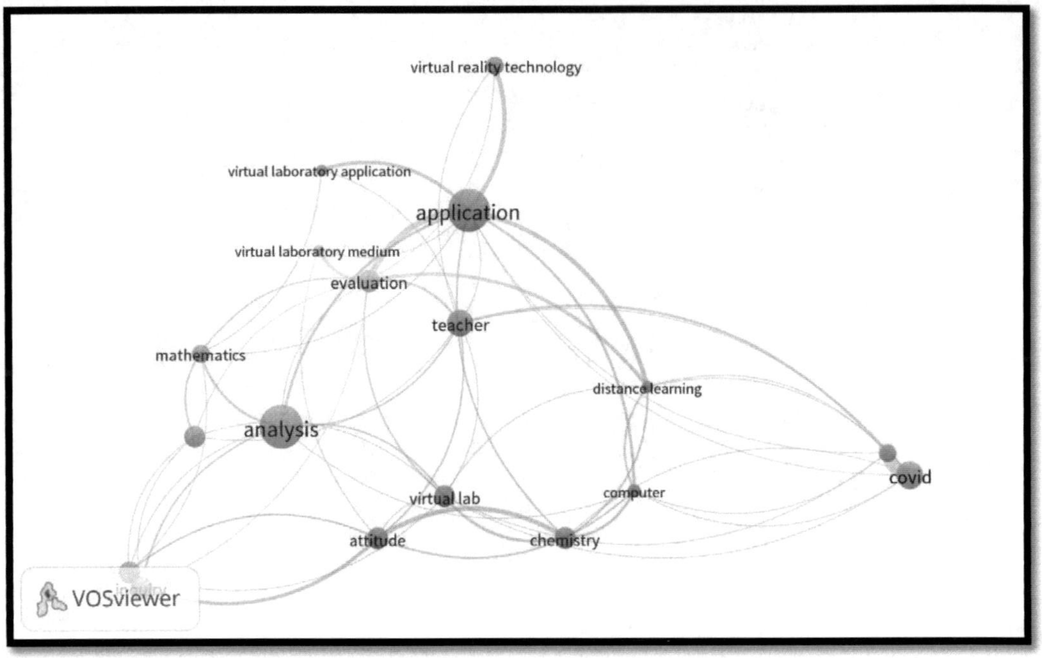

Figure 1. The Visualization of the network in google scholar

Figure 1 shows a visualization of the relationship between keywords in the publications contained in the Google Scholar database. The color difference in Figure 1 shows that there are five keyword clusters, namely blue, yellow, green, purple, and red color clusters. Keywords in closely related publications will usually be located close to each other and are indicated by the same color line. Based on Figure 1, the virtual laboratory keywords are shown in red and are closely related to publications related to attitude, chemistry, computers, and distance learning.

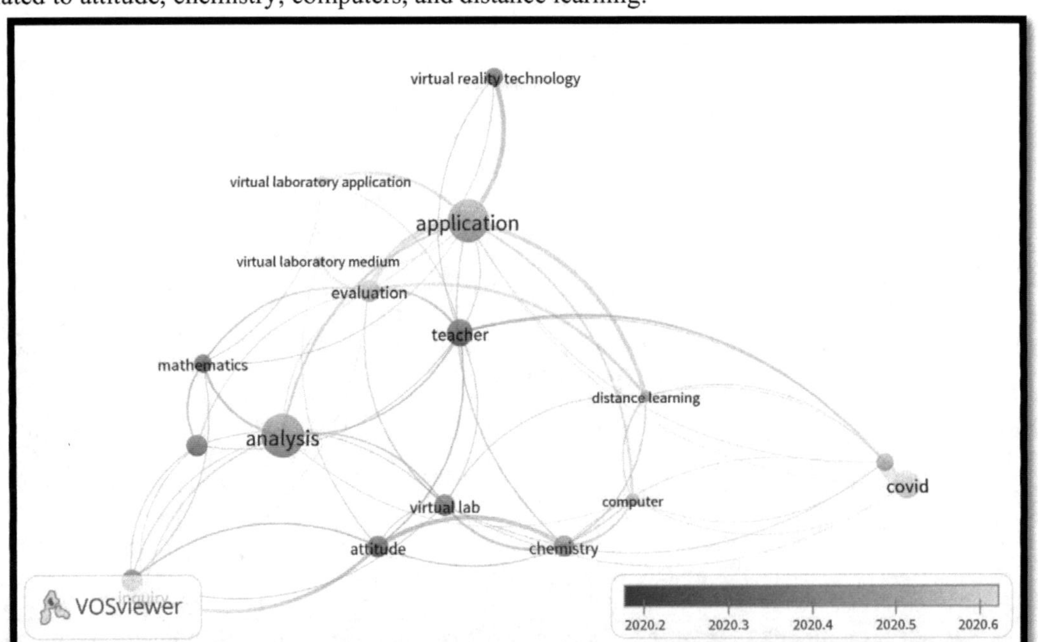

Figure 2. The visualization of the google scholar databased overlay

Figure 2 shows the level of novelty of the publication keywords through the color change from purple to yellow. Table 2 keywords for the latest publications are covid, evaluation, inquiry, virtual laboratory medium, and virtual laboratory application. However, when viewed from the relationship with virtual laboratories, the keywords covid and evaluation of publications have the closest relationship.

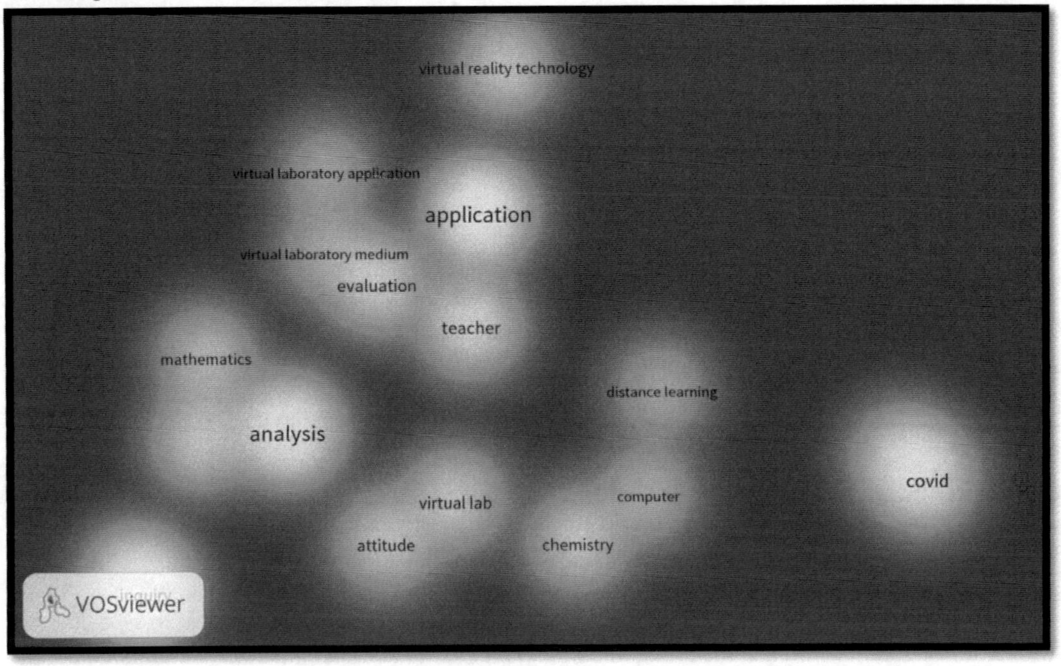

Figure 3. The Visualization of the density of the google scholar database

Figure 3 shows the density of relationships between keywords covered in the same region. The bold color indicates a very close relationship between these keywords. The information obtained from Figure 3 is that the keyword attitude has the most intimate relationship with the virtual laboratory. This result shows that the discussion regarding the implementation of virtual laboratories during the COVID-19 pandemic did not dampen the teacher's intention to provide learning that does not just convey material but still pays attention to student attitudes.

The information that can be summarized based on the visualization of images 1, 2, and 3 is the latest trend of publications related to virtual laboratories during the covid 19 pandemic, related to the keywords of covid publications and evaluation. However, Figure 3 shows that many publications still focus on the relationship between virtual laboratories and attitude.

Evaluation is a procedure used to determine whether the subject (i.e., students) meets the predetermined criteria [23]. Discussions on evaluation in the implementation of virtual laboratories become more dominant during the performance of distance learning. The initial concept of implementing a virtual laboratory in science learning is as a solution to science problems. The problem is related to the difficulty of teaching abstract science concepts, which causes students to be less interested in learning science. The idea of virtual laboratories has now shifted to including didactic aspects in virtual laboratories and how to evaluate them [8].

The novelty of this research lies in the success of providing an overview of publication trends related to virtual laboratories during the COVID-19 pandemic using the Google Scholar database. However, the drawback of this study is that it has not revealed the publication trend in terms of the author and the author's country of origin, so that it has not been able to provide an overview of the publication trend related to virtual laboratories during the COVID-19 pandemic as a whole.

4. Conclusion

Overall, articles with the keyword virtual laboratory published in the late 2020-2021 period focus on its relationship with Covid 19 and evaluation. The results of this study can be used as a reference to carry out further research related to the development of learning tools in virtual laboratories and the implementation of these learning tools. During the past year, the focus of research has been on evaluating learning through virtual laboratories. Some suggestions related to this research are: (1) the need for an analysis of the relationship between the author and the country of origin, (2) changing the database for collecting publication articles other than Google Scholar, and (3) the need for research on how to evaluate knowledge, attitudes, and skills in the implementation of virtual laboratories.

References

[1] Wolf T 2010 *IEEE Trans. Educ.* **53** 216–22
[2] Jiang R, Li C, Huang X, Sung S and Xie C 2020 *Sci. Teach.* **88** 63–71
[3] Mubarok H, Suprapto N and Adam A S 2019 *J. Phys. Conf. Ser.* **1171** 1–6
[4] Alexiou A, Bouras C and Giannaka E 2005 *IFIP Adv. Inf. Commun. Technol.* **171** 19–28
[5] Hatherly P A, Jordan S E and Cayless A 2009 *Eur. J. Phys.* **30** 751–62
[6] Okoyeigbo O, Agboje E, Omuabor E, Samson U A and Orimogunje A 2020 *Int. J. Electr. Comput. Eng.* **10** 5883–90
[7] Tatli Z and Ayas A 2010 *Procedia - Soc. Behav. Sci.* **9** 938–42
[8] Budai T and Kuczmann M 2018 *Acta Polytech. Hungarica* **15** 191–204
[9] Mackenzie A H 2021 *Sci. Teach.* **88** 6–7
[10] Annetta L, Klesath M and Meyer J 2009 *J. Coll. Sci. Teach.* **39** 27–33
[11] Merigó J M and Yang J-B 2017 *Omega* **73** 37–48
[12] Tomaszewska E J and Florea A 2018 *Eng. Manag. Prod. Serv.* **10** 41–56
[13] Kapilan N, Vidhya P and Gao X Z 2021 *High. Educ. for the Futur* ***8*** 31-46
[14] Radhamani R, Kumar D, Nizar N, Achuthan K and ... 2021 *Educ. and Inf. Technol.* 1-19
[15] García-Vela M, Zambrano J L, Falquez D A and ... 2020 *Proc. of ICERI 2020* **9**
[16] Yap W H, Teoh M L, Tang Y Q and Goh B H 2021 *Biochem. Mol. Biol. Educ.* **49** 685–91
[17] Sherrer S M 2020 *Biochem. Mol. Biol. Educ.* **48** 659–61
[18] Vasiliadou R 2020 *Biochem. Mol. Biol. Educ.* ***48*** 482-483
[19] Bunyamin E M, Siregar E S and Kusumawardani D K 2021 *J. Educ. Res. Eval.* **5**
[20] Mafudi I and Handhika J 2021 *Impuls. J. Res. Innov. Phys. Educ.* **1** 42-49
[21] Lohrasbi M 2020 *Chemistry Training with the help of Virtual Laboratory in the Corona Era* (sid.ir)
[22] Tamura N, Taira H and Umemura K 2021 *J. Phys. Conf. Ser.* **1893** 012017
[23] Eck N J Van and Waltman L 2016 Text Mining and Visualization *Text Min. Vis.* 1–5

Profile of students' scientific literacy on particle dynamics

S Mahtari[1], M Wati[1], S Rizky[1], D Dewantara[1] and B K Prahani[2]

[1]Department of Physics Education, Universitas Lambung Mangkurat, Banjarmasin,
[2]Department of Physics, Universitas Negeri Surabaya, Surabaya

saiyidah_pfis@ulm.ac.id

Abstract. This study measures students' ability to complete scientific literacy instruments on particle dynamics material analyzed using the Rasch model. This study uses a quantitative descriptive research design with a survey method. The research subjects were 29 students of class XI IPA 2 at State Senior High School 6 Banjarmasin. The data collection technique was carried out using a scientific literacy test on particle dynamics material in the form of reasoned multiple choice questions as many as ten questions. The research data were analyzed using the Rasch model with the Winstep program. The results showed that the MNSQ outfit was 0.94, Outfit ZSTD was 0.05, person reliability was 0.58 in the weak category, Cronbach's alpha reliability was 0.92, the separation value was 1.17 with two groups. The students' scientific literacy ability is quite good but still needs to be improved and retrained.

1. Introduction

PISA (Program for International Student Assessment) is an international student assessment program conducted every three years by the Organization for Economic Co-operation and Development (OECD) or economic cooperation and development based in Paris, France. The PISA survey was conducted to measure students' abilities in reading, math, and science. The essential knowledge that is the main focus tested in the PISA survey is literacy or reading. In science, scientific literacy is defined as the ability to explain a scientific phenomenon, evaluate and design scientific questions, and interpret data and evidence to devote attention to scientific topics or ideas as a form of reflection [1].

Indonesia has been following the PISA study since 2000. One of the main reasons is understanding student achievement in Indonesia compared to international standards and other countries facing the same challenges [2]. The Indonesian state PISA study results in science that Indonesia's scientific literacy is still very low [3]. Indonesia's scientific literacy score is still far from the international average score and consistently ranks in the top 10 from the bottom, decreasing. In addition, the results of the PISA study also show that most of the students in Indonesia are at level 1, where students have limited knowledge of science that can only be applied to situations or problems that they are familiar with [4].

Scientific literacy is critical to be mastered by students because it relates to the way students understand their environment, health, economy, and other problems faced by modern society, which is very dependent on progress and technology, and the development of science [5]. Therefore, it is essential to survey the field to determine the students' scientific literacy ability. Thus, this study aims to measure students' ability to complete scientific literacy instruments on particle dynamics material analyzed using the Rasch model.

2. Method

The method used in this research is the descriptive method using the survey technique. The subjects of this study were students of class XI IPA 2 at State Senior High School 6 Banjarmasin, totaling 29 people who had studied particle dynamics material in the previous semester. The instrument used to measure scientific literacy skills is in the form of multiple-choice questions based on scientific literacy as many as ten questions that are done online via google records with one day processing time. The students' test results were then used as data in this study and then analyzed using the Rasch model in the Winstep program to determine the logit value.

The output in this software is in the form of table measuring items (item fit order), size of people (person fit order), variable maps (variable maps), reliability, and summary statistics (summary statistics). According to Suminto and Widhiarso, dividing items into four categories [6]. The criteria are used to check the suitability of the items [7]. Outfit means-square (MNSQ) value received: 0.5 < MNSQ < 1.5. Outfit Z-standard (ZSTD) value accepted: -2.0 < ZSTD < +2. Point Measure Correlation (Pt Mean Corr) value : 0.4 < Pt Mean Corr < 0.85. An item is considered biased if the probability value is < 5% (<0.05).

3. Result and Discussion

Based on Table 1, It is found that the logit or measure value of the person measure is 0.11 and the item measure is 0.00, which means that the student's ability is higher than the level of difficulty of the questions. So that it allows all the questions to be answered correctly by students, and students who have high abilities can answer complex questions correctly. In addition, it is obtained from Table 1. that person reliability is 0.58. The reliability is included in the weak category (<0.67), while item reliability is 0.85, which is included in the excellent category (0.8-0.9). The value of Cronbach's Alpha (measuring the reliability of interaction between person and item as a whole) is 0.71, which is included in the good category. The separation value of the person is two, and the thing is 4, which means that for the person, there are two groups, capable and unable or high ability and low ability. The item groups are divided into four groups: difficult, medium, easy, and very easy.

Table 1. Statistic summary

	Description	Score
Logit	Person	0.11
	Item	0.00
Reliability	Person Reliability	0.58
	Item Reliability	0.85
	Alpha Cronbach	0.71
Separation	Person	1.17
	Item	2.42
Outfit MNSQ	Person	0.94
	Item	0.94
Outfit ZSTD	Person	0.05
	Item	0.05

Another quantity shown in table 1 is the value of the MNSQ outfit, both person and item, of 0.98, which is included in the fit criteria if it lies between 0.5 < MNSQ < 1.5, meaning that the test questions used are appropriate to measure students' scientific literacy skills. In addition, the ZSTD outfit value obtained was 0.05, which is between the interval -2 < ZSTD < 2, meaning the data has a possible rational value or in other words, the question as a whole is by the Rasch model and can be used as a test question on particle dynamics material to measure the ability student scientific literacy.

The variable maps are one of the outputs of the winstep program. The variable map shows student ability and item difficulty distribution on the same logit scale [8]. Students' abilities are listed on the left side of the map, while item difficulty levels are on the right side of the map. Logits 0 is set as the average of test items [9]. Based on Figure 1, a map of variables shows that item number 10 is the most challenging question and number 2 is a straightforward question. In addition, 19 students can solve scientific literacy problems with logit > 0.00, while ten students get logit scores < 0.00. The students with codes 05P, 24L, and 27P have the highest ability or ability, while those with codes 08L, 12p, 25L, and 17P have the lowest abilities. Therefore, the students' scientific literacy skills are pretty good but still need improvement and retraining. Items 3, 6, and 9 are on the same line. It shows that the questions have the same level of difficulty according to the program, so that from these three items, only one item can be selected as well as items 4 and 5 which are on the same line.

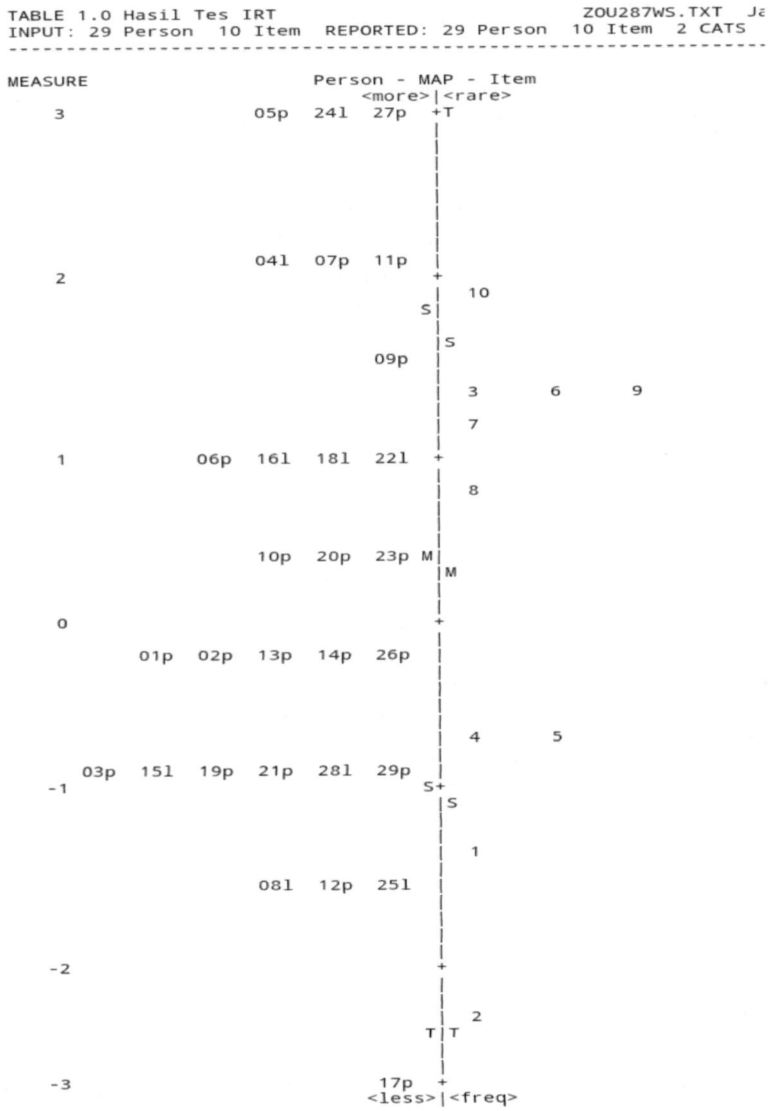

Figure 1. Variable Maps

More frequent students can improve students' scientific literacy with scientific literacy questions [10]. It can also be done by using textbooks and learning patterns oriented to scientific literacy [11]. In addition, students' scientific literacy skills can be improved by using innovative learning models, modules, student worksheets, media, and evaluation questions that can improve students' scientific literacy skills [12].

4. Conclusion

Based on the results and discussion, it can be concluded that the scientific literacy instrument made can measure students' scientific literacy skills, and the results obtained are pretty good for students' abilities but still need to be improved based on the logit value of 0.11. Thus, it is necessary to have an essential role in learning and providing examples in everyday life to enhance students' scientific literacy skills.

References

[1] OECD 2019 *The Pisa 2015 Assessment and Analytical Framework* (Paris: OECD Publishing)
[2] Kemendikbud 2019 *Pendidikan di Indonesia Belajar dari Hasil PISA 2018* (Jakarta: Balitbang Kemendikbud)
[3] Hewi L and Shaleh M 2020 *J. Golden Age* **4** 34–5
[4] Fatmawati I N and Utari S 2015 *Edusains* **7** 151–9
[5] Asyhari A and Hartati R 2015 *J. Ilm. Pendidik. Fis. Al-Biruni* **4** 179
[6] Sumintono B and Widhiarso W 2015 *Aplikasi Pemodelan Rasch pada Assesment Pendidikan* (Cimahi: Trim Komunikata)
[7] Widyaningsih S R and Yusuf I 2018 *Gravity* **4** 36
[8] Wati M, Mahtari S, Hartini S and Amalia H 2019 *Int. J. Interact. Mob. Technol.* **13**
[9] Wati M, Mahtari S, Azmi A N, Salam A and Mastuang M 2021 *J. Phys. Conf. Ser.* **1832**
[10] Zainab, Wati M and Miriam S 2017 *J. Ilm. Pendidik. Fis.* **3** 113–25
[11] Budiningsih T Y, Rusilowati A and Marwoto P 2015 *J. Innov. sciensce Educ.* **4** 39
[12] Mahtari S, Misbah, Hartini S and Suryati 2019 *Kasuari Phys. Educ. J.* **2** 11–6

The effect of physics learning using ardouno uno based media on higher-order thinking skills

Matsun, Boisandi, I N Sari, S Hadiati and D F Saputri

Department of Physics Education IKIP PGRI Pontianak

matsunpfisika@ikippgriptk.ac.id

Abstract. The purpose of this study was to determine the effectiveness of contextual media with Arduino Uno-based media on students' higher-order thinking skills on density material. The population of this research is class XII IPA MAN 1 Pontianak totaling six classes, with samples of class XII IPA A and class XII IPA B. This study is quasi-experimental with a pretest-posttest group design—sampling with a simple random technique to determine the experimental and control classes. The data collection technique used is a test in the form of pretest and post-test questions. The data analysis technique used is the one-party t-test based on the sig value and the normalized gain (N-gain). The results of research data analysis showed that contextual learning media with media based on Arduino Uno had a positive effect and improved students' higher-order thinking skills than students who used conventional media. It is indicated by the results of the t-test with a value of sig. The post-test score is 0.001, and the results of the improvement test with the N-gain of the experimental class are in the medium category, higher than the control class category, namely low. So that learning physics using learning media based on Ardouno Uno has a positive effect and can improve students' Higher Order Thinking skills.

1. Introduction

Physics learning is a branch of natural science that determines a process that emphasizes a fact, concept, principle, theory, or law. The ideas that exist in physics have the basis of grouping several non-verbal presentations, which ultimately have an abstract tendency that requires the ability to describe them [1]. Learning physics in the classroom must be supported by media that can change students' perceptions of learning physics. Students' perception in the learning process in class is that the concept of physics is complex. The learning process is related to the transfer of concepts, skills, and values between teachers and students [2]. A teacher has a vital role in the learning process. A teacher must have an exciting learning innovation that makes students interested in learning physics. Students' low physics learning outcomes are caused by factors including the material in the book, which is felt to be too theoretical to be understood, is still less effective in learning media, the learning media used and selected by the teacher is not correct, the students are not optimal to be involved in a learning process and class activities that are not optimal dominated by teachers or educators [3]. The renewal of learning in the 2013 curriculum sees the need for higher-order thinking skills to run the objectives of achieving the curriculum learning [4].

Information obtained from students of MAN 1 Pontianak that learning physics is considered difficult due to the lack of understanding of the concepts being taught. The difficult understanding obtained by MAN students occurs in Density material when observing irregular objects; students assume that the density of uniform objects and irregular objects is the same. In addition, the students of MAN 1 Pontianak have difficulty in doing a practicum. Practical tools available in schools are still limited in terms of the number and quality of devices. The practicum tools used still use simple tools and do not have automated tools. The problem that exists at the school is that there are no Arduino Uno-based teaching aids with sensor-assisted or technology-based tools, one of which is to use an Arduino Uno Microcontroller-based automation laboratory tool. The learning media based on the Arduino Uno microcontroller is quite effective and easy to understand in the teaching and learning process according to the characteristics of the supporting media. The research and learning process results using the Arduino Uno microcontroller can provide motivation and improve students' High Order Thinking Skills. The students are said to handle problems if they can examine problems by combining their knowledge in new situations. This ability can be known as High Order Thinking Skills. Skills that allow students to connect, manipulate and change understanding with their experience critically and creatively in making decisions to solve problems. There are several realms of thinking from Bloom with the categories of analyzing, evaluating, and creating/creating [5].

Higher-order thinking skills are students' active abilities when facing unusual problems, uncertainties, questions, and dilemmas. This ability continues to develop forward, providing valid results in accordance with knowledge and experience [6]. Meaningful experiences are beneficial for students' learning so that students' high-level abilities are realized. Higher-order thinking skills include cognitive skills in the realm of analyzing (C4), evaluating (C5), and creating (C6) [7]. Students must be able to explore the questions given, assess the problems that occur, and create the problems they face. Indicators to measure higher-order thinking skills include [8]: 1) Analyze incoming information and divide or structure information into smaller parts to identify patterns or relationships. Able to recognize and distinguish the causes and effects of a complex scenario. Identify or formulate questions. 2). Evaluating provides an assessment of solutions, ideas, and methodologies using suitable criteria or existing standards to ensure the value of effectiveness or benefits and creating hypotheses, criticizing, and testing. Accept or reject a statement based on predetermined criteria. 3) Creating is making generalizations of an idea or perspective on something. Design a way to solve the problem. Organize elements or parts into a new structure that has never existed before. Humans have several abilities that can be developed through experience. The experience occurs because of human interaction with both physical and social environments [9]. Physics learning requires learning media that can make students understand the material from virtual to real.

Media Contextual Teaching and Learning is a learning concept that helps teachers relate the material being taught to real-world students and encourages students to connect their knowledge and its application in real-life [10]. Arduino Uno-based learning media is a media that is developed automatically. In the experiment in the actual laboratory, the calculations are done manually. Using the density practicum tool based on the Arduino Uno microcontroller, the calculation results can be observed directly on the LCD screen. The learning media measures density using an Arduino Uno microcontroller with the help of ultrasonic sensors and load cells. Ultrasonic sensors function to determine the distance of the object's surface to measure the object's volume. Wh le cell functions as a sensor to find the mass of objects. So that the density of objects can be calculated automatically, both regular and irregular solids. The learning media based on the Arduino Uno microcontroller applies scientific procedures that students can do. Scientific procedures are an effort to train students' higher-order thinking skills. Changes that arise because the learning process is effective, namely meaningful learning. That is, these changes bring specific effects, meanings, and benefits for students [11]. It also determines the effectiveness of learning by looking at the impact on students. Students who take part in direct learning will have the ability to formulate and find concepts [12]. This ability is used to solve problems and problems faced by students in real life.

The effectiveness of this research can be measured from student learning outcomes which become a minimal statement regarding student abilities [13]. Suppose the learning process is carried out with the correct and appropriate stages. In that case, the explanation description can achieve the research objectives related to the effectiveness of contextual media with Ardouno Uno-based media on students' higher-order thinking skills on density material.

2. Method

This study was conducted using a quasi-experimental design with a pretest-posttest control group design [14]. The population of this research is class XII IPA MAN 1 Pontianak with six classes. Sampling was carried out using a more straightforward random sampling method which required a homogeneous population [15]. A homogeneous population means that each class has the same condition or no difference between one class and another. Determination of homogeneous population or not using homogeneity test, normality test, and one-way ANOVA test. The data collection technique used a test with pretest and post-test questions in the form of multiple choices. The homogeneity test (variance) uses Levene's test to find out the research sample is homogeneous. Normality test with Kolmogorov-Smirnov to show the data in a normally distributed state. The data analysis technique uses parametric statistics which data is typically distributed and homogeneous. The parametric statistic uses the right-hand t one-tailed test to determine the effect and the Ngain test to determine the improvement of students' higher-order thinking skills. The one-tailed t-test uses SPSS 18 software assistance in the data analysis process. Test the hypothesis by looking at the value of sig. obtained from the analysis through the SPSS 18 program. The category of improving students' higher-order thinking skills is seen by using the N-gain equation [16] as follows:

$$N - gain = \frac{posttest\ score - priest\ score}{maximum\ possible\ score - pretest\ score}$$

Classification of normalized N-gain values can be seen in table 1. Below [17].

Tabel 1. Classification score N-gain

Score N-gain	Classification
0,70 < N-gain ≤ 1,00	High
0,30 < N-gain ≤ 0,70	Medium
N-gain ≤ 0,30	Low

3. Result and Discussion

3.1 *Result*

Learning physics using Arduino Uno during the COVID-19 pandemic towards the new normal period is very effective for students. Learning media based on the Arduino Uno microcontroller is a learning media that instill concepts and stages in implementing the practicum. Students can calculate the density of objects, both regular and irregular objects, so effective learning is carried out. T-Test Results Posttest and Control Class Scores can be seen in Table 2.

Table 2. N-gain of Experiment Class and Control Class

Class	N	Average Post-test	Average Pretest	N-gain	Classification
ExperimentExperiment	40	18,70	10,231	0,375	Medium
Control	40	15,54	8,891	0,251	Low

Determination of the sample is viewed by looking at the homogeneity of the population. Homogeneity in the population was seen by Levene's test, normality test, and one-way ANOVA test. The student's initial abilities were obtained from the pretest scores before the two groups received treatment. Analysis of the pretest scores showed that the value of sig. (2-tailed) of 0.140, which means it is greater than the significance level of a. So it can be seen that the initial abilities of the experimental class and control class students before receiving treatment were the same.

The difference in learning outcomes of higher-order thinking skills in experimental class and control class students is seen from the post-test scores. The post-test scores were tested statistically using a two-party t-test using SPSS 18. There is no difference in the average post-test scores of the experimental class and the control class seen from the sig value. (2-tailed) which is compared with a significance level of 5% or 0.05, it can be seen that the value of sig. (2-tailed) is 0.0021, which means it is smaller than the significance level of a. So it can be seen that there are differences in learning outcomes of higher-order thinking skills between experimental class and control class students. Furthermore, to determine the effect and effectiveness of the treatment given in the experimental class, a follow-up test was carried out using the right-hand one t-test. A one-sided t-test was carried out through the SPSS 18 program. Based on table 3, it can be seen that the value of sig. (1-tailed) of 0.0015, which means it is smaller than the significance level of 5% or 0.05. So it can be concluded that the treatment (learning using contextual media) with learning media based on the Arduino Uno microcontroller) given in the experimental class has a positive effect on students' higher-order thinking on density material.

The improvement of students' higher-order thinking ability learning outcomes can be seen from the N gain test scores of pretest and post-test between the experimental and control classes. The results of the analysis are presented in Table 2 below. Based on table 2, it is known that the N-gain classification of the experimental class and the control class is different. The experimental class N-gain value is 0.375 and is included in the medium category. The N-gain value for the control class is 0.251 and is classified as low. So it can be concluded that the treatment (learning using contextual media) with the Arduino Uno microcontroller-based media given in the experimental class can improve students' higher-order thinking skills on density material.

The difference in treatment given to the two samples has affected students' higher-order thinking skills. Based on the hypothesis test results, the difference in the average post-test scores of the experimental and control groups, there is a difference indicated by the value of sig. (2-tailed) 0.0021 < 0.05 significance level. Previously, it was known that the initial abilities of the students of the two groups were the same (the results of the two-tailed t-test, the pretest score was sig. 0.140 > sig. a 0.05). Then a further test to determine the effect of treatment is known to be the value of sig. (1-tailed) 0.001 < 0.05 significance level, meaning that the treatment given in the experimental class has a positive effect on students' higher-order thinking skills. The results of the N-gain test show that the experimental class N-gain classification is greater than

Low control class N-gain classification. The N-gain values indicate it for the experimental and control classes, 0.375 and 0.251, respectively. Based on the hypothesis test results, it can be concluded that the treatment given in the experimental class in the form of learning using contextual media with media based on the Arduino Uno microcontroller is more effective in achieving students' higher-order thinking skills on density material.

3.2 Discussion

Measuring the density of an irregular object uses a container filled with water. The whole system is conditioned to zero, so the density is the same. The object to be measured the density is inserted into a container filled with water. The ultrasonic sensor will read the increase in water height due to new objects entering the container. The increase in water will be read as the total volume of the object. The Ardouno Uno microcontroller will recalculate the density according to the command given. The LCDs are the readings ordered from the microcontroller to display the mass, volume, and density of the object being measured. The measurement of the density of irregular objects is carried out using natural stone. The volume will increase for each additional mass of natural stone, but the density will always be the same. This is due to the similar characteristics of natural stone. Table 3 below shows the results of natural stone density measurements measured using Ardouno Uno microcontroller along with manual measurements:

From table 3, it can be seen that the results of the density test using the Ardouno Uno microcontroller have measurement results that are close to the results of manual calculations. With a measurement error using this tool of 1.44%. By testing the error of the Ardouno Uno microcontroller tool compared to the manual, it has a small error so that the tool used can improve students' Higher Order Thinking skills.

Table 3. Testing the Density of Natural Stone

Testing	Density (g/cm^3)	
	Manual	Tool
1	2.66	2.67
2	2.65	2.67
3	2.65	2.68
4	2.67	2.68
5	2.67	2.67

Learning that involves higher-order thinking skills requires clarity in conveying facts to eliminate ambiguity and confusion to improve students' attitudes in thinking [18]. The learning media Arduino Uno provides facts related to the value of the density of objects that are read automatically so that students do not feel ambiguous. The opinion follows the results of students' higher-order thinking skills as seen from students' answers in working on questions [19]. The results of the improvement analysis are presented in Table 4.

Table 3 provides information that students have succeeded in practicing higher-order thinking skills during the learning process. It can be seen from the results of increasing student answers after treatment. The improvement result of students in the experimental class is higher than that of students in the experimental class. However, the indicators for evaluating experimental and control class students experienced the same increase. Some indicators of higher-order thinking ability starting from the initial ability are zero, meaning that students don't know about it. The treatment given has given a significant increase to a value of 45. The increase in experimental class students who take learning with contextual media and Arduino-based learning media is higher than students in the control class. Students' initial knowledge has been constructed through physics learning, which applies to digging students' prior knowledge into complete knowledge through hands-on practical experience using Arduino Uno-based practicum tools [20]. Students find, process, analyze existing data and then compare them with existing physical concepts. These results follow previous research conducted [21], which stated that the mastery of concepts of students who studied with the guided inquiry was higher. The higher-order thinking abilities of students were well trained. The improvement of students' higher-order thinking skills can be seen from the pattern of students' answers to higher-order thinking skills.

Tabel 4. Improving Students' Higher Order Thinking Ability

Indicator	Experiment			Control		
	Pre	Post	N-gain	Pre	Post	N-gain
Analyze incoming information and share information into smaller parts to recognize the relationship.	22	41	0,405	47	31	-0,865
Able to recognize and distinguish factors cause and effect of a scenario	6	80	0,575	3	26	0,181
Formulating questions	109	115	0,157	93	97	0,126
Provide an assessment of the solution, ideas, and methodologies by using criteria appropriate	90	130	0,478	58	111	0,567
Make a hypothesis, and do the test	2	15	0,118	8	23	0,176
Accept or reject a statement based on predetermined criteria.	3	84	0,678	0	75	0,570
Generalize an idea	53	67	0,714	45	62	0,594
Designing a way to solve the problem.	30	41	0,087	6	38	0,241
Organizing elements or parts Becomes structure new	0	45	0,317	0	3	0,018

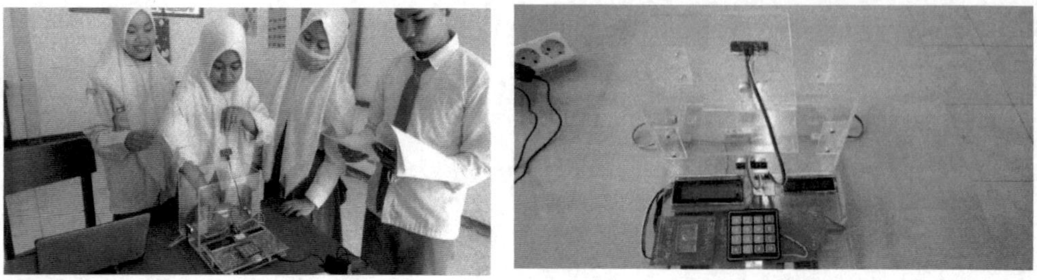

Figure 1. The use of the Arduino Uno tool by students and pictures of the density tool

The pattern of student answers leads to problem-solving by applying higher-order thinking aspects. Students make assumptions (hypotheses) on the problems that exist in the problem. The hypothesis is tested by analyzing the calculation results in a way that they understand. Students give reasons to support the hypothesis according to the calculations' results [22]. The process of students in solving problems has proven that learning using media based on the Arduino Uno microcontroller is successful in training students' higher-order thinking skills. These results are consistent with the statement that students' success in practicing higher-order thinking skills depends on the ability of students to apply, reorganize or rearrange, and embellish their knowledge in situations where students think following existing circumstances. [23]. Students' thinking levels at the level of analyzing, evaluating, and creating are considered higher-order thinking abilities. The success of students in practicing higher-order thinking skills affects the success of student learning. Treatment needs to be given to students to change thinking patterns, see the problem as a whole, solve problems, and then concretize abstract knowledge because it is an important thing that students need to have.

4. Conclusion

Based on the results of the data analysis of the experimental and control class post-test scores, obtained in the medium category and the N-gain value for the control class in the low category. The results of this research analysis can be concluded that learning physics using learning media based on Ardouno Uno has a positive effect and can improve higher-order thinking skills. Based on this, the ardouno uno learning media effectively achieves students' higher-order thinking skills on density material. The treatment given in the experimental class in the form of learning physics using learning media based on the Arduino Uno microcontroller has a positive impact on students' higher-order thinking skills. The influence and improvement are also supported by the learning process that has trained students' higher-order thinking skills.

References

[1] Sarah F, Khaldun I and Gani A 2021 *J. Pendidik. Sains* **9** 51
[2] Stahre Wästberg B, Eriksson T, Karlsson G, Sunnerstam M, Axelsson M and Billger M 2019 *Educ. Inf. Technol.* **24** 2059–80
[3] Matsun M, Ramadhani D and Lestari I 2018 *J. Pendidik. Mat. dan IPA*
[4] Matsun M, Darmawan H and Fitriyanti L 2019 *J. Pendidik. Mat. dan IPA*
[5] Suparman S, Juandi D and Tamur M 2021 *ACM Int. Conf. Proceeding Ser.* 44–51
[6] Amin D I and Ikhsan J 2021 *Eur. J. Educ. Res.* **10** 261–74
[7] Setiawan J, Sudrajat A, Aman and Kumalasari D 2021 *Int. J. Eval. Res. Educ.* **10** 545–52
[8] Krause U, Béneker T, van Tartwijk J and Maier V 2021 *London Rev. Educ.* **19**
[9] Putri A 2021 *J. Inov. dan Pembelajar. Fis.* **8** 1–13
[10] H N, Arnidah A and Hasfat H 2021 *J. Educ. Sci. Technol.* **7** 84
[11] Salar R 2021 *Revista Cubana de Física* **38** 38–42
[12] Ansori Z M, Anifah L and Buditjahjanto 2021 *J.Pendidik. Teknik Elektro* **11** 69-78
[13] Yurtyapan E and Kandemir N 2021 *Particip. Educ. Res.* **8** 62–87
[14] Arikunto S 2013 *Prosedur Penelitian Suatu Pendekatan Praktik* (Jakarta: PT Rineka Cipta)
[15] Sudijono A *Pengantar Statistik* (Jakarta: Raja Grafindo Persada)
[16] Arifin Z 2012 *Evaluasi Pembelajaran* (Jakarta: Direktorat Jenderal Pendidikan Islam Kementrian Agama RI)

[17] Hake R R 1999 *Educational Research Association* (Indiana University: Dept. of Physics)
[18] Fayakun M and Joko P 2015 *J. Pendidik. Fis. Indones.* **11** 49–58
[19] Payadnya I P A A and Wibawa K A 2021 *J. Phys. Conf. Ser.* **1957**
[20] Rozi A, Khoiri A, Dewi Mutia Farida R, Sunarsi D, Jasmani, Paeno, Munawaroh and Iswadi U 2021 *J. Phys. Conf. Ser.* **1764** 0–18
[21] Görgülü Arı A and Meço G 2021 *Biology (Basel).* **10** 506
[22] Ngurahrai A. H., Farmayanti S. D. N 2019 *Radiasi: J. Berk. Pendidik. Fis.* **12** 76–83
[23] Latifah Z, Ikhsan J and Sugiyarto K H 2019 *J. Phys. Conf. Ser.* **1156** 6–12

E-module on elasticity of solids topic through cooperative learning to improve learning outcomes and motivation: Validity aspects

N Muhammad[1], F Hamid[1], M Misbah[2], and D Dewantara[2]

[1]Physics Education Study Program, Faculty of Teacher Training and Education, Universitas Khairun, Indonesia
[2]Physics Education Study Program, Faculty of Teacher Training and Education, Universitas Lambung Mangkurat, Indonesia

Nurlaelamuhammad83@gmail.com

Abstract. This study aims to describe the validity of the electronic module on the elasticity of solids topic through the cooperative learning model to improve students' learning outcomes and motivation. This research is a research and development with the ADDIE development model. This electronic module was assessed by using the electronic module's validation sheet. The validators of this e-module consisted of two academics and one practitioner. Based on the validation results, it was obtained that the validity of the electronic module was 3.49 with a valid category. It shows that the electronic module on the elasticity of solids topic through cooperative learning models can be used in the classroom learning process to improve student's learning outcomes and motivation.

1. Introduction

The Industrial Revolution 4.0 is the integration of technology in the learning process [1]. It still makes learning easier even though we are currently experiencing a covid-19 pandemic[2]. The current condition, where most students study from home or can be called distance learning, can be done online. So teachers need to innovate in the use of teaching materials [3]. Teachers in the industrial revolution 4.0 era act as facilitators and motivators in the learning process [4].

Digital technology and online media function to find teaching materials that can assist in understanding learning materials[4]. The use of technology-integrated teaching materials such as electronic modules is needed in current conditions [5,6]. E-module are teaching materials that can help students learn subject matter independently using electronic media. As the electronic module can help students learn individually/independently in the classroom, the teacher learns only as a facilitator [7]. E-module is the right choice to make it easier for teachers and students [8].

An electronic module presents self-study materials that are systematically arranged into the smallest learning units to achieve specific learning objectives, which are presented in an electronic format [9]. This e-module can be accessed anytime and anywhere. The advantages of e-module compared to print modules are that they are interactive, easy to navigate, allow displaying/loading of images, audio, video, and animation and are equipped with formative tests/quizzes that allow immediate automatic feedback [8].

The electronic module using Flip PDF Corporate Edition can be used without the need to download the application. First, it is effectively used in the learning process [10,11] to improve student learning outcomes. The use of this e-module can increase student motivation. Motivation is an internal process that is the main factor in the success and achievement of learning. High motivation to learn will provide a clear goal in a person to absorb information and understand the concept of the material. Learning outcomes are obtained from a person's knowledge, skills, and attitudes after doing learning activities [12]. Motivation is an essential factor in the success of learning outcomes. The higher the motivation in oneself, the higher the process of absorbing information and understanding material concepts [13,14].

The E module in this study uses the Flip PDF Professional application. Several studies have shown that e-module using Flip PDF Professional can improve students' learning outcomes [15,16]. Electronic modules can help students improve their cognitive understanding. With this interactive electronic module, the learning process involves audio-visual displays, sound, movies, and others. The program is easy to understand to be used as a suitable learning medium [17]. Based on the results of interviews with physics teachers during the preliminary study and needs analysis, the teaching materials used in schools were rarely technology-based. In addition, no one has developed materials regarding the elasticity of solids. Learning outcomes were also categorized as moderate, and learning motivation was still in the poor category.

Several studies have been carried out on the development of elasticity of solids e-module [18,19]. However, the study has not explicitly mentioned the model used in learning to use the e-module. Research on learning models at the time of learning shows that cooperative learning models can improve student learning outcomes [20–22] and student motivation in learning [23–25]. It will support the success of the e-module in learning. Therefore, research on the development of electronic modules on the elasticity of solids was carried out through cooperative learning models to improve learning outcomes and student motivation. This study aims to describe the validity of the electronic module on the elasticity of solids to improve learning outcomes and student motivation.

2. Method

This research is a research and development study with the ADDIE model. This electronic module was assessed using the electronic module's validation sheet. The validators of this e-module consisted of 2 academics and one practitioner. Data validation was calculated with the average value of the total score for each aspect of the assessment. The calculation results were adjusted to the assessment criteria [26] in Table 1.

Table 1 Electronic Module Validation Criteria

No	Equation	Average (x)	Category
1	$X > \bar{X}_i + 1.8 \times S_{b_i}$	$\bar{x} > 3.4$	Very good
2	$\bar{X}_i + 0.6 \times S_{b_i} < X \leq \bar{X}_i + 1.8 \times S_{b_i}$	$2.8 < \bar{x} \leq 3.4$	Good
3	$\bar{X}_i - 0.6 \times S_{b_i} < X \leq \bar{X}_i + 0.6 \times S_{b_i}$	$2.2 < \bar{x} \leq 2.8$	Quite good
4	$\bar{X}_i - 1.8 \times S_{b_i} < X \leq \bar{X}_i - 0.6 \times S_{b_i}$	$1.6 < \bar{x} \leq 2.2$	Poor
5	$X \leq \bar{X}_i - 1.8 \times S_{b_i}$	$\bar{x} \leq 1.6$	Not Good

Where:
$\bar{X}_i$ = Ideal average = ½ (ideal maximum score + ideal minimum score)
S_{b_i} = Ideal standard deviation = $1/6$ (ideal max score - ideal min score) = Empirical score

The reliability of the validation results was calculated using the Alpha Crombach equation. To determine the reliability of the validated teaching materials assessment, it used the reliability criteria r [27].

3. Result and Discussion

The result of the developed e-physics module with a cooperative learning model was used to assist the learning process following the characteristics of the elasticity of solids topic and the characteristics of students in class XI SMAN 3 Ternate City. The developed e-module was equipped with experimental Student Worksheet (LKPD) guidance, aiming to improve student learning outcomes. E-module is one of the teaching materials that can improve students' mastery of concepts and think independently [28,29]. The e-module can be accessed at the following link **http://elastisitaszatpadat.online/mobile/index.html.** The following is an example of the appearance of the solid elasticity e-module developed, which can be seen in Figure 1.

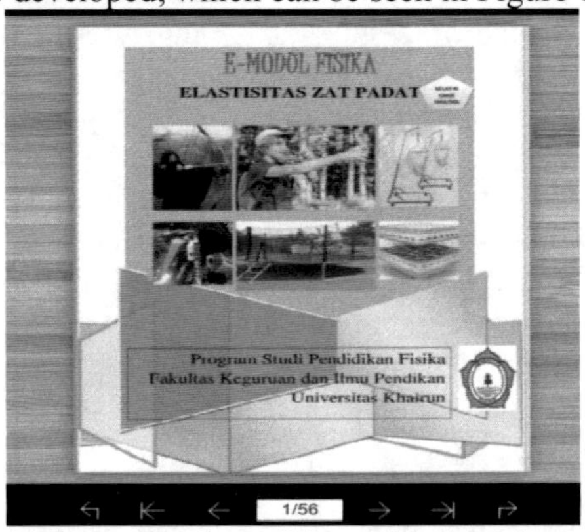

Figure 1 The cover view of the solid elasticity e-module

The developed e-module contained the elasticity of solids topic. It consisted of the cover, introduction, table of contents, instructions for using the module, concept map, keywords, chapter titles, motivations related to life, student worksheet with PhET Simulation media, material descriptions, scientists' physics, physics applications, physics info, examples of thinking question, notes on important formulas, summaries, independent questions, glossary, answer keys, and bibliography. The elasticity of solids topic was divided into three meetings. Each meeting contained a problem equipped with student worksheets to conduct experiments through PhET Simulation media and questions of strengthening thinking in the e-module. It was to assist students in analyzing experiments [30,31]. Then, in the e-module, there were examples related to life. The material and application in life contained physics info, notes on important formulas, physics applications, a summary of each sub-chapter, and a competency test used to determine students' understanding in each meeting. The update of the developed module compared to existing books is presented in Table 2.

Table 2 Update of modules developed by researchers

No	Module Update
1	E-module with Flip Builder
2	The module was equipped with a PhET media on Student Worksheet (LKPD) for virtual experiments to be carried out and given questions of strengthening thinking
3	The initial motivation was linked to daily life's examples in each lesson, such as a child pulling a slingshot, playing with plasticine, spring swings, motorized vehicles that have shock breakers
4	Equipped with examples of thinking related to everyday life problems
5	Given a column for physics info, physics scientists, physics applications and notes on important formulas for each sub-section of the material, video links/video barcodes

The e-module of the elasticity of solids in this study used the Flip PDF Professional application. The hallmark of the e-module was that it could be used anywhere in the form of a professional Flip PDF. There was a video link to add insight and practicum media using PhET simulation, which helped children interactively conduct virtual experiments. The development of the module was divided into three meetings on the elasticity of solids, namely the modulus of elasticity, Hooke's law, and the arrangement of springs (series and parallels).

The e-module developed was reviewed based on the results of the assessment of 3 validators from media experts and material experts, both from academics and practitioners. E-module validation was based on aspects of e-module component assessment such as validation of student teaching material formats, language validation, content validation of student teaching materials, presentation validation, validation of the benefits/usability of teaching materials on electronic modules. The results of the assessment of the validity of the e-module elasticity of solids are presented in Table 3.

Table 3. The results of the validity and reliability of the module

Aspect	**Validator**			**Average**	**Criteria**
	1	2	3		
Module Format	3.00	3.67	3.67	3.44	Valid
Module Contents	2.80	3.90	4.00	3.57	Very Valid
Presentation	2.87	3.73	4.00	3.53	Very Valid
Benefits/usage	3.00	4.00	4.00	3.67	Very Valid
Language	2.20	3.68	3.78	3.22	Valid
			Average	3.49	Valid
			Module validity category		Very Valid
			Average overall reliability		0.99
			Category overall reliability		Very High

Suggestions from reviewers on the developed e-module can be seen in Table 4.

Table 4 Suggestions from reviewers on the module

Many misspell, such as missing letters
Inconsistent spacing
Example questions are reproduced with answers
More practice questions
Add a game section to overcome student boredom in learning
If possible, tuck in the corner of character education and national culture

The following changes to the contents of the e-module based on suggestions from the validator are presented in Table 5.

Table 5 e-module content changes

Before	**After being validated**
No games yet	Add a game section to overcome student boredom in learning
No tucked character education corner and the nation's culture yet	Tucked in the corner of character education and national culture

Table 2 shows the results of e-module validation with a value of 3.49 in the valid category. The reliability is 0.99, with a very high degree of reliability. It shows that the e-module format already contains learning objectives, guidelines in student worksheets, examples of questions related to everyday life, information, and a summary of the material. The e-module can stand alone, the numbering system is clear, the design is clear. The e-module is attractive, easy to use, and communicative. The validated e-module was then revised according to the suggestions from the three validators.

The characteristics of the e-module were self-contained; if all learning materials are contained in the e-module, it can stand alone. The developed e-module did not depend on other media or teaching materials, and the format of the E-module corresponded to the characteristics of the e-module. The developed e-module helped students by providing convenience for students in learning [32]. Furthermore, in table 2, the results of the calculation of the validation of the contents of the e-module, which consists of three aspects of the assessment, namely aspects of material coverage, aspects of the material description, and aspects of up to date. Validation of the contents of this e-module is included in a very valid category. Obtaining a score with a very valid category indicates that the content of the e-module is extensive and deep, the content of the e-module is accurate to facts, principles, theories, terms and concepts of physics, and the content of the e-module follows the development of science.

It corresponds to the characteristics of the e-module [32] that a good e-module is one of them with adaptive characteristics, namely the developed e-module adapted to knowledge and technology. The validation of the presentation of the e-module contained in Table 2 consists of four aspects of assessment, namely aspects of presentation techniques, aspects of supporting material presentation, aspects of presenting learning in e-module, and aspects of presenting student worksheets in e-module. Obtaining a score with a very valid category indicates that the presentation of the e-module is systematic, the balance of substance between the chapters in the chapter, the concept is coherent, has been centred on students, fosters curiosity. The presentation of student worksheets on the e-module is appropriate. Flip PDF Professional is different from the pdf that is usually used. In terms of appearance, Flip PDF Professional is like an e-book display that can be flipped while reading it [33,34].

Validation of the e-module's benefits/usability in table 2, which consists of aspects of the benefits containing three items, is considered a very valid category. Obtaining a score with a very valid category indicates that the e-module can be used as a teacher's guide in learning. The e-module can be used as a guide for students in independent learning, and the e-module can be used in learning to solve problems. By using the e-module, students can learn independently and focus on their abilities [32,35]. Language validation in table 2, which consists of two aspects, obtained a very valid category. Obtaining a score with a very valid category indicates that the language in the e-module already contains proper grammar and spelling.

Therefore, the results of the e-module validation were reviewed from aspects of format, content, presentation, and benefits/usability and language aspects. In Table 2, it is found that the developed e-module was good but still needed a minor revision to be better. Besides, the developed physics e-module was feasible and could be used in the trial phase at school, and the e-module is reliable. Finally, this e-module was revised so that it was feasible to be implemented in classroom learning.

4. Conclusion

The conclusion from the results of this study is that the electronic module on the elasticity of solids topic through cooperative learning models can be used in the learning process in the classroom to improve learning outcomes and student motivation. This research implies that the module on solid elasticity topic is valid for use in schools. Thus, this e-module can be implemented in classroom learning to review further the effectiveness and the student's responses to the module.

References

[1] Wiyono K and Zakiyah S 2019 *Pros. Semin. Nas. Pendidik. Progr. Stud. Pendidik. Fis. FKIP ULM*
[2] Cahyani A, Listiana I D and Larasati S P D 2020 *IQ (Ilmu Al-qur'an) J. Pendidik. Islam* **3** 123–40
[3] Matsun M and Saputri D F 2020 *ORBITA J. Kajian, Inov. dan Apl. Pendidik. Fis.* **6** 213–20
[4] Yuniani A, Ardianty D I and Rahmadani W A 2019 Era revolusi industri 4.0: peran media sosial dalam proses pembelajaran fisika di SMA *GRAVITASI J. Pendidik. Fis. dan Sains* **2** 18–24
[5] Dewantara D, Wati M, Misbah M, Mahtari S and Haryandi S 2020 *J. Phys. Conf. Ser.* **1491**

[6] Saregar A, Jamaludin W and Septiani R 2019 *J. Phys. Conf. Ser.* **1155**
[7] Asrial A, Syahrial S, Maison M, Kurniawan D A and Piyana S O 2020 *JPI (Jurnal Pendidik. Indones.* **9** 30
[8] Prihatiningtyas S and Sholihah F N 2020 *J. Pendidik. Fis.* **8** 223–34
[9] Puspitasari A D 2019 *J. Pendidik. Fis.* **7** 17–25
[10] Susanti E D and Sholihah U 2021 *range J. Pendidik. Mat.* **3** 37–46
[11] Darwis D, Fitriani E and Styariyani D 2020 *JRPK J. Ris. Pendidik. Kim.* **10** 130–8
[12] Rijal S and Bachtiar S 2015 *J. Bioedukatika* **3** 15
[13] Yuliastini L I, Wiyasa I K N and Manuaba I S 2020 *Mimb. Ilmu* **25** 11–9
[14] Sardiman 2011 *Interaksi & Motivasi Belajar-Mengajar* (Jakarta: PT Raja Grafindo Persada)
[15] Komikesari H, Mutoharoh M, Dewi P S, Utami G N, Anggraini W and Himmah E F 2020 *J. Phys. Conf. Ser.* **1572**
[16] Seruni R, Munawaroh S, Kurniadewi F and Nurjayadi M 2020 *J. Phys. Conf. Ser.* **1521**
[17] Sugianto D, Abdullah A G, Elvyanti S and Muladi Y 2013 *invotec* **9** 101–16
[18] Muzijah R, Wati M and Mahtari S 2020 *J. Ilm. Pendidik. Fis.*
[19] Puspitasari R, Hamdani D and Risdianto E 2020 *J. Kumparan Fis.* **3** 247–54
[20] Yunita Y and Makiyah Y S 2021 *J. Pendidik. Fis. dan Teknol.* **7** 71–4
[21] Djajadi M 2020 *J. Sipatokkong BPSDM sulsel* **1** 252–61
[22] Loko I M 2021 *Suluh Pendidik.* **19** 85–98
[23] Qudsyi H, Indriaty L, Herawaty Y, Khaliq I and Setiawan J 2021 *Proyeksi J. Psikol.* **6** 34–49
[24] Fitriani F, Hala Y and Taiyeb A M 2016 *J. Nalar Pendidik.* **4**
[25] Suparta D G, Lasmawan I W and Marhaeni A N 2015 *Pendasi J. Pendidik. Dasar Indones.* **5**
[26] Widoyoko E P 2016 *Evaluasi Program Pembelajaran* (Yogyakarta: Pustaka Belajar)
[27] Arikunto S 2012 *Dasar-Dasar Evaluasi Pendidikan* (Jakarta: Bumi Aksara)
[28] Kairmo G A, Ratu T, Hermansyah, Yahya F and Fitriyanto S 2021 *J. Sci. Instr. Technol.* **1** 29–33
[29] Misbah M, Khairunnisa Y, Amrita P D, Dewantara D, Mahtari S, Syahidi K, Muhammad N, Prahani B K and Deta U A 2021 *J. Phys. Conf. Ser.* **1760**
[30] Mahtari S, Wati M, Hartini S, Misbah M and Dewantara D 2020 *J. Phys. Conf. Ser.* **1422** 1–7
[31] Maulani R N, Wati M, Misbah, Dewantara D and Mahtari S 2018 *Semin. Nas. Fis.*
[32] Daryanto D and Dwicahyono A 2014 *Pengembangan Perangkat Pembelajaran (Silabus, RPP, PHB, Bahan Ajar)* (Yogyakarta: Gava Media)
[33] Bagas 2018 Flip pdf professional
[34] Putra R R and Fitrihidajati H 2021 *Berk. Ilm. Pendidik. Biol.* **11** 116–26
[35] Kurnia R, Ayu F and Fauzi A 2020 *J. Penelit. Pembelajaran Fis.* **6** 73–80

Journal of Physics: Conference Series **2104** (2021) 012016 doi:10.1088/1742-6596/2104/1/012016

The development of electronic modules with science literature through direct instruction of impulse and momentum materials

M Arifuddin, N P Hidayah and S Mahtari

Department of Physics Education, Universitas Lambung Mangkurat, Banjarmasin, Indonesia

nitapurnamahidayah@gmail.com

Abstract. This research aims to produce an electronic module with scientific literacy through direct instruction on impulse and momentum materials and describe its feasibility. The specific objective of this study is to describe the validity, practicality and effectiveness of the electronic module. The research method used is research and development using the ADDIE model (Analyze, Design, Develop, Implement, Evaluate). The test subjects in this study were ten students at one of the senior high schools in Banjarmasin. The data were obtained through the electronic module validity sheet instrument, student response questionnaire sheets, and student learning outcomes tests. The results show that: (1) the electronic module is declared valid with a very high category, (2) the electronic module is declared practical in the good category, (3) the electronic module is declared effective with a gain score of 0.64 in the moderate category. In this study, it can be concluded that the electronic module with scientific literacy through direct instruction on impulse and momentum material is declared fit for use in learning.

1. Introduction

Science education is challenged to prepare quality human resources, which are proficient in science and technology, can think logically, critically, and creatively, and have scientific literacy to solve various daily life problems. PISA is an examination system initiated by the Organization for Economic Cooperation and Development (OECD) to evaluate the education systems of 72 countries around the world. Every three years, 15-year-old students are randomly selected to take a test of three essential competencies, namely reading, math and science. PISA measures what a student knows and what he does (application) with his knowledge. The latest results from PISA (Program for International Student Assessment) 2018 Indonesia were ranked 73rd out of 78 participating countries that participated and got a score of 371 to aspects of literacy, 379 to aspects of mathematics, and a score of 396 to aspects of science. However, it is still low compared with the score OECD average is 487 for a score of literacy and 489 for the score in mathematics and science.

Based on the results of PISA, it can be seen that Indonesia is low in literacy aspects. The research results of the Research and Development Institute of the Ministry of Education and Culture in researching the level of students' scientific literacy, many factors affect students' lack of scientific literacy. One of them is the lack of teaching materials in Indonesia compared to other countries in developing students' scientific literacy [1]. Similar to the OECD, one of the international institutions that evaluates student achievement is the IEA (International Association for the Evaluation of

Education Achievement) through the TIMMS study (Trends of International Mathematics and Science Study), surveying fourth-grade elementary school students and eighth-grade, junior high school students. This survey is conducted every four years. This survey was first conducted in 1999 and last conducted in 2019.

Based on the results of the preliminary studies, it was found that 63.24% of students did not know about scientific literacy, 77.94% of students were not interested and enthusiastic in learning, and 63.24% of students preferred electronic books to printed books. Obtained test results of class X students at one of the senior high schools in Banjarmasin that is 51.43% have not met the Minimum Completeness Criteria applicable at the school. It shows that the test of student learning outcomes is low. One of the reasons for this low level is students' incomplete understanding of learning when receiving information from the teacher. Everyone must have scientific literacy skills, including scientific knowledge, science process skills, and scientific attitudes [2]. Based on statistical analysis results, cognitive learning outcomes only affect 2% of scientific literacy (not significant) [3]. In this study, cognitive learning outcomes and scientific literacy stand alone. There is no significant relationship. Influence each other because each is influenced by several internal and external factors from the environment.

Many studies related to scientific literacy have been carried out, such as scientific literacy comics [4], scientific literacy in contextual-flipped classroom learning [5]. However, so far, there has been no scientific literacy research on impulse and momentum material. The material of impulse and momentum requires students to apply the concepts of momentum and impulse and the law of conservation of momentum in everyday life. This study uses electronic modules that can be accessed easily on various networks (wifi, 3G, and 4G), accompanied by videos, images, and links.

2. Method

The type of research conducted is research and development. Development research is a research method used to produce specific products and test the effectiveness of these products [6]. The purpose of developing the product is to solve the problem in the problem formulation. The research and development carried out is the development of e-modules for high school students. The steps of this research are to use the ADDIE model. ADDIE Model is a systematic instructional design model that includes five phases (Analyze, Design, Development, Implementation, Evaluation).

Table 1. Stages of *ADDIE* Development Model

Research Stage	Information
Analyze	The analysis phase consists of two stages, namely, performance analysis and needs analysis. Performance analysis identifies and clarifies whether performance problems require solutions to organize training programs or improve work management. The second stage is to conduct a needs analysis. At this stage, the researcher considers a need to research and develop e-modules containing scientific literacy at one of the senior high schools in Banjarmasin.
Design	The design phase is focused on dealing with and resolving the problems described in the previous analysis phase. The e-module is designed in such a way according to the previous problem analysis. At this stage, the design is based on scientific literacy relating to momentum and impulse material to make students more active in learning and interested in participating in learning activities, deepening material, adding to understanding and insight, and improving student literacy.
Development	At this stage, researchers develop e-modules to foster scientific literacy. E-modules were developed covering the material concept of impulse and momentum, the law of conservation of momentum, and other types of collisions. At this stage, the e-module was also validated by three validators before being implemented. The aim was to determine the validity of the developed e-module.

Research Stage	Information
Implementation	At this stage, research and development product trials are conducted on a small scale to conduct a series of trials of the learning process. The implementation subjects in this research and development were ten class X MIPA, one of senior high school in Banjarmasin. The trial design used in this study compares the conditions before and after using the developed scientific literacy e- module (before-after).
Evaluation	This evaluation stage includes the feasibility of the e-module. The e-module is declared feasible if it has gone through validity, effectiveness, and practicality tests. Validity testing determines the level of conformity of the e-module with the concept of the material being taught, effectiveness to determine the extent to which the e-module can achieve the expected goals, and practicality testing to determine the extent of ease of use of the e-module.

The trial subjects related to the feasibility of the e-module being developed were ten students of class X MIPA 5, one senior high school in Banjarmasin, for the 2019/2020 school year. The subject of this research is the feasibility of the developed e-module. Feasibility e-module is the suitability of the e-module obtained from the validity, practicability and effectiveness developed. The validity of the e-module is the level of validity of the e-module, which is validated by three validators using a validation sheet. The practicality of the developed scientific literacy e-module is measured based on the student response questionnaires distributed when the e- module learning has been completed. The learning outcomes test (summative evaluation) is used to measure the achievement of student learning outcomes that indicate the level of effectiveness of the developed product and measure the achievement of learning outcomes in the form of mastery of the material. Data on test results on students' learning outcomes were then analyzed by calculating the difference between post-test and pre-test (gain normality test)[7].

3. Results and Discussion

The developed electronic module contains scientific literacy on impulse and momentum material using the Flip PDF Professional application to present interesting electronic modules accompanied by videos, images and links to help students understand more easily in studying physics material. There are four categories for scientific literacy indicators in e-modules: science as a body of knowledge, science as a way of investigating), science as a way of thinking, and the interaction between science, technology, and society [8].

The validation of the Electronic Module containing scientific literacy through direct instruction on impulse and momentum material was carried out by three validators. The results of the electronic module validation consist of three assessment aspects, content, display and scientific literacy. The recapitulation of the validation results of the electronic module containing scientific literacy through direct instruction on impulse and momentum material is presented in Table 2. The results of electronic module validation are 86.29%, with a very valid category.

Table 2. The results of the validation of the electronic module for each aspect

No	Aspect	Total score	Total score maximum	Percentage Validity
1	Contents	158	180	87.78%
2	Appearance	134	156	85.89%
3	Science Literacy	184	216	85.19%
Category all aspects				Very Valid

Based on the validation results of the electronic module, the validator consists of three aspects: review the content, look, and scientific literacy in the category of very valid to the category of reliability is very high. The characteristics of the module are self-instructional, self-contained, stand-alone, adaptive, user friendly, and consistent. The first characteristic of the electronic module is self-instructional, meaning that students can learn on their own, not depending on other parties. The second characteristic is self-contained, meaning that the content in the module contains all material (material, worksheets, evaluations) of one competency that students must learn. The third characteristic is stand alone. The electronic module is said to be stand-alone. In the use of the electronic module, it can be used alone as a complete medium without using other media as a compliment. The fourth characteristic is adaptive, meaning that the electronic module is adapted to the characteristics of students. The fifth characteristic is user friendly. The electronic module meets the rules of being friendly to the user. The last characteristic of the electronic module is consistency. It is consistent in using fonts, spacing, and layout and must be the same and balanced. Electronic modules provide opportunities for students to be active in the process of teaching and learning activities.

The assessment of the scientific literacy aspect in the electronic module is categorized as very feasible. It is because it has fulfilled the aspects of science as a body, science as a way to investigate, science as a way of thinking, the interaction between science, technology and society. Based on scientific literacy, the electronic module presents facts, concepts, principles, laws, theories, models, describing how science is used in technology for society. The presentation of the contents of the electronic module is adjusted in the setting of the direct instruction model.

Direct instruction is a teacher-centred teaching model. The direct instruction model is a learning model to support student learning with the teacher as the centre of the learning process [9]. In addition, this learning model is also shown to help students learn basic skills and obtain information that can be taught step by step. So that this electronic module can be tested in the classroom because the developed electronic module is suitable for use as a guide for learning tools for students.

Table 3. Results of student response questionnaires

Aspect	Average Score
Benefit	3.38
Efficiency	3.25
Convenience	3.30
Science Literacy	3.03

The practicality of the module was carried out by calculating the average score of the student response questionnaire distributed when learning with the electronic module had been completed [10]. A product is said to have high practicability if the product is practical and easy to administer [11]. There are two types of statements in these four aspects: positive and negative statements. Twenty-five statements exist in the questionnaire responses, both positive and negative statements in every aspect. Once the students learn the material contents contained in the electronic module, the learners complete a questionnaire response online using google form.

The convenience aspect based on Table 3 shows that the electronic module is in the practical category. It means that using an electronic module containing scientific literacy through direct instruction on impulse and momentum material is easy. This developed electronic module makes it easier for students in the learning process because it uses language, words, and phrases. Sentences, and appropriate paragraphs, this electronic module presents animations, videos, images and links that help make it easier for students to understand learning material. This electronic module is more accessible anytime and anywhere than printed books and is easily accessible on various networks. Research conducted in line with Mujizah, Wati, & Mahtari states that students can access electronic modules to be flexible to study independently [12]. The electronic module could be used anywhere because it combines print and computer media to present information in a structured, attractive manner high level of interactivity [13]. Students can teach scientific literacy skills through science subjects [14].

The learning outcomes test is used to measure the achievement of student learning outcomes that indicate the level of effectiveness of the developed product and measure the achievement of learning outcomes in the form of mastery of the material. Data on student learning outcomes test results were then analyzed by calculating the difference between post-test and pre-test (gain normality test). The effectiveness of the e- module is determined from the suitability of the gain score obtained. The results of the calculation of the normality gain test of the student are 23.25 for pre-test average, 72,7 for post-test average and 0.64 for N-gain score with medium category

The effectiveness of electronic modules is calculated by the N-gain test, which consists of three categories, namely high, medium, and low. The calculation for the cognitive learning outcomes test on the value of each student was carried out by the N-Gain test, from ten students whose data was used in the study and took the pre-test and post-test. The pre-test is a test carried out before being given an electronic module. The Post-test is a test that is carried out after being given an electronic module. The learning outcomes test contains ten multiple-choice questions and six essays that students do. The average post-test scores of students after studying the developed electronic module give relatively good results. It indicates that there is an increase in the ability of students to solve physics problems. It is in line with the research of Zulkarnain, Kadaritna, & Tania, which states that academic ability is a factor from within students in understanding the material [15]. It shows with the results student's response after using module electronics in the process defender's horse. In addition to ability, motivation also affects the higher the motivation, the higher the academic achievement. From the results of the average value of the pre-test and post-test, the results of the N-gain test scores are in the medium category. These results show that the electronic module containing scientific literacy through direct instruction on impulse and momentum material is effectively used during the learning process. It is supported by research that uses the Normalized gain score to measure the effectiveness of the developed product [16].

Overall, students' cognitive abilities have increased compared to before being taught using the developed electronic module, based on the average pre-test and post-test scores. Analysis of learning outcomes tests that have been carried out shows that learning using this electronic module is effective because it can help students achieve learning objectives that direct students to undergo the learning process. The effectiveness is related to how product development can achieve specific goals [17]. This electronic module developed is effective because it is influenced by several factors in which it contains scientific literacy by the 2013 curriculum—appropriate links for strengthening learning materials.

4. Conclusion

This research produces an electronic module suitable for learning physics on Impulse and Momentum material with valid, practical, and effective ways to improve students' science literacy skills.

References

[1] Kemendikbud 2016 *Peringkat dan Capaian PISA Indonesia Mengalami Peningkatan*
[2] Ogunleye B O 2012 International Journal of Educational Leadership **4** 41-8
[3] Dhaniaputri R, Amin M and Al-Muhdar M 2019 *Indonesian Journal of Natural Science Indonesia*
[4] Aisyah R, Zakiyah I A, Farida I and Ramdhani M A *Journal of Physics: Conference Series* **895**
[5] Paristiowati M, Hadinugrahaningsih T, Purwanto A and Karyadi P A 2019 *Journal of Physics: Conference Series* **1156**
[6] Hake R R 1998 *Am. J. Phys.* **66** 64–74
[7] Sugiyono 2012 *Metode Penelitian Pendidikan* (Bandung: Alfabeta)
[8] Chiapetta E L, Koballa Jr T R and Collette A T 1999 *Chapter 1: The Nature of Science. Science Instruction in the Middle and Secondary Schools*
[9] Cobern W W, Schuster D, Adams B, Applegate B, Skjold B, Undreiu A, Loving C C and Gobert J D 2010 *Research in Science & Technological Education* **28** 81-96

[10] Syafri F S 2019 *Pengembangan Modul Pembelajaran Aljabar Elementer di Program Studi Tadris Matematika IAIN Bengkulu* (Bengkulu: CV Zigie Utama)
[11] Widoyoko E 2016 *Evaluasi Program Pembelajaran* (Yogyakarta: Pustaka Belajar)
[12] Mujizah R, Wati M and Mahtari S 2020 *Jurnal Ilmiah Pendidikan Fisik*a **4** 89-98
[13] Sugianto D, Abdullah A G, Elvyanti S and Muladi Y 2013 *INVOTEC* **9** 101-16
[14] Andriani N, Saparini and Akhsan H 2018 *Berkala Ilmiah Pendidikan Fisika* **6** 278-91
[15] Zulkarnain A, Kadaritna N and Tania L 2015 *Jurnal Pendidikan dan Pembelajaran Kimia* 222-35
[16] Coletta V P and Steinert J J 2020 *Physical Review Physics Education Research* **16**
[17] Tegeh 2014 *Model Penelitian Pengembangan* (Yogyakarta: Graha Ilmu)

Learning how to plan a science lesson: An exploration of pre-service science teacher reflection in online microteaching

Z Haryanto[1], N F Sulaeman[1], A Nuryadin[1,2], P D A Putra[3], S A Putri[1], A Z Rahmawati[1]

[1]Physics Education Study Program, Faculty of Teacher Training and Education, Mulawarman University, Indonesia
[2]Graduate School of Sciences and Technology for Innovation, Yamaguchi University, Japan
[3]Science Education Study Program, Faculty of Teacher Training and Education, University of Jember, Indonesia

zeni.haryanto@fkip.unmul.ac.id

Abstract. As a compulsory approach in shaping pre-service teachers for Technological Pedagogical Content Knowledge (TPACK), microteaching is either essential and challenging. During the COVID-19 pandemic, the shift from offline to the online microteaching course had occurred. This major shift drives the change of planning science lessons. Our research examined how pre-service science teachers designed and reflected their planning on a 30-minute microteaching unit. The participants of our study were four groups of pre-service science teachers that consisted of five students. The participants were pre-service science teachers from two national universities in Indonesia that were chosen by purposive sampling. The data was collected through the lesson plan, self-reflection to the lesson plan, and group discussion. We conducted both qualitative and quantitative analysis on research data for content analysis. The qualitative analysis dealt with the themes and patterns of the theme, while quantitative analysis dealt with the duration and frequency of the theme. Based on the analysis, the single domains such as CK, PK, TK were shown in the lesson plan. However, the complex domain of TPACK was rarely seen. The group discussions revealed that teacher educators need to become specific and explicit about which technology is being infused to support learning. It is significant to enhance the TPACK during planning in online microteaching classes and conduct further study of the implementation of this planning in the lesson.

1. Introduction

During the 4-year program of pre-service science teachers, they learn how to develop the specific teacher's professional ability that builds their identity[1]. The discussion of science teacher preparation focuses on how to prepare science teachers connected to the vision of the national curriculum and the goals for science teacher education[2]. The teaching ability is a complex terminology that is well-known as Technological Pedagogical Content Knowledge (TPACK). This terminology could be traced from the terminology of Pedagogical Content Knowledge (PCK) in 1987[3], its implication also needs the new element of technology that strongly influence the ability of teaching [4]. Recently, the discussion of the definition of TPACK has been explored deeply and became a concern of researchers in teacher education. TPACK consists of seven components: Technology Knowledge (TK), Pedagogical Knowledge (PK), Content Knowledge (CK), Technological and Content Knowledge

(TCK), Technological and Pedagogical Knowledge (TPK), Pedagogical and Content Knowledge (PCK), and Technological Pedagogical Content Knowledge (TPACK)[5]. These domains were interconnected [6].

On the other hand, pre-service science teachers are a unique group at the university level. They learn science subjects as well as learn how to teach the subjects. Even though the development of TPACK started from their first semester, the activities in the microteaching course are suitable to be the course that is blending all the knowledge into observable skills. However, the COVID-19 pandemic changes science teacher education in many ways [7,8]. Moreover, in the era of science and technology innovation, the role of a science teacher became crucial to bring an understanding of the integration of science, technology, mathematics, and engineering (STEM) concepts[9]. The need to fulfill the role of a science teacher in the future highlights the requirement to explore the development of TPACK during distance learning [10]. The planning phase of microteaching has been given considerably less attention in both practice and research[11]. A deeper exploration of this issue is needed to address such education shortcomings in preparing future teachers who can plan online science lessons. This paper reported the findings from a study designed to explore the development of TPACK from pre-service science teachers during their activities in planning the science lesson.

2. Method

2.1. Data collection

This study is part of a larger collaborative project between science education teacher preparation programs in Indonesia that mainly focused on designing and redesigning the microteaching course. This study presents results from the initial cycle of what took place during the planning phase of the microteaching unit. The authors had introduced the research project to the pre-service science teachers two weeks in advance and acquired permission to conduct the recordings. The participants were pre-service science teachers in two national universities in Indonesia who were in their 6th semester and taking the online microteaching course. All participants joined two online meetings (@ 60 minutes) where authors 2 and 3 described the microteaching class and planned the science lesson. Then they developed their lesson plan. Moreover, the last meeting was discussion time for reflecting on their lesson plan. The data was collected through the lesson plan and group discussion.

Table 1. Participants Demography

Institution	Participants	Gender	
A	10	Male	2
		Female	8
B	10	Male	3
		Female	7

2.2. Data analysis

Data were then carefully read and reread several times by two authors in line with our analytic approach [10]. We conducted both qualitative and quantitative analysis on research data for content analysis. The qualitative analysis dealt with the themes and antecedent consequent patterns of the theme, while quantitative analysis was related to the duration and frequency of the theme. The theme in qualitative analysis followed the seven elements in TPACK. The observation sheet used for observing the lesson plan could be seen in Table 2.

Table 2. TPACK Observation Sheet

No	Aspects	Classification		Description in the lesson plan
		Existed (1)	**Cannot found (0)**	
1	Technology Knowledge (TK)			
2	Pedagogical Knowledge (PK)			
3	Content Knowledge (CK)			
4	Technological and Content			

	Knowledge (TCK)
5	Technological and Pedagogical Knowledge (TPK)
6	Pedagogical and Content Knowledge (PCK)
7	Technological Pedagogical Content Knowledge (TPACK)

Authors 5 and 6 became the raters for the lesson plan, and inter-rater reliability is estimated with the following equation [12]:

$$\text{Inter} - \text{rater agreement} = \frac{\text{Number of cases with the same score from two raters}}{\text{Number of cases}} \text{x}100 \quad (1)$$

3. Result and Discussion

The analysis and estimation of the inter-rater reliability were conducted on the collected data with SPSS software. The result showed that our instrument had a coefficient of reliability of 0.7, which was categorized as sufficient [13]. From the lesson plan data analysis, the appearance of seven components of TPACK could be identified. Moreover, the identification of strength and vulnerable components were conducted. The confirmation of those findings was conducted through classroom discussions.

3.1. The strength in basic components (CK, PK, TK)

The lesson plan analysis showed that Content Knowledge (CK), Pedagogical Knowledge (PK), and Technological Knowledge (TK) were the three components that were clearly stated. During the planning phase of microteaching sessions, either online or offline, it is commonly found that the pre-service teacher firstly decides the content part. Therefore, the kins of science content to be delivered in the lesson plan always becomes a priority. The class discussions revealed that the CK domain was relatively easy to decide at the beginning of the debate.

Student A: During the group meeting, my group discussed what content each student should have. Then every student made their lesson plan and put the content in it.

Moreover, the PK domain was seen through the opening and closing part of the lesson plan. PK refers to the methods and processes of teaching and includes knowledge in classroom management [4]. This result is similar to the result in in-service physics teachers [14]. Specifically, during the online microteaching class, the students were urged to decide which technology or platform would be used during their lesson. This situation drives the TK domain to be strongly addressed in the lesson plan. For example, the platform for online meetings or supporting their presentation.

3.2. The vulnerable in more integrated component (TPACK)

Although the participants could show their understanding of a single component in the lesson plan, the formation of the complex component remained a challenge for them. The result showed that the more complex domains such as TPACK became more vulnerable components. There was a tendency that the lesson plan for the online microteaching had more briefs than the offline lesson plan. Therefore, there was a lack of evidence of the TPACK domain found in the lesson plan. TPACK is a framework that introduces the relationships and the complexities between all three basic components of knowledge (technology, pedagogy, and content) [5]. These findings confirmed that teacher educators need to become very specific and explicit about which technology to be infused to support learning [15]. Understanding this domain is more complex than understanding the six single components.

Student C: I was thinking of using a digital platform during my lesson; the content was about mechanical spring. I did not mention it clearly in the lesson plan. It was challenging to decide which technology that suitable for the topic. It would be great if there were more examples from the lectures.

The formation of TPACK needs a constant bidirectional process among the single domains [16]. The method of TPACK includes an integrated understanding of how to represent science concepts with technologies, pedagogical techniques that use technologies in constructive ways to teach science, and knowledge of what makes concepts easy to learn. More examples of TPACK from the teacher educators during the microteaching course are required [17]. Digital literacy for teachers also became a significant concern for the online teaching process [18,19]. Our result highlights the need for teacher preparation programs to foster the development of complex domains on TPACK, bringing the ability to enhance the lesson plan.

4. Conclusion

The results conclude that the single domains such as CK, PK, TK were shown in the lesson plan developed by the participants. However, the complex domain of TPACK was rarely seen. The group discussions revealed that teacher educators need to become very specific and explicit about which technology to be infused to support learning. It is significant to enhance the TPACK during planning in online microteaching classes. Although the planning phase is essential, the implementation of this lesson plan was unexplored by this research. Therefore, it is beneficial to conduct further study on pre-service science teachers' TPACK performance.

Acknowledgments

The work described in this article was fully supported by a research grand from PIU IsDB Mulawarman University, Indonesia (Grand number 405/UN17.11/PL/2021).

References

[1] Munfaridah N, Avraamidou L and Goedhart M 2021 *Res. Sci. Educ.* 1–17
[2] Olson J K 2017 *Sci. Educ.* 523–37
[3] Schneider R M and Plasman K 2011 *Rev. Educ. Res.* **81** 530–65
[4] Schmidt D A, Baran E, Thompson A D, Mishra P, Koehler M J and Shin T S 2009 *J. Res. Technol. Educ.* **42** 123–49
[5] Mishra P and Koehler M J 2008 *Pap. Present. Annu. Meet. Am. Educ. Res. Assoc.* 1–16
[6] Bayram-Jacobs D, Henze I, Evagorou M, Shwartz Y, Aschim E L, Alcaraz-Dominguez S, Barajas M and Dagan E 2019 *J. Res. Sci. Teach.* **56** 1207–33
[7] Campbell T, Melville W, Verma G and Park B Y 2021 *J. Sci. Teacher Educ.* **32** 1–6
[8] Erduran S 2020 Science Education in the Era of a Pandemic *Sci. Educ.* **29** 233–5
[9] Qadar R, Mahdiannur M and Sulaeman N F 2021 *Cypriot J. Educ. Sci.* **16** 1154–74
[10] Efwinda S and Mannan M N 2021 *J. Phys. Conf. Ser.* **1806**
[11] Karlström M and Hamza K 2019 *J. Sci. Teacher Educ.* **30** 44–62
[12] Retnawati H 2016 *Analisis Kuantitatif Instrumen Penelitian* (Yogyakarta: Parama Publishing)
[13] Murphy K R and Davidsofter C . 2005 *Physicological Testing: Principle and Testing* (New Jersey: Upper Sadler River)
[14] Qadar R and Haryanto Z 2019 *J. Phys. Conf. Ser.* **1185** 012130
[15] Zhou G, Xu J and Martinovic D 2017 *Eurasia J. Math. Sci. Technol. Educ.* **13** 85–103
[16] Sperandeo-Mineo R M, Fazio C and Tarantino G 2006 *Pedagogical content knowledge development and pre-service physics teacher education: A case study* vol 36
[17] Elias S K 2018 *Open J. Soc. Sci.* **06** 205–24
[18] Aditya D S 2015 *J. Technol. Sci. Educ.* **5** 184–93
[19] Abidah A, Hidaayatullaah H N, Simamora R M, Fehabutar D and Mutakinati L 2020 *Stud. Philos. Sci. Educ.* **1** 38–49

Electronic thermodynamics teaching materials based on authentic learning to practice students' problem-solving skills: Aspects of validity

M Rizki, M Wati, and M Misbah
Physics Education, Faculty of Teacher Training and Education, Universitas Lambung Mangkurat, Indonesia

rizkifisikaulm@gmail.com

Abstract. The purpose of this study was to describe the validity of authentic learning-based electronic thermodynamic teaching materials to train students' problem-solving skills during the covid-19 pandemic. The model used in this study was the ASSURE development model, which only reached the third stage, namely Selection of Media and Materials. The validation of teaching materials was assessed by three validators consisting of two physics education academics and one practitioner. Based on the validation results, it was found that the authentic learning-based electronic thermodynamic teaching material for training students' problem-solving skills was declared valid with a value of 3.14. Appropriate teaching materials can then be used in schools' teaching and learning processes and tested for practicality and effectiveness.

1. Introduction

National education standards in Indonesia are expected to improve students' skills, one of which is problem-solving skills. Schools are expected to be able to create students who have good problem-solving skills [1,2]. This skill is essential to train students because it is beneficial for students in dealing with problems, both in physics problems and in real-world issues [3,4].

The results of an interview with one of the physics teachers at a high school in the city of Banjarmasin explained that the learning resources during the COVID-19 pandemic were student worksheets and videos on YouTube. The existing learning media have not been optimized in facilitating students to learn independently. The material in the student worksheets tends to display general concepts and abstract concepts but cannot be applied to learning in the real world. Teaching materials used in schools are also less helpful for students to solve analytical problems. Because of this, students tend to have less problem-solving skills when viewed from the problem-solving skill indicators. This can be seen after the researchers test students by giving physics questions in essays that practice problem-solving skills. As a result, students answer questions by describing the problem or what is known about the issue without telling or visualizing the problem encountered.

The problems above need exceptional innovations so that the ordinary skills of students are not lost. One solution is to develop authentic learning-based thermodynamic electronic teaching materials to train students' problem-solving skills during the COVID-19 pandemic. Authentic Learning focuses on real-world situations, problem-based activities, complex problems and solutions, case studies, and individual and group practice [5]. Authentic learning is learning that can provide a real learning experience [6–8]. Authentic learning is designed to connect students' issues, problems, and applications in the real world [9–11]. Submission of material in the teaching and learning process, in physics learning, will be more meaningful if it is associated with the surrounding environment [12–14]. The surrounding environment is closely related to the concepts of physics lessons [15–17]. Authentic learning is more inclined to a constructivist learning model that can create meaning for students [18]. Authentic learning can develop students' skills for the better [19]. Authentic learning allows learners to engage in practical tasks using real-world resources and tools and provides opportunities for learners to learn with intention by thinking and acting like professionals as they tackle real problems [20]. Authentic learning can improve understanding, comparing, proving, suggesting, and reflecting on the problem-solving process [21].

The subject used in the development of authentic learning-based electronic teaching materials is the subject of thermodynamics. Thermodynamics material is relatively difficult to complete by students, especially in an online learning system like this. It will complicate material that is already relatively difficult. The application of thermodynamics is also relatively often encountered by students in everyday life or the real world. This thermodynamic material is easier to relate to real-world problems that can provide more value in the learning process. Examples of materials in thermodynamics that can be loaded in authentic learning are thermodynamic processes, heat engines, and cooling engines. In addition, it can also be applied to the toy boat game; to make the toy boat move, there is a thermodynamic process in it, it takes oil in the body of the boat which will be given fire in it, with the heat from the fire that remains burning as long as the oil is still hot it will rotate propeller under the boat that causes the ship to move. Thermodynamics teaching materials are expected to help and facilitate students in solving complex problems.

The results of previous studies stated that authentic learning-based teaching materials could improve student learning outcomes and improve students' problem-solving skills [22–24]. However, the weakness is that students still struggle to remember the sequence of problem-solving skills [25]. Previous research still uses printed teaching materials in the teaching and learning process with the face-to-face method. As for this study, teaching materials are made in such a way as to adapt to the conditions that occurred in Indonesia, namely COVID-19. These teaching materials are made in the electronic form to make it easier for students to access anywhere and anytime. This electronic teaching material also adapts to the needs of students who have difficulty getting access to textbooks. Authentic learning must also adapt to students' circumstances, for example, by adjusting online learning to meet the needs of students well [26].

This research produced a learning product, namely authentic learning-based thermodynamic electronic teaching materials, to train students' problem-solving skills during the covid-19 pandemic. The focus of this research was to describe the validity of original learning-based thermodynamics electronic teaching materials to teach students problem-solving skills.

2. Method

This research was part of the development research with the ASSURE model. However, this article was only up to the fourth stage, namely the Utilization of instructional materials. The following are the stages of the ASSURE model.

The first step is student analysis, the characteristics in one of the senior high schools in Banjarmasin class XI range from 16 to 17 years old. Based on Piaget's theory of cognitive development level, the age of senior high school students in Banjarmasin belongs to the formal operational category. At this age, students are considered capable of communicating socially with their friends, applying concepts, solving problems, and assessing work. Learning activities must emphasize abstract and analytical thinking processes to find answers to a pain in the learning process. However, students still do not understand how to relate learning to everyday life, so problem-solving skills are still lacking.

The second step is to formulate learning objectives. The teaching material used in the research was thermodynamics for the high school of class XI. Learning objectives were taken based on essential competencies in the revised 2013 (K-13) curriculum.

The third step is choosing media, and learning packages, the material used was thermodynamics based on authentic learning. The fourth step is using learning materials; at this stage, validation will be carried out.

Validity measures a teaching material that can or cannot be used to measure what should be measured [27]. According to Ridho [22], validity is an activity to validate a device based on the suitability of the criteria needed in electronic teaching materials. Approval, in this case, is approved by the validator, namely from academics and practitioners. The requirements for the validity of teaching materials are listed in Table 1. Meanwhile, the reliability is calculated using the Cronbach's Alpha equation.

Table 1. Criteria for Validity of Electronic Teaching Materials

No	Mean Score	Category
1	$x > 3{,}4$	Very Valid
2	$2{,}8 < x \leq 3{,}4$	Valid
3	$2{,}2 < x \leq 2{,}8$	Adequately Valid
4	$1{,}6 < x \leq 2{,}2$	Less Valid
5	$x \leq 1{,}6$	Very Less Valid

3. Result and Discussion

The result of this research was electronic thermodynamics teaching material based on authentic learning. There are eight characteristics of authentic learning [28], which some of them are contained in the making of this teaching material, including (1) learning is centered on authentic tasks that are interesting to the learner, (2) students are encouraged to explore the world around them and inquire, (3) students become motivated in more complex tasks, and higher-order thinking, (4) the teacher as a facilitator who helps in learning, and (5) students carry out the basic framework of techniques.

Electronic teaching materials that authentic learning included teaching materials that contained life problems, such as the inclusion of specific regional game objects and nearby tools into teaching materials. This is expected to increase students' motivation to learn so that it is easy to understand the material presented by the teacher. In addition, students can understand the relationship between physics material in everyday life. This electronic teaching material is also equipped with pictures and videos to make it easier for students to understand electronic teaching materials individually. The electronic teaching materials produced have gone through various stages, one of which was validation by two academic validators and one practitioner validator. Two validators are physics education lecturers and one practitioner from a state high school in Banjarmasin. The following suggestions given by the three validators can be seen in Table 2.

Table 2. Validator's Recommendation on Electronic Thermodynamics Teaching Materials

No	Recommendation		
	Validator 1	**Validator 2**	**Validator 3**
1	This teaching material is in accordance with current conditions and helps students learn online independently	Improvements to the table of contents, concept definitions that must use university standard books, science corners must be consistent; learning objectives are replaced with learning indicators, italicized sentences are only for foreign languages, sample questions must be given hands, formula numbers should not be blocked, image forms must be proportional, add a glossary, and pay attention to writing at the beginning of each meeting.	Pay attention to the writing aspects of teaching materials such as punctuation, spaces, italics, formula writing, etc.
2	The electronic teaching materials	Electronic teaching materials are also expected to increase students' motivation to learn.	Clarify authentic learning indicators in the teaching materials,

presented are adjusted to the syllabus		multiply practice questions related to problem-solving.

The following display of electronic teaching materials before and after being validated can be seen in Table 3.

Table 3. Electronic teaching materials before and after being validated

Before	After
	Cover

	Preliminary of authentic learning
None	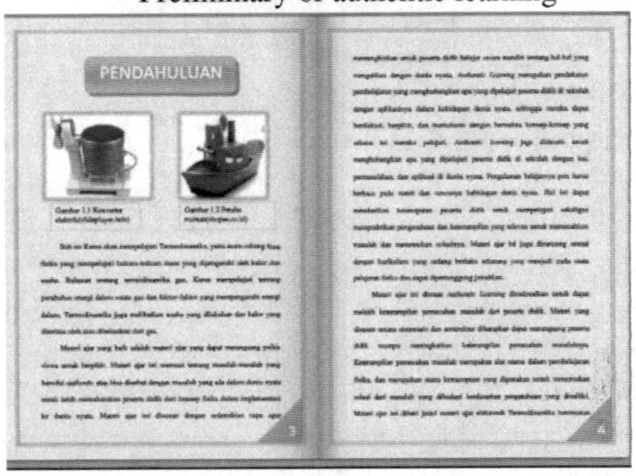
	Example questions with steps to solve the problem
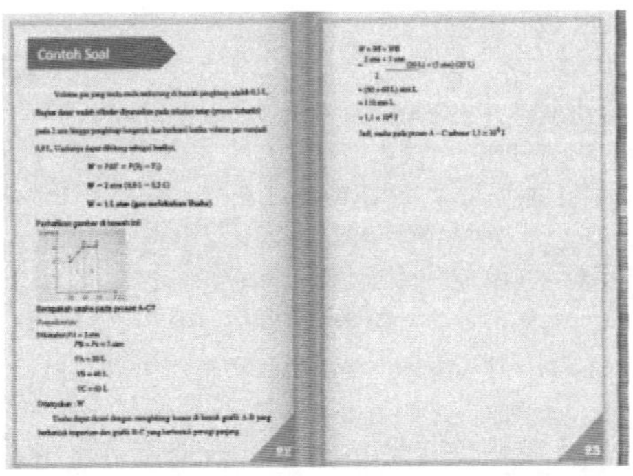	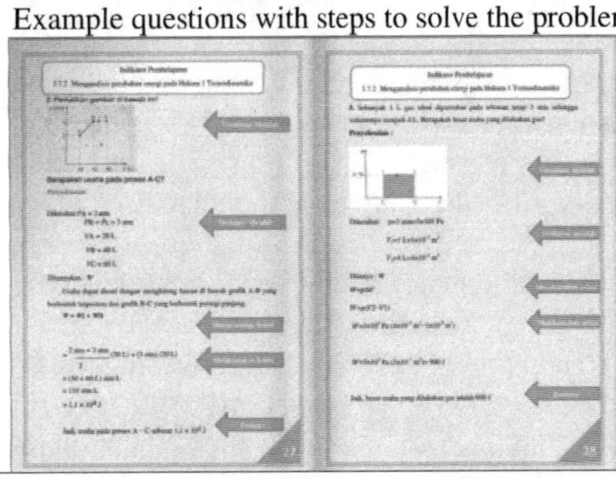

The validity of electronic teaching materials consists of aspects of format, language, content, presentation, and usability. After being assessed by the validator and then analyzed by the author, the validation results were obtained. The results of the validity of electronic teaching materials for each aspect can be seen in Table 4.

Table 4. The results of the validation of teaching materials

Aspect	V1	Category	V2	Category	V3	Category
Format	3,17	Valid	3,33	Valid	3,08	Valid
Language	3,26	Valid	3,21	Vaid	2,95	Valid
Content	3,00	Valid	3,30	Valid	3,00	Valid
Delivery	3,17	Valid	3,29	Valid	2,83	Valid
Usage	3,50	Very valid	3,00	Valid	3,00	Valid

Table 4 shows that the validation results of electronic teaching materials in all categories are reasonable or valid. First, the language aspect of the teaching material is categorized as accurate. It means that the language aspect used in the electronic teaching material is in accordance with the rules of good and correct Indonesian. Second, the aspect of the content of electronic teaching materials is categorized as valid. This also shows that the contents of electronic teaching materials are in accordance with national standards or competency standards that should be adapted to the curriculum currently circulating during the COVID-19 pandemic. The aspect of presenting the material is categorized as valid. It also shows that the presentation delivered in electronic teaching materials is good enough to be used in the teaching and learning process in the classroom. Finally, the aspects of the benefits and uses of electronic teaching materials are categorized as valid. This also shows that the developed teaching materials will have good benefits and uses for their users later during the teaching and learning process. The four aspects in the validation show that this electronic teaching material is categorized as valid and ready to be used in authentic learning. This indicates that the developed teaching materials can be used for further trials. This electronic teaching material, after being validated, was then revised according to the validator's suggestions. This was done to obtain better electronic teaching materials than before. In addition, teaching materials that are declared valid indication that the teaching materials can measure precisely what will be measured, in this case, students' problem-solving abilities [29–31].

This electronic teaching material was designed to adapt to the current situation, making electronic teaching materials not in printed form to make it easier for students and teachers to access teaching materials anytime and anywhere. Electronic teaching materials are equipped with sample questions with problem-solving skill methods so that students can quickly obtain electronic teaching materials that can improve their problem-solving skills. Electronic teaching materials that are declared valid can be suitable and used in the learning process [22,32]. Appropriate teaching materials according to Prastowo [33], have several criteria such as: (1) titles or materials that are prepared based on essential competencies or basic materials that students must achieve; (2) the design is clear and attractive, the order of contents must be easy, the title is short, the cognitive structure is clear, and contains a table of contents, summaries, and student assignments; (3) language that is easy to understand, seen from the flowing vocabulary, clear sentences and relationships between sentences, and not long-winded sentences; (4) the content of the material that can be used to test understanding, meaning that teachers can assess students' understanding through teaching materials; (5) contains stimulants, this concerns the content of the material that encourages students to think; (6) easy to read, seen from the eye-friendly teaching materials, the letters are not too small and easy to read, and the order of the text is structured; and (7) instructional materials, meaning that the contents of the teaching materials contain clear instructions and instructions to make it easier for students to learn.

Suitable teaching materials must contain relevant facts, concepts, principles, and procedures and are written in points according to the formulation of the competency indicators [34]. Electronic teaching materials are developed to train students' problem-solving skills. This teaching material contains authentic learning knowledge, motivations, sample questions with problem-solving skills, practice questions, and summaries. The advantages of authentic learning are that students will gain new understanding and experience a level of change in making perceptions [35]. The benefits of Authentic Learning are included in education that can occur anywhere so that students do not feel bored with learning; students have more extraordinary skills in analyzing problems that arise in real life, students get a qualified learning experience in interacting with the surrounding environment, learning is student-centered so that students can understand their material in depth [36].

In general, authentic learning-based thermodynamic electronic teaching materials are declared valid. It shows that these electronic teaching materials can be used to train students' problem-solving skills. In addition, this teaching material can be used as a reference during the COVID-19 pandemic. Furthermore, this teaching material needs to be implemented in the teaching and learning process to measure its practicality and effectiveness.

4. Conclusion

The authentic learning-based thermodynamic electronic teaching material developed is categorized as valid with an average value of 3.14. Teaching materials that are said to be helpful can then be used in the class trial stage. As for suggestions for further research, further research is needed to test the validated electronic teaching materials.

Acknowledge

Researchers would like to thank the Universitas Lambung Mangkurat for funding PNBP University in 2021, in the program “Dosen Wajib Meneliti”, with contract No: 010. 24/UN8.2/PL/2021.

References

[1] Fahrina F, Jamal A and Salam A 2018 *Berk. Ilm. Pendidik. Fis.* **6** 98–117
[2] Thersia V, Arifuddin M and Misbah M 2019 *Berk. Ilm. Pendidik. Fis.* **7**
[3] Yuberti Y, Latifah S, Anugrah A, Saregar A, Misbah M and Jermsittiparsert K 2019 *Eur. J. Educ. Res.* **8** 1217–27
[4] Hafizah E, Misbah M and Annur S 2018 *Momentum Phys. Educ. J.* **2** 72–8
[5] Lombardi M M 2007 *Educ. Learn. Initiat.* **1** 1–12
[6] Astuti E R and Baysha M H 2018 *J. Ilm. Mandala Educ.* **4** 231–237
[7] Nicaise M, Gibney T and Crane M 2000 *J. Sci. Educ. Technol.* **9** 79–94
[8] Wati M, Sutiniasih N K, Misbah M, Mahtari S, Annur S and Mastuang M 2020 *J. Phys. Conf. Ser.* **1567** 1–10
[9] Pearce S 2016 *E-teaching; Manag. Strateg. Classr.* **10**
[10] Har L B 2013 Authentic Learning (Hong Kong: The Hong Kong Institute of Education)
[11] Wati M, Safiah S and Misbah M 2021 *J. Phys. Conf. Ser.* **1760**
[12] Hasani N L, Hartini S and An’Nur S 2019 *J. Ilm. Pendidik. Fis.* **3** 65–76
[13] Hartini S, Isnanda M F, Wati M, Misbah M, S A and Mahtari S 2018 *J. Phys. Conf. Ser.* **1088**
[14] Wati M, Putri M R, Misbah M, Hartini S, and Mahtari S *J. Phys. Conf. Ser.* **1422**
[15] Hartini S, Firdausi S, Misbah and Sulaeman N F 2018 *J. Pendidik. IPA Indones.* **7** 130–7
[16] Hartini S, Misbah M, Helda H and Dewantara D 2017 *AIP Conference Proceedings* **1868**
[17] Misbah M, Hirani M, Annur S, Sulaeman N F and Ibrahim M A 2020 T *JIPF (Jurnal Ilmu Pendidik. Fis.* **5** 1
[18] Roach K, Tilley E and Mitchell J 2018 *High. Educ. Pedagog.* **3** 495–509
[19] Ornellas A, Falkner K and Edman Stalbrandt E 2019 *High. Educ. Ski. Work. Learn.* **9** 107–120
[20] Herrington J, Parker J and Boase-Jelinek D 2014 *Aust. J. Educ.* **58** 23–35
[21] Dolapcioglu S and Doğanay A 2020 *Int. J. Math. Educ. Sci. Technol.* 1–24
[22] Ridho M, H ., Wati M, Misbah M and Mahtari S 2020 *J. Teach. Learn. Phys.* **5** 2 5 87–98
[23] Arsyad Z, Wati M and Suyidno S 2020 *SEJ (Science Educ. J.* **4** 113–128
[24] Triuitami R R and Ruwanto B 2017 *J. Pendidik. Fis.* **4** 368–378
[25] Supriyadi S 2020 Pengembangan Materi Ajar Gerak Lurus Berbasis Authentic Learning untuk Melatihkan Keterampilan Pemecahan Masalah Peserta Didik (Skripsi FKIP ULM Banjarmasin. Tidak Dipublikasikan)

[26] Chiu P S, Pu Y H, Kao C C, Wu T T and Huang Y M 2018 *Innov. Educ. Teach. Int.* **55** 336–347
[27] Sugiyono S 2017 Metode penelitian kuantitatif, kualitatif dan R&D (Bandung: Alfabeta)
[28] Donovan M S, Bransford J D and Pallegrino J W 1999 How people learn: Bridging Research and Practice (Washington, DC: National Academy)
[29] Arikunto S 2012 Dasar-dasar Evaluasi Pendidikan (Jakarta: Bumi Aksara)
[30] Widoyoko E P 2016 Evaluasi program pembelajaran (Yogyakarta: Pustaka Pelajar)
[31] Yusup F 2018 *J. Ilm. Kependidikan* **7**
[32] Aini N, Zainuddin Z and Mahardika A I 2018 *Berk. Ilm. Pendidik. Fis.* **6**
[33] Prastowo A 2014 Panduan kreatif membuat perangkat pembelajaran inovatif (Yogyakarta: Diva Press)
[34] Suyidno S and Jamal M . 2012 Strategi Belajar Mengajar Pegangan bagi Pembelajar Kreatif, Inovatif dan Berkarakter (Banjarmasin: Microteaching FKIP ULM)
[35] Lee H 2020 Encycl. Glob. Archaeol 8253-8256.
[36] Mukson M 2014 Peningkatan Hasil Belajar Ilmu Pengetahuan Sosial (IPS) Melalui Model Pembelajaran Authentic Learning Peserta Didik Kelas 4 MIN Pucung Ngantru Tulungagung (Tulungagung: IAIN Tulungagung. Skripsi: Tidak Dipublikasikan)

Validity of teaching materials for environmentally friendly technology products using STEM-based guided inquiry to improve students' scientific literacy competence

Syahmani[1], M Kusasi[1], I Najmiati[2]

[1] Chemistry Education Program Study, Faculty of Teacher Training and Education, Lambung Mangkurat University, Banjarmasin, Indonesia
[2] Science Education Program Study, Faculty of Teacher Training and Education, Lambung Mangkurat University, Banjarmasin, Indonesia

syahmani_kimia@ulm.ac.id

Abstract. Students' scientific literacy skills need to be improved because students in the 5.0 society era must have various skills. The availability of relevant student books has limitations in presenting the material. In overcoming this problem, science teaching materials that contain the scientific literacy referred to the 21st-century situation are needed. Therefore, research was conducted on developing integrated science teaching materials with STEM-based Guided Inquiry (STEM-based GI) to improve students' scientific literacy competence. This study aims to determine the validity of teaching materials (integrated science modules, lesson plans, and learning outcomes tests) for STEM-based GI on the topic of Environmentally Friendly Technology (EFT) products to improve students' scientific literacy competence. The research and development procedure used was the formative evaluation model. This research procedure was divided into three stages: self-evaluation, prototyping, and evaluation of field tests. However, this research only reached the stage of expert validation (prototyping). The results showed that the average value of the three validators was 94.33% which was included in the very valid category. The coefficient of agreement between validators was (R) 87.10%, included in the high category. Therefore, the resulting teaching material was valid and improved scientific literacy in junior high school science learning.

1. Introduction

Globalization has resulted in changes in the overall life of society, including the education sector. The 21st-century life skills are the answer to the Industrial Revolution 4.0 and the Society 5.0 challenges. Basic literacy skills and other competencies such as critical thinking, reasoning, creativity, communication, collaboration, and problem-solving abilities are essential.

Science education can be implemented in various ways, one of which is through science learning [1]. Science learning itself is a learning that can grow and instill scientific literacy in students through learning about science concepts, doing practical work, and solving problems scientifically. Based on the 2018 PISA tests and evaluations in mathematics, science, and reading, Indonesia was ranked 71st out of 77 countries with an average score of 382 [2].

This situation proves that students in Indonesia, compared to other countries, are still very low in scientific literacy skills. The factors that influence this are the education system and curriculum, the selection of learning models and methods, learning facilities, teaching materials, learning resources, and other factors.

Student learning resources are a factor that directly influences student learning activities that affect the low level of students' scientific literacy skills. Books are still the primary source of student learning in schools. The book used by the teacher in science learning does not contain complete scientific literacy. The material presented is also too broad, so students have difficulty understanding the science concepts presented in the book[3].

Teaching materials are one of the learning tools that help students understand the subject matter. Rizki [4] states that teaching materials categorized as good are attractive in terms of appearance and content and must adapt to students' needs. The teaching materials can be in the form of modules. Modules are teaching materials that are arranged systematically and made by using language that students easily understand. Modules are dynamic and self-contained. The module also has the advantages that can be adjusted to the students' age and knowledge level to learn independently through the instructions in the module. To be successful in the learning process, the teacher must be precise in choosing the approach used during the learning process.

Science, technology, engineering, and mathematics (STEM) education is crucial in today's educational trends. STEM is an alternative approach to education in the 21st century that realizes the fundamental goal of growing knowledge and expertise (IPTEK). Kelley [5] states that the STEM approach used during the learning process can train students to implement their knowledge to arrange something as a form of problem-solving related to their living environment by utilizing technology. The integrated STEM approach uses real-world contexts to investigate authentic problems using contextual and active learning approaches [6][7]. This approach helps students to be problem solvers, independents, innovators, inventors, creators, logical thinkers, science, and technology literate [8-12].

The approach enhances higher-order thinking skills, motivation, and achievement [13,14]. In addition, it improves students' cognitive skills, such as listening, thinking, problem-solving, and decision making, and also triggers the capability of self-expression, observation, cooperation, collaboration, and communication [15-18].

The STEM-based GI learning model can develop conceptual understanding[19], meta-cognitive knowledge, and skills [20,21]. It also improves student learning, achievement, motivation, and interest [22-24], as well as directs students to identify problems, find solutions, formulate questions, conduct experiments, analyze and draw conclusions [25]. Through inquiry, students engage in discovery, ask real questions about academic topics, and are interested in learning more and sharing with others[27]. A well-designed inquiry-based approach is effective in promoting positive learning outcomes such as deep thinking, application of knowledge, and logical reasoning [28,29].

One of the integrated science materials taken is Eco-Friendly Technology or Environmentally-Friendly Technology (EFT) products material for Junior High School class IX, based on the syllabus and basic competencies. EFT often involve some of the following: (a) recycled, recyclable, and/or biodegradable content, (b) plant-based materials, (c) reduction of polluting substances, (d) reduction of greenhouse gas emissions, (e) renewable energy, (f) energy-efficiency, (g) multi-functionality, (h) low-impact manufacturing. EFT is any technology that aims to make our lives and environment better [30]. This material can be related to STEM-based GI learning because it contains the four pillars of science and scientific literacy processes in everyday life. Water purification, composting, and biogases are examples of processes that involve the simplest scientific literacy and manufacture through a series of processes that use natural materials and simple tools. However, the existing textbooks do not contain elements of STEM-based GI, so there is still a need to develop teaching materials, especially on EFT product materials, to improve students' scientific literacy competence. Researchers hope that with the development of these teaching materials, they can make it easier for teachers and students in the learning process and easier to understand and master the material of EFT products.

2. Method

The research and development design was the formative evaluation model adapted from Plomp & Nieveen [31], which included the stages of self-evaluation, prototyping, and field test evaluation. The research object was an integrated science learning material for EFT using STEM-based GI, which is relevant to clean technology to produce environmentally friendly products [31]. The research was conducted from February to August 2020.

In the self-evaluation stage, the product developed went through the analysis and design stages related to relevant journal articles and STEM-based GI learning. The EFT module using STEM-based GI presented: (1) Determination and principles of EFT, (2) Application of EFT, and (3) Energy-saving behavior. The first meeting was about the lack of clean water as a problem, and the solution to this was manufacturing simple water filtration products. The filtration method in water treatment was simple by using bottles filled with gravel, fine sand, stones, fibers, zeolite, and activated carbon from coconut shells. These materials were arranged in bottles as dirty water filters to get clearer water. Students were assigned to find out how the composition and arrangement of these ingredients produced the clearest water. At the second meeting, the produced product was a simple technology for making biogas and organic fertilizer by solving the problem of animal manure or household organic waste. Then, the third meeting resolved the problem of the scarcity of fuel oil as an energy source. The EFT was produced in the form of bio-ethanol and bio-diesel. At the last meeting, the issue discussed was the lack of awareness in terms of energy saving. Then, a poster/mock-up of an energy-efficient home was made as a product.

Next is the prototyping stage, consisting of an expert review, one-to-one evaluation, and a small group. At the expert review stage, the product developed was validated by three experts and proceeded to the one-to-one evaluation stage, where product trials were conducted on a small scale of about three students. Then in the small group stage, the product was tested on a larger scale than the previous one. After that, the product was tested on a large scale at the last stage, named field test evaluation. However, this research was only carried out until the expert review stage. The following are the stages of development in this research:

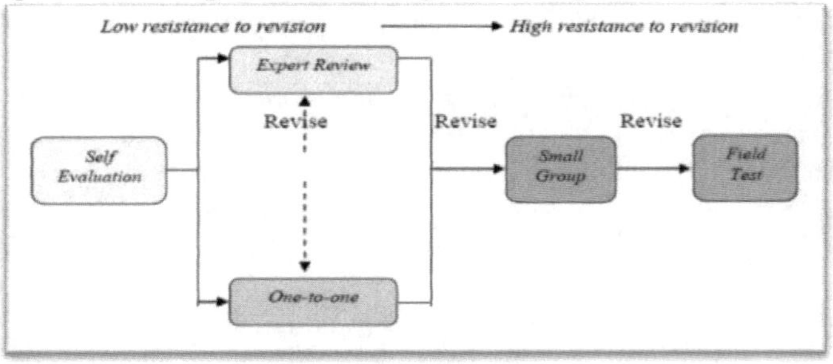

Figure 1. Stages of development module STEM-based GI in the topic EFT product

Data was collected by the questionnaire method. Questionnaire forms were given to 3 validators. The score of the validators used a Likert scale of 1 – 4. The interpretation of the validity of the module, lesson plans and literacy tests were determined based on the following criteria: 85 < x 100 (very valid), 70 < x < 85 (valid), 50 < x < 70 (less valid), and x < 50 (invalid) [7]. The reliability of the validation results of the STEM-based GI learning tool was based on the statistical analysis of the percentage of agreement (R) [32], with $R = \left[1 - \left(\frac{A-B}{A+B}\right)\right] X100\%$. Description: A = the highest score , B = the lowest score, wherein the value of R is the percentage of agreement (consistency between validators). The consistency of the validation results is included in the reliable category if the R coefficient is above 75%. The criteria was adapted from Fraenkel et al. [33], which is: 90 < x 100 (very good reliability), 70 < x 90 (high reliability), 50 < x 70 (medium reliability), and x 50 (low reliability).

3. Result and Discussion

Teaching materials developed in this study were in the form of modules, lesson plans, and literacy test instruments. The feasibility of the developed teaching materials can be seen from the validation results at the stage expert review in Table 1.

Table 1. Results of teaching materials validation

Aspect of	Validity Result		Coefficient of agreement between validators	
Assessment	Percentage	Criteria	R (%)	Criteria
Module	94.00	Very Valid	86.00	High consistency
Lesson plan	95.00	Very Valid	86.00	High consistency
Literacy test	94.00	Very Valid	89.50	High consistency
Average	94.33	Very Valid	87.10	High consistency

3.1 Validation Module

The e-modules that have been compiled have various features to improve scientific literacy, namely: (1) Did you know, (2) Do it with STEM integration, and (3) *Let's* practice. The features to improve scientific literacy competence are "STEM integration with EFT products" and "*let's* practice" because of the integration and interaction of STEM aspects in solving social-scientific societal issues.

Validation results conducted by three experts were grouped into five reviews covering design aspects, format aspects, material aspects, linguistic aspects, and presentation aspects. Based on the five aspects of the validation assessment, the validation percentage of the module developed was 94%, classified as very valid; and the module reliability was 86% with very reliable criteria. So, the STEM-based GI module is feasible to be used as a support for science learning in schools. Students discover and explain natural phenomena and scientific concepts. This module also contains problem-solving and equips students with social skills through individual and group assignments, discussions, delegating responsibilities, and collaborating with their peers. This module also supports student-centered learning and the interdisciplinary nature of science, involving students in constructing their knowledge based on personal and meaningful experiences.

This developed module has advantages on the materials in KI and KD related to STEM. This developed module contains four pillars of STEM science using a guided inquiry model. Therefore, students can more quickly master the topic related to everyday life as a learning guide to independence. The module is designed with a more attractive appearance and equipped with pictures related to the subject. The module is attractive to foster curiosity and enthusiasm for learning in students. The weakness in this module is that the material developed is only limited to EFT products, so it needs to be developed more broadly in other materials. This module consists of study instructions, achievement competencies, subject materials, student worksheets, question sheets, answer keys, evaluation, and feedback on the evaluation results. The environmental problem and linkage of the material in the module are shown in Figure 2-5.

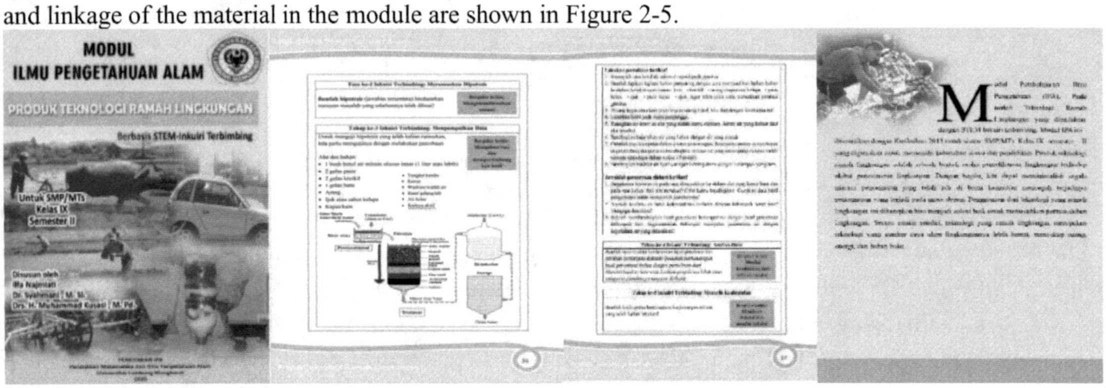

Figure 2. EFT module layout using STEM-based GI model

There are three problems for EFT using STEM-based GI learning presented below.

Problem 1:

"Consuming polluted water will cause various diseases such as typhoid, diarrhea, cholera, dysentery, tapeworm, hepatitis, polio, itching, diarrhea, and scaly skin [34,35], which can directly or indirectly undermine human health. Indeed, many people underestimate these diseases, but most are not aware that these diseases can also cause death. This water treatment aims to kill microorganisms in the water. How do you process this dirty water into clean and healthy drinking water?"

SCIENCE
Factual: Removes chemicals, biological contaminants, solid particles, and gases from contaminated or dirty water.
Conceptual: Water purification is the processing of dirty water into clean water.
Procedural: Procedure for making water purification correct
ENGINEERING: Designing tools and materials for making water purification
TECHNOLOGY: A simple tool used to support the manufacture of water purification.
MATHEMATICS: Calculate the speed of the filtration process.

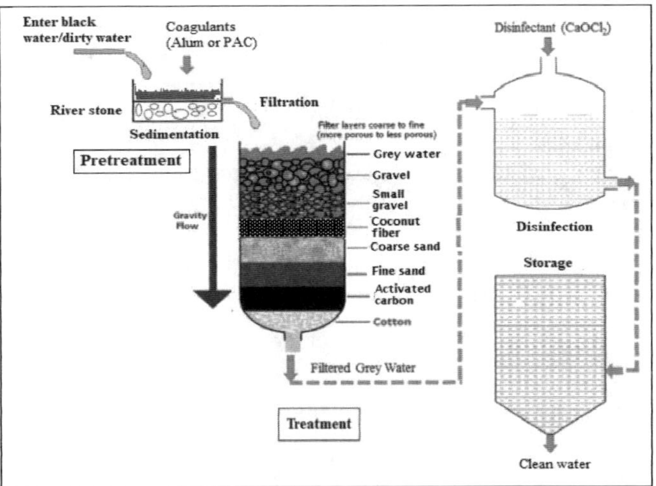

Figure 3. The relationship between water purification and STEM

Several methods can be used to purify or treat raw water, such as coagulation, flocculation, sedimentation, filtration, adsorption, and chemical disinfection [36] . The adsorbent for decolorization and the removal of organic pollutants in wastewaters are activated carbon, alumina, chitin, chitosan [38,39]. However, it is necessary to find and choose an effective and low-cost household water treatment method.

Problem 2:

"Energy is vital for human life because any form of human action or activity requires energy. We are now developing alternative energy other than fossil fuels, whose quantity is becoming increasingly limited. Waste from food production can be processed into biogas. This technology is useful for producing alternative energy sources, but it can also keep the environment clean. Organic waste from livestock and the food manufacturing industry is not simply thrown away, but used as biogas."

SCIENCE
Fact: Anaerobic bacteria convert organic matter into methane gas (CH_4) by 75% and other gases such as carbon dioxide, hydrogen, and hydrogen sulfide.
Conceptual: Biogas is gas produced by anaerobic activity or fermentation of organic materials.
Procedural: Procedure for making biogas

ENGINEERING
1. Designing procedures for making biogas
2. Designing tools and materials for making biogas.

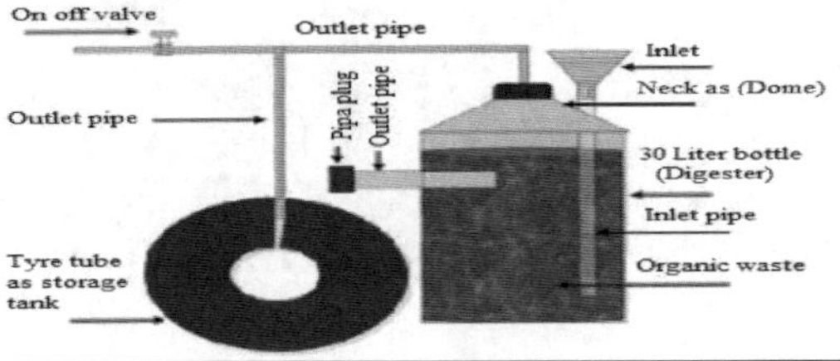

TECHNOLOGY
Simple tools used to support biogas production presented by Re-cyclopedia.

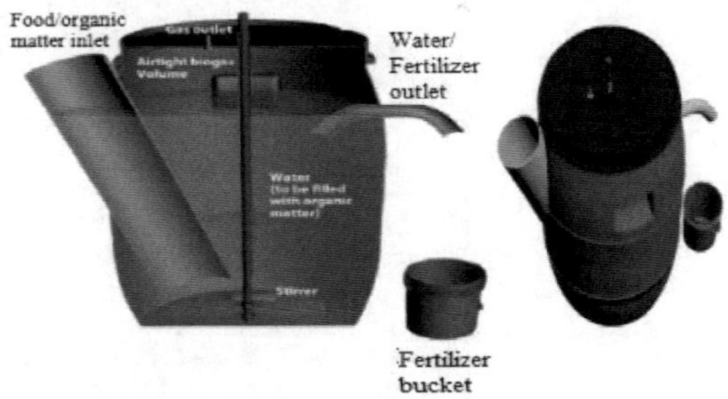

MATHEMATICS
1. Measure the diameter and length of the hose used.
2. Measure the volume of methane (biogas) produced.

Figure 4. The link between biogas and STEM

Biogas is a product of the anaerobic digestion of organic matter by methanogenic bacteria. Biogas technology utilizes organic agricultural waste and converts it into fuel and fertilizer. The direct impact of biogas is saving firewood and kerosene energy. Biogas technology also solves the problem of environmental pollution and improves household or communal sanitation.

Energy conservation is the first step to start the energy-saving movement to prevent an energy crisis. Current technological advances have created a variety of energy-efficient products that are much more environmentally friendly, such as bio-fuel technologies that can also reduce the use of fossil fuels for

vehicles. Through a simple energy conservation movement, we can save the Earth from an energy crisis and save the environment from the damage caused by excessive energy use. Let us start the energy conservation movement!
Regarding the last issue, which discussed energy efficiency and energy-saving behavior.

Problem 3:

"Energy efficient homes are an important concern for current and future urban development. Housing contributes about 37.5% to significant energy consumption and carbon emissions [40]. What makes a home energy-efficient and energy-saving? The teacher instilled in students' awareness in terms of saving energy. Make a mock-up of an energy-efficient home made as a product. For example, designing a miniature house that saves electricity."

SCIENCE **Factual:** the lamp will turn on if it is connected to a current source. **Conceptual:** with the same current, the voltage in the series circuit will be divided into several lamps, while in a parallel circuit, the voltage on each electrical load is the same as the source voltage. **Procedural:** how to assemble open, closed, series, and parallel electrical circuits. **Metacognitive:** save electricity consumption	**TECHNOLOGY** 1. Energy efficient homes. 2. Electrical installation at home. 3. Internet to find information related to energy-efficient homes and electrical circuits. 4. Computer for making reports.
ENGINEERING 1. Design a miniature energy-efficient house 2. Make a miniature energy-efficient house 3. Design electrical installations in energy-efficient miniature houses. 4. Make electrical installations in energy-efficient miniature houses. 5. Evaluate the work.	**MATHEMATICS** 1. Using a scale, calculate the length, width, and height of the house. 2. Prove the Pythagorean theorem. 3. Determine the surface area and volume of the flat figure! 4. Using social arithmetic, calculate the amount of electric power, the price, and the number of materials used.

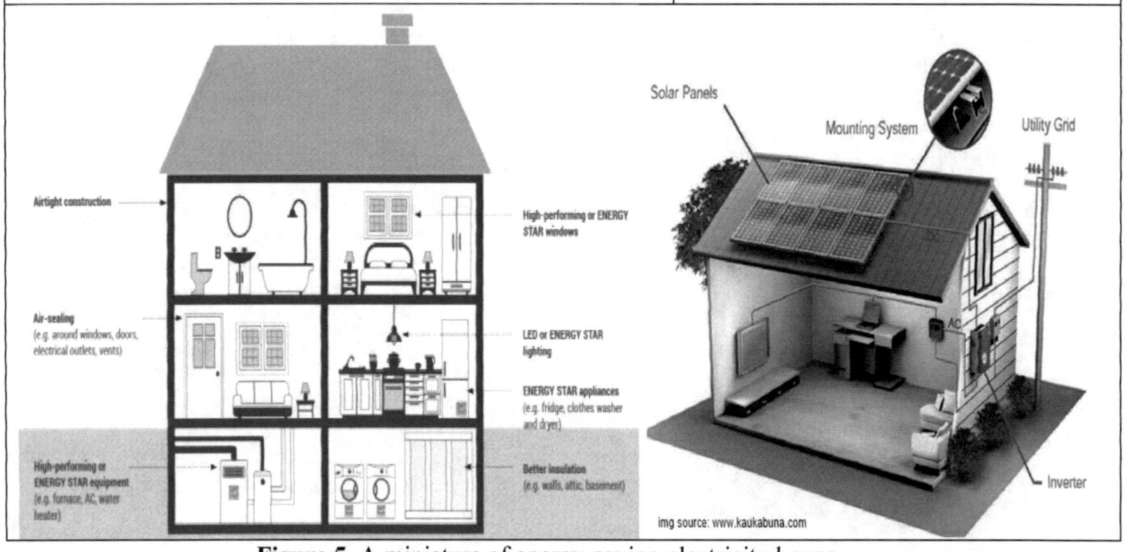

Figure 5. A miniature of energy-saving electricity house

Figure 5 shows a description of the technical energy saving at home. Actions to be taken to increase energy efficiency at home include using compact fluorescent lamps (CFLs) or LED lamps. Turn off the lights, Air Conditioner, and heater when nobody is home. Using better insulation, high-performing or energy star windows, and equipment. Using photovoltaic (PV) solar panels at home is a good option if there is available electricity.

Based on research made by Islamyah [41], the implementation of learning with a STEM-based GI model can support students to find and explore a science concept more easily. It also motivates students to apply the knowledge they understand into their daily lives. Additionally, it can also improve students' scientific literacy competence. The application of modules using the STEM approach for high school students has a positive effect [42].

3.2 Validation of Lesson Plan

The standard lesson plan elements consist of title, objectives, standards, explanation of the warm-up activity, assignments, materials, assessment, and questioning [43,44]. The results of the lesson plan analyzed by three validators include aspects of format, language, and content. Based on these three aspects, the percentage of lesson plan validation was 95%. Meanwhile, the result of the reliability analysis was 86%. From the assessment range determined, the value of 86% shows that the lesson plan is high consistency (reliability).

The lesson plan validation results with valid categories in each aspect indicate that the components building up the lesson plan were correctly fulfilled. Therefore, it can be concluded that the lesson plan developed is very valid and feasible to use in learning and can be used as a guide for teachers to improve students' competence in the learning process following the teaching materials used. Instructional planning is a vital aspect of any teacher's work to give students meaningful learning experiences [45]. The advantage of developing this learning implementation plan is that it applies a guided inquiry model that makes students more proficient in the learning process. The learning implementation plan with the STEM approach with Engineering Design Process (EDP) is interrelated with the module, which is expected to make it easier for students to understand the learning provided by the teacher. The EDP has seven stages: define the problem, plan possible solutions, choose the possible solution, design, test, redesign, and communicate [46]. It is simplified into six stages for junior and senior high school, as presented in Table 2.

The STEM-based GI learning model is implemented to improve scientific literacy through scientific investigation or scientific methods because the three are related (Table 2).

Table 2. The link between STEM-based GI and scientific literacy

Guided inquiry syntax[25][26]	STEM (EDP stages)[47]	Scientific literacy indicator [48]
Initiation. Formulate a well-formed investigation research question	Define the problem	Explain phenomena scientifically
Invention. Generate competitive alternative hypotheses and predictions	Plan	Design scientific investigations
Investigation. Design and carry out experiments for data collecting	Carry out investigations Design	Design scientific investigations
Interpretation. Analyze the data		Interpret data and evidence scientifically
Instruction. Conclude, evaluate, and reflection	Evaluate and Redesign	Evaluate scientific investigation

The GI learning had five phases [25], and the STEM integration was done as follows. Two aspects of STEM: science and technology, were integrated into the first phase (Initiation) and the second phase (Invention). Three aspects of STEM: science, mathematics, and engineering, were integrated into the third phase (Investigation). All aspects in STEM: science, technology, engineering, and mathematics, were integrated into the fourth phase (Interpretation) and the fifth phase (Instruction). In STEM, students' activities were primarily dominated in the engineering aspect. The EDP had six steps (define the problem, plan, carry out investigations, design, evaluate, and redesign). Through STEM-based GI learning, students can practice acquiring knowledge or concepts and conduct experiments. Therefore, they can learn to find various solutions to a particular task or problem. STEM education can help people practice their literacy skills in technology and science [5]. The integration of STEM aspects, especially engineering aspects in GI learning, is that applying mathematics and science concepts helps students understand certain phenomena.

The scientific processes, such as problem analysis, original investigation, communication with peers, formulation of solutions, and interpretation of results, with the help of mathematics, are essential in operating data [49]. In technology, students are given the skills to understand the tools used to simplify all existing

Journal of Physics: Conference Series **2104** (2021) 012019 doi:10.1088/1742-6596/2104/1/012019

problems. Students are also be provided with how to make these tools and get ideas to find out a problem that can be done more efficiently. The engineering aspect helps transform understanding knowledge into practical technologies that can help to solve similar problems further [50]. The STEM-based GI learning can develop the skills and abilities necessary for work and daily life in the 21st century [51, 52], develop scientific literacy and problem-solving skills[53], processes, and student's academic success [54].

3.3 Validation of Literacy Test

The OECD Programme for International Student Assessment (PISA) [44] describes three competencies needed in scientific literacy, namely: (1) explain phenomena scientifically, (2) evaluate and design scientific inquiry, and (3) interpret data and evidence scientifically. Indicators of scientific literacy competency are presented in Table 3.

Table 3. Scientific literacy competency with the indicators

Scientific Literacy Competency	Indicators
Explain Phenomena Scientifically	• recall and apply appropriate scientific knowledge • identify, use and generate explanatory models and representations • make and justify appropriate predictions • offer explanatory hypotheses • explain the potential implications of scientific knowledge for society
Evaluate and Design Scientific Enquiry	• identify the question explored in a given scientific study • distinguish questions that could be investigated scientifically • propose a way of exploring a given question scientifically • evaluate ways of exploring a given question scientifically • describe and evaluate how scientists ensure the reliability of data, and the objectivity and generalisability of explanations.
Interpret data and evidence scientifically	• transform data from one representation to another • analyse and interpret data and draw appropriate conclusions • identify the assumptions, evidence and reasoning in science-related texts • distinguish between arguments that are based on scientific evidence and theory and those based on other considerations • evaluate scientific arguments and evidence from different sources (e.g. newspapers, the Internet, journals)

The learning outcomes test analyzed by three validators include several aspects: format, questions following scientific literacy indicators, clarity of questions' purposes, question sentences preventing multiple interpretations, questions using clear interrogative sentences or commands, clear and functioning images, using correct Indonesian rules, and using language that is easy to understand. In each aspect, various validation scores were obtained. The presentation of the test validation of learning outcomes was 94.00% with very valid criteria and a reliability value of 89.50% with very reliable criteria, showing that the questions for the learning outcomes test developed by the researcher are feasible to be used as learning tools. According to [19], learning tools are effective if they provide results following the goals that have been set. The guided inquiry learning model is very effective in managing various types of the learning atmosphere in the classroom and can facilitate the improvement of students' scientific literacy [55].

The advantages of the developed learning outcomes test questions refer to indicators of scientific literacy achievement. The increase in students' scientific literacy can be measured using questions that researchers have developed. The scientific literacy indicators used are for questions (1) Analyzing and assessing objective and reliable information relevant to energy issues (2) Applying appropriate and environmentally friendly alternative energy methods (3) Distinguishing scientific questions that enable scientific investigations in the manufacture of simple biogas; (4) Analyzing precise data on water pollution; (5) Identify, use and produce illustrative examples and illustrations; (6) Determining the correct order of ingredients in water purification; (7) Distinguishing scientific questions that allow for scientific investigation in the manufacture of simple water purifiers; (8) Explain the implications of energy-saving technology for society; (9) Evaluating the lifestyle that contributes to solving global energy problems. (10) Interpreting the data and scientific evidence of energy efficiency and draw the appropriate conclusion. The learning outcomes test that has been developed has the advantage that there are ten valid and reliable items with indicators of scientific literacy and learning objectives. According to Turner [56], a literacy test is a handy tool in providing natural science. Therefore, this learning outcomes test provides multiple-choice questions that can measure scientific literacy levels and students' knowledge.

The EFT teaching materials using STEM-based GI provide students with a rich learning context in which students learn new knowledge and learn how to discover facts, concepts, and principles. If students are endeavored to think and act wisely towards the environment, the learning objectives oriented towards the environment will be achieved. Therefore, one thing that can be done to overcome the problem of environmental damage is through education [57].

4. Conclusion
Based on the results of expert validation, it was concluded that the EFT teaching materials using STEM-based GI in the form of modules, lesson plans, and literacy test instruments developed were very valid, with the coefficient of agreement between the validators (R) in the high category. Teaching materials are appropriate to be used as alternative learning resources and guidelines for teachers in EFT learning.

References

[1] Rahayu R and Laksono E W 2015 *J. Kependidikan Penelit. Inov. Pembelajaran* **45** 29–43
[2] OECD 2019 *PISA 2018 Results*
[3] Leksono S M, Syachruroji A, Marianingsih P 2015 *J. Kependidikan* **45** 168–83
[4] Rizki S and Linuhung N 2017 *AKSIOMA J. Math. Educ.* **5** 137
[5] Kelley T R and Knowles J G 2016 *Int. J. STEM Educ.* **3**
[6] Hernandez P, Ibrahim, B, Bodin R, Rambo-Hernandez K E 2014 *Int. J. Technol. Des. Educ.* **24** 107–20
[7] Yunita A, Suyidno S, and Syahmani S 2021 *J Phys Conf. Ser* **1760**
[8] Morrison J 2006 *TIES (Teaching Inst. Excel. STEM)*
[9] Kennedy T J and Odell M R L 2014 *Sci. Educ. Int.* **25**246–58
[10] English L D 2017 *Int. J. Sci. Math. Educ* **15** 5–24
[11] Rusyati R, Permanasari A, and Ardianto D 2019 *J. Sci. Educ. Pract.* **2** 10–22
[12] Syahmani, Sauqina, and Hafizah E 2021 *IOP Conf. Ser. Earth Environ. Sci.* **758**
[13] Cunningham C M and LaChapelle C P 2014 *Eng Pre-College Settings Synth Res Policy Pract* 117–40
[14] English L D and King D T 2015 STEM learning through engineering design: fourth-grade students' investigations in aerospace *Int. J. STEM Educ.* **2**
[15] Taylor B 2016 *Electronic Int. J. of Educ. Arts and Sci.* **3** 1-22
[16] Chu S K W, Chow K, and Tse S K 2011 *Library & Information Sci. Research* **33** 132-43
[17] Liu X 2013 *J. Am. Soc. Inf. Sci. Technol* **64** 1852–63
[18] Chu S, Tavares N J, Chu D, Yee H S, Chou K, Siu F, and Wong M 2012 *Developing upper primary students' 21st century skills: inquiry learning through collaborative teaching and web 2.0 technology* (Hong Kong: Centre for Information Technology in Education, Faculty of Education, The University of Hong Kong)
[19] Anisah A, Wati M, and Mahardika A I 2016 *Berk. Ilm. Pendidik Fis* **4**
[20] Masnah, Syahmani, and Kusasi M 2018 *J. Chem. Educ.* **1** 225–36
[21] Leny, Syahmani, Ningsih F, and Sanjaya R E 2020 *J. Phys. Conf. Ser.* **1422**
[22] Riskowski J L, Todd C D, Wee B, Dark M, and Harbor J 2009 *Int. J. Eng. Educ.* **25**181–95
[23] Becker K and Park K 2011 *J. STEM Educ.* **12** 23–38
[24] Barcelona K 2014 21st *Am. J. Educ. Res.* **2** 862–75
[25] Syahmani, Saadi P, Clarita D, and Sholahuddin A 2021 *J. Phys. Conf. Ser.* **1760**
[26] Seraphin K D, Philippoff J, Kaupp L, and Vallin L M 2012 *Sci. Educ. Int.* **23** 366–82
[27] Maniotes L K and Kuhlthau C C 2014 *Knowl. Quest* **43** 8–17
[28] Hmelo-Silver C E, Duncan R G, and Chinn C A 2007 *Educ. Psychol.* **42** 99–107
[29] Hu S, Kuh G D, and Li S 2008 *Innov. High Educ.* **33** 71–81
[30] Sabharwal M *Masters Int. J. of Management Research and Development* **I** 76–85
[31] Plomp T and Nieveen N 2010 *Proceedings of the seminar conducted at the East China Normal University Shanghai (PR China) November 23-26, 2007*

[32] Pujari D 2006 *Technovation* **26** 76–85
[33] Borich G D 2003 *Observation's skills for effective teaching* 4th ed (Pearson Education Ltd)
[34] Fraenkel J R, Wallen N E, and Hyun H H 2012 *How to design and evaluate research in education* 8th ed (Mc. Graw Hill)
[35] Ahmad A and Azam T 2019 *Water Purification Technologies* (Elsevier Inc)
[36] Syahmani, Leny, Iriani R, and Sanjaya R E 2018 *Adv. in Social Sci. Educ. and Humanities Research* **147** 316–21
[37] Bain R, Wright J, Yang H, Gundry S, Pedley S, and Bartram J 2014 *Glob Water Issues Insights*
[38] Syahmani 2016 *The Proceeding of International Seminar on Ethnopedagogy* 489-502.
[39] Allen S J, Koumanova B 2005 *J. of the University of Chemical Technology and Metallurgy* **40** 175-92
[40] Zakiah A 2020 *DIMENSI (J. of Architecture and Built Environment)* **47** 11–8
[41] Islamyah D G, Yasa P, and Rachmawati D O 2019 *Jppf* **8** 86–94
[42] Afriana J, Permanasari A, and Fitriani A2016 *J. Pendidik. IPA Indonesia* **5** 261–267
[43] Shimmel T M and Columba L 2016 *US-China Edu Rev A* **6** 513–28
[44] Bin-Hady W R A and Abdulsafi A S T 2019 *SSRN Electron. J.* 7 275-89
[45] Dunne R and Wragg T 2005 *Effective teaching Taylor & Francis e-Library*
[46] Mesutoglu C and Baran E 2020 *Int. J. of Sci. and Math. Educ.*, 1-21
[47] Lie R, Aranda M L, Guzey S S, and Moore T J 2019 *Research in Science Education* 1-21
[48] OECD 2017 *PISA for Development Assessment and Analytical Framework: Reading, Mathematics and Science* (Preliminary Version Paris: OECD Publishing)
[49] Hanover Research 2011 *Successful K-12 STEM Education: Identifying Effective Approaches in Science, Technology, Engineering, and Mathematics* (Washington, DC: National Academies Press)
[50] Bligh A 2015 *Towards a 10-year Plan for Science, Technology, Engineering and Mathematics (STEM) education and skills in Queensland* (Queensland: Departemen of Education, Training and the Arts)
[51] FitzGerald L 2010 The twin purposes of Guided Inquiry: Guiding student inquiry and evidence-based practice *IASL Annu. Conf. Proc.* 1–18
[52] Opara J A and Oguzor N S 2011 *Curr. Res. J. Soc. Sci.* **3** 188–98
[53] Yuliati L, Parno, Hapsari A A, Nurhidayah F and Halim L 2018 *J. Phys. Conf. Ser.* **1108**
[54] Gucluer E and Kesercioglu T 2012 *Online Submiss.* **1** 8–13
[55] Shellawati S and Sunarti T 2018 *Inov. Pendidik. Fis* **7** 407–12
[56] Turner J 2016 *the Assessment of General Education natural Science Program* **63** 1–14
[57] Kılınç A, Stannis Street M, and Boyes E 2011 *Intern. J. of Environm. & Sci. Educ.* **3** 89 – 98

Journal of Physics: Conference Series **2104** (2021) 012020 doi:10.1088/1742-6596/2104/1/012020

The implementation of guided discovery learning model to improve students' characters of responsibility and academic skills

F Ahmad, M Mastuang, and M Misbah

Physics Education Study Program, Faculty of Teacher Training and Education, Universitas Lambung Mangkurat, Indonesia

Mastuang_pfis@ulm.ac.id

Abstract. The low level of responsibility and academic skills can hinder students in developing all their potentials to obtain more meaningful information and stay on one's memory. Thus, research was conducted to improve the character of responsibility and the academic skills of Grade XI Science 1 students at SMA Negeri 12 Banjarmasin. This type of research is a class action model adapted from Kemmis and Mc Taggart, consisting of 2 cycles. Each cycle includes planning, implementation/observation, and reflection. The research subjects involved were 23 pupils. Observation and tests were conducted to collect the data needed. The instruments used were observation sheets on the implementation of the Learning Implementation Plan. Observation sheets were utilized to analyze student responsibility characters, science process skills tests, and cognitive ability tests. Data were analyzed descriptively. The findings of the research instruments from cycle I to cycle II stated: (1) the implementation of the Learning Implementation Plan in the first cycle of the first meeting obtained a score of 79.43% (good), and the second meeting received 85.36% (very good). In the second cycle, the first meeting got 91.73% (excellent) and the second meeting achieved 93.17% (very good); (2) The average level of student responsibility in cycle I and cycle II increased from 65.81% (high) to 81.42% (very high); (3) Students' science process skills in cycle I and cycle II also increased from 0.58 (quite skilled) to 0.92 (very professional); (4) The classical cognitive ability of students in cycle I and cycle II increased from 78.26% (completed) to 100% (completed). It was concluded that the guided discovery learning model helped improve students' responsibility and academic skills.

1. Introduction

Learning must be comprehensive, covering all aspects of affective, psychomotor, and cognitive. Responsibility is one branch of the affective element that schools must teach [1–3]. Humans who have a responsible character can be aware of their behavior and actions, whether intentional or unintentional [4,5]. Therefore, the nature of responsibility is a person's attitude and efforts to do what should be done for oneself, society, environment, the State, and God Almighty. Hence, the character of responsibility is needed by every individual in everyday life [6].

One aspect directly related to the learning process that greatly determines student success in school is cognitive development. Cognitive development relates to the perception, way of thinking, memory, language skills, and information processing which all enable a person to acquire knowledge, solve problems, and plan for the future- in other words; it is all psychological processes related to how individuals learn, pay attention, observe, imagine, estimate, assess, and think about the environment [7].

Learning in schools is not only taught by delivering information from teachers or other learning resources, such as books. However, it must also be integrated through activities that emphasize actual activities, including observations, experiments, or trials. These activities are a means of transferring information/knowledge and training various process skills. Science process skills are used to understand what phenomena have occurred. These process skills are needed to acquire, discover, develop, and apply scientific concepts, principles, laws, or theories [8–10].

In general, the primary source of information during the learning process in the classroom tends to be dominated by the teacher, which results in students becoming less actively involved in the learning process [11]. As a result, students have difficulty understanding the concepts and principles of physics (cognitive domain). In addition, students' lack of courage to take the opportunity to ask questions is also a problem that causes them to have difficulties in understanding the lesson. It can be seen from the results of the physics tests given to students of grade XI Science 1 SMA Negeri 12 Banjarmasin in the 2016/2017 academic year during the odd semester. The tests showed that as many as 82.60% of students obtained scores below the minimum completeness criteria that have been set, which was 75. In addition, the average final exam scores of the 23 students in grade XI Science 1 SMA Negeri 12 Banjarmasin was 58.76. These results indicated that students' cognitive abilities in XI Science 1 SMA Negeri 12 Banjarmasin are still relatively low.

The absence of attention on character education (affective domain) in learning activities raises concerns regarding the long-term impact on students. This is supported by the results of the researcher's observations on the level of student responsibility when the teacher teaches. The data indicated that: 2 people were categorized in the very high category, four people were in the high category, six people were included in the reasonably high category, nine people in the low category, and two were categorized in the deficient category. This indicates that the character of student responsibility at the time of learning is still relatively low. Therefore, the researchers aim to increase student responsibility during the learning activities.

Another problem in learning physics that was not focused on was the weakness of students' science process skills. This was reflected in the results of TIMSS (Third International in Mathematics and Science Study), which indicated the ability of Indonesian students in the science field to rank 45th out of 48 countries [12]. One of the reasons is that science learning has been more oriented to the content of science materially rather than the process of science itself. Based on the initial data collection results carried out in grade XI Science 1, the students' science process skills were very low, with an average of only 20%. Therefore, this research also aimed to improve students' science process skills. Thus, to overcome the problems above, an appropriate learning model is needed to develop character and improve science process skills and students' cognitive abilities. One of the learning models that can be used is the guided discovery learning model.

Several existing studies have proven that discovery learning models could develop responsibility, effectively improve students' cognitive learning outcomes, and improve students' science process skills [13–15]. Therefore, research was conducted on applying the guided discovery learning model to enhance students' character of responsibility and academic skills. This study aimed to describe the application of the guided discovery learning model to improve students' character of responsibility and academic skills.

2. Method

The type of this research is Classroom Action Research which was adapted from the Kemmis and Mc Taggart model. The model consists of 2 cycles, in which each cycle is carried out in 2 meetings. Each cycle consists of three stages, namely planning, implementing and observing, and reflecting.

The research subjects involved in this study were 23 students of grade XI Science 1 SMA Negeri 12 Banjarmasin. The techniques used to collect data in this study were observation and tests. Observations were conducted to obtain data on the implementation of guided discovery learning model lesson plans and the character of students' responsibility. Partner teachers and one research colleague carried out observations of the performance of the lesson plans. Three research colleagues carried out words of the character of responsibility. The test was conducted to collect data on science process skills and cognitive abilities.

The tools and instruments used in this research are Learning Implementation Plans, teaching materials, Student Activity Sheets, observation sheets, science process skills tests, and cognitive ability tests.

The lesson plan implementation data were analyzed descriptively quantitatively with the percentage technique. The percentages obtained are further categorized according to the criteria [16] shown in Table 1. To calculate the reliability of the lesson plan implementation between two observers, Cronbach's Alpha formula is used (Ratumanan and Laurens, 2011).

Table 1. Lesson Plan Implementation Criteria

No.	Percentage	Criteria
1.	$P \leq 20$	Very Poor
2.	$20 < P \leq 40$	Poor
3.	$40 < P \leq 60$	Fair
4.	$60 < P \leq 80$	Good
5.	$P > 80$	Very Good

The assessment of the level of student responsibility observed during the learning process is expressed in a percentage. The percentages obtained are further categorized according to the criteria shown in Table 1.

The data of students' science process skills were analyzed descriptively quantitatively. The data were expressed in the form of scores. The science process skill scores obtained were further categorized according to the criteria [16] in Table 2.

Table 2. Science Process Skills Criteria

No.	Score	Criteria
1.	$P \leq 0.2$	Poor Skilled
2.	$0.2 < P \leq 0.4$	Low Skilled
3.	$0.4 < P \leq 0.6$	Quite Skilled
4.	$0.6 < P \leq 0.8$	Skilled
5.	$P > 0.8$	Highly Skilled

Cognitive ability test data were analyzed individually and classically. Mastery learning individually seen from the achievement of scores obtained according to the Minimum Completeness Criteria set by SMA Negeri 12 Banjarmasin was ≥75.

This action research is considered successful if the implementation of the lesson plans at least results in the excellent category, the character of the minimum responsibility are categorized in the reasonably high category, the science process skills are at least included in the quite skilled in the category, and the classical completeness of the cognitive ability test is at least 75%.

3. Result and Discussion

3.1 The Implementation of the Lesson Plan

The data recapitulation of the lesson plan implementation for each meeting is based on the observation sheet as shown in Figure 1.

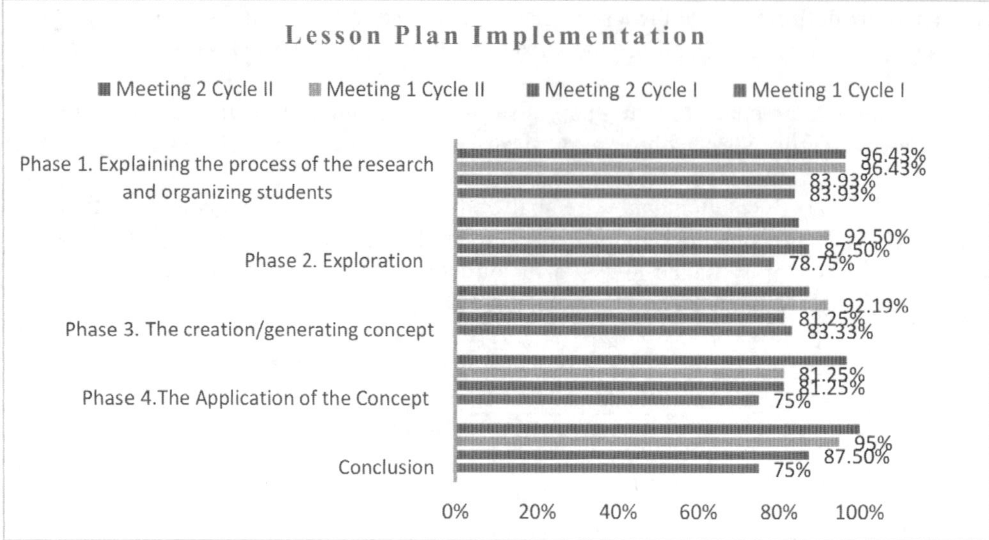

Figure 1. Implementation of the Learning Implementation Plan for each meeting

Overall, the implementation of the lesson plan has increased from the good category to the outstanding category. Although each phase has been carried out under the very good category, in phase 1, during Stage 3 and 4, Meeting 1, which focused on motivating students and reminding students of previous learning, was still categorized as good. This was due to the lack of preparation made before the learning activities took place. In addition, phases 2 and 3 were not following the time allocation previously determined in the Lesson Plan. The teacher took too much time in phases 2 and 3 to guide students who were still having difficulties collecting data, analyzing data based on experimental data obtained, and making conclusions based on the results of data analysis. Students were still in the adjustment phase with learning activities that integrated the guided discovery learning model, which was the first time it was applied in the class. Actions taken by the researchers as an effort to improve in cycle II, namely preparing as well as possible, and during the time of motivating and reminding students of the previous meeting, researchers should pose problems related to the material discussed through simple questions related to everyday life. In addition, the researchers should try to improve their time management, primarily during phases 2 and 3. Each stage in the Lesson Plan could be carried out by the time allocation that was previously planned in the Lesson Plan and enable the learning activities to run smoothly.

Overall, the average score of the implementation of the lesson plans in cycle II increased from 91.72% (meeting 1) to 93.16% (meeting 2), which was categorized in the outstanding category. All phases in cycle II were carried out under the scale of the outstanding category, following the time allocation specified in the lesson plan, and met the indicators of success. The average implementation of the lesson plans as a whole in cycle I and cycle II have increased. Thus, the efforts made by researchers according to the results of the reflection of the cycle I succeeded in expanding the implementation of the lesson plan cycle II as a whole, and the performance of all phases in the lesson plan in cycle II had met the indicators of success.

The efforts made by the researchers in preparing the lesson plans in cycle II were based on the results of the reflection conducted during the process I. It resulted in an increase in the implementation of each phase in the lesson plan and helped to achieve the established success indicators. This was in accordance with the statement of Suprihatiningrum [17], in which the learning process would run smoothly, learning objectives could be achieved optimally, and the success of learning itself could be described if it is planned as well as possible. Sumantri [18] also stated that good learning planning would result in a good learning process. The description above showed that the implementation of the lesson plans always increases at each meeting in the research cycle. In addition, in cycles, I and II, all phases in the lesson plan had been implemented in an excellent category and met the indicators of research success.

3.2 Character of Responsibility
The data recapitulation based on the student responsibility level observation sheet in each cycle is shown in Figure 2.

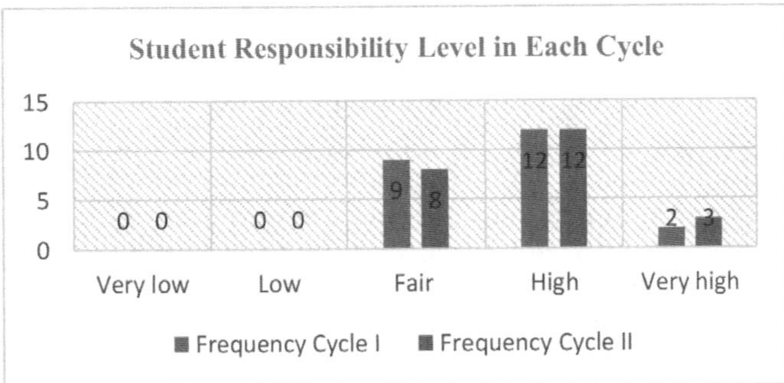

Figure 2. The Student Responsibility Level in Each Cycle

The recapitulation of the achievement of each responsibility criterion in cycles I and II is shown in Table 3.

Table 3. Recapitulation of Student Achievement of Each Responsibility Characteristic Criteria

Responsibility Characteristic Criteria	Cycle		Cycle 2	
	Percentage	Category	Percentage	Category
Self-discipline	100.00%	Very high	100.00%	Very High
Compliance	57.61%	Quite High	78.26%	High
Persistent	100.00%	Very High	100.00%	Very High
Giving the best effort	44.93%	Quite High	55.07%	Quite High
Applying Self-control	76.09%	High	97.83%	Very High
Thinking before acting	91.30%	Very High	100.00%	Very High
Responsible towards choices	78.26%	High	91.30%	Very High

Based on Table 3, it appears that the achievement of all responsibility criteria meets the indicators of success. In addition, the description above shows that learning by applying the guided discovery learning model has succeeded in increasing the character of students' responsibility, following several research results that state that using physics learning tools oriented to the guided discovery learning approach is proven to develop the level of commitment [14,19,20]. In addition, this was also by the opinion of Carin [21], which stated, "...through discovery learning students are guided and encouraged to become more independent, self-directed, and responsible for their learning...".

3.3 Students' science process skills
The recapitulation of the students' science process skills achievement in cycles I and II for each category is shown in Table 4.

Table 4. Recapitulation of the achievement of students' science process skills aspects

Students' Science Process Skills	Cycle I Percentage (%) Categories					Cycle II Percentage (%) Categories				
	HS	S	FS	LS	PS	HS	S	FS	LS	PS
Formulating Hypotheses	4.35	69.56	26.09	0.00	0.00	100.00	0.00	0.00	0.00	0.00
Identifying Variable	30.43	43.48	26.09	0.00	0.00	100.00	0.00	0.00	0.00	0.00
Analyzing Data	0.00	0.00	0.00	100.00	0.00	17.39	78.26	4.35	0.00	0.00
Drawing Conclusion	47.83	8.69	0.00	43.48	0.00	95.65	4.35	0.00	0.00	0.00
Formulating Hypotheses	4.35	69.56	26.09	0.00	0.00	100.00	0.00	0.00	0.00	0.00

Note: HS = Highly Skilled; S = Skilled; FS = Fairly Skilled; LS = Low Skilled; PS = Poorly Skilled

Based on Table 4 above, it appears that the achievement of all aspects of students' science process skills in cycle I for each category has met the indicators of success, namely at least categorized in the category of quite skilled, except for the skills of analyzing data and making conclusions. There are still skills classified as less experienced, namely analyzing data and making conclusions, which received 100% and 43.48%, respectively.

Based on the deficiencies that occur, the efforts made by researchers as improvements in cycle II are needed to improve students' abilities in analyzing data and concluding experimental data. These efforts provided guidance and emphasized the steps that students must consider and carry out before studying and completing experimental data. Skills in analyzing data, the teacher said, was what needs to be done to investigate the practical data table, describe the experimental data into sentences that explain the influence of the manipulation variable on the response variable following the problem formulation, hypothesis formulation, and guiding questions that have been provided, relate the data description using the appropriate mathematical equation, and explain the mathematical equation while connecting it to the manipulation, response, and control variables. Skills in concluding data, the teacher conveys, were that the data is completed to data analysis and the formulation of hypotheses. It must answer the formulation of the problem.

Aspects of science process skills that have met the indicators of success need to be further improved. Therefore, in cycle II, the researcher reduces the guidance when students formulate hypotheses and identify variables. It aims to make students more proficient in developing hypotheses and identifying variables.

Based on Table 4, the achievement of all aspects of students' science process skills in cycle II for each category has met the indicators of success. In addition, the percentage of achievement of all aspects of students' science process skills in cycle II for each category also increased.

Based on the results obtained, it indicates that the efforts made by the researchers by trying to guide the form of direction and emphasis on the steps that must be taken by students before analyzing data and making conclusions inflicted a positive impact, namely the achievement of aspects of data analysis skills and making conclusions, that have finally met the indicators of success. Biggs [22], which stat if teachers want students to understand the information they have learned, teachers must help students build their meanings from the information they have learned. This is following the results of research, which states that through discovery learning, students are encouraged to order to be able to find information or knowledge based on the materials that have been provided. The teacher acts as a facilitator to guide until the students can use the ideas, ideas, and skills learned to find new information or knowledge [23].

Efforts made by researchers by reducing guidance when students formulate hypotheses and identify variables also positively impacted students. Students become more proficient in developing ideas and identifying variables, as seen from the increase in the percentage of skill aspect achievement. This is following Lev Vygotsky's constructivist learning theory, namely scaffolding, which describes that students are given a large amount of guidance and assistance during the early stages of learning, then reduce the advice and assistance to provide opportunities for students to take on greater responsibility as soon as they can do something. The task independently [24].

Referring to the above explanations, it can be concluded that the application of the guided discovery learning model has succeeded in improving students' science process skills. This is aligned with the research results, which states that the application of physics learning tools oriented to guided discovery learning has proven to improve students' science process skills [13,25–28].

3.4 Student cognitive abilities

The recapitulation of student cognitive abilities for each cycle is shown in Figure 3.

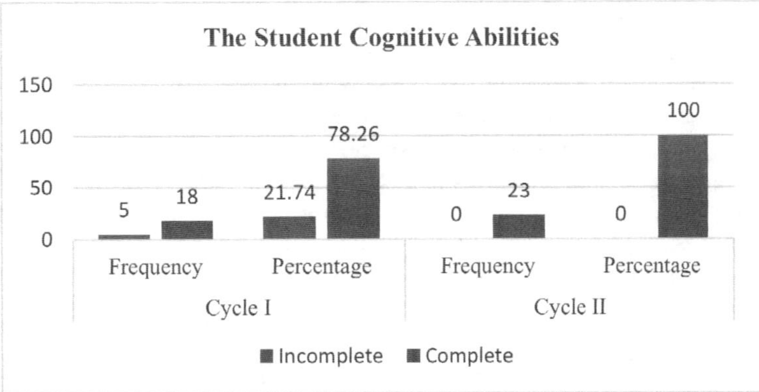

Figure 3. The Student Cognitive Abilities for Each Cycle

The mastery of students' cognitive abilities in the first cycle can be seen in Figure 3. The cognitivStudents' cognitive abilities cycle have met the indicators of success with the percentage of students who have completed with a score of 78.26%. Based on the data in Figure 3, the cognitive abilities of students in cycle II have met the indicators of success with the percentage of students who completed classically at 100%. The classical students' cognitive skills in cycle I and cycle II increased from 78.26% (completed) to 100% (completed). Thus, the application of the guided discovery learning model succeeded in improving students' cognitive abilities. This follows Bruner's opinion, which states that through discovery learning, a person has broad opportunities to build their knowledge so that their intellectual potential will increase. Bruner claims that students are not mere listeners. He believes that children learn best when they are actively involved in hands-on and mind-on activities. The discovery approach provides broad opportunities for students to organize and manage the information found through their investigations so that the information obtained or the concepts built are more meaningful, which ultimately affects students' cognitive abilities [21].

Thus, through discovery learning, students are allowed to learn deeper so that the facts/concepts/principles/theories/laws learned through the discovery process become more meaningful and can be understood in-depth to ensure that they can last a long time in the students' memories. The learning process using the guided discovery learning model effectively improved cognitive learning outcomes [15]. Several other studies also show that the application of discovery learning impacts student learning outcomes that are getting better [29–32].

The researchers' efforts to immediately share the results of cognitive ability tests to students, provide feedback and reinforcement, and distribute prizes to students who have passed contribute positively to improving students' cognitive ability test results. In accordance with the results of research by Dwiantoro and Kartiko [33]) which states that gift-giving has a significant effect on student learning outcomes. In addition, Hosnan [34] suggests that informing student learning outcomes and giving awards in the form of prizes or praise is one of the efforts to increase student learning motivation.

Efforts to provide feedback and reinforcement are intended to acknowledge and remember which steps are proper and which actions are wrong to avoid making the same mistakes in cycle II. According to Sabriani [35], giving feedback aims to make students realize the strengths and weaknesses of their tasks. In addition, Sulistyaningsih and Cahyani [36] stated that reinforcement seeks to make students follow or not follow specific behavior.

4. Conclusion

It can be concluded that the application of the guided discovery learning model can be used to improve the character of responsibility and academic skills of students in Grade XI Science 1 SMA Negeri 12 Banjarmasin. This statement was supported by the findings of the research on the application of guided discovery learning models in Grade XI Science 1 SMA Negeri 12 Banjarmasin as follows: (1) The implementation of the guided discovery learning model lesson plans at the first meeting of the first cycle was carried out with the score of 79.73% which belonged in the excellent category, the second meeting of the first cycle was carried out at 85.36% in the very good category, the first meeting in the second cycle was 91.72% in the very good category, and the second meeting in the second cycle was 93.16% in the very good category; (2) The level of student responsibility after the guided discovery learning model was applied had met the indicators of success and increased from cycle I to cycle II; (3) Students' science process skills applied by guided discovery learning model met the indicators of success and increased from cycle I to cycle II; and (4) classical students' cognitive abilities in cycle I and cycle II increased from 78.26% to 100%.

References

[1] Malinda H, Mwania J and Maithya R 2017 *African Educ. Res. J* **5** 64–74
[2] Suyidno S, Susilowati E, Arifuddin M, Misbah M, Sunarti T and Dwikoranto D 2019 *J. Penelit. Fis. dan Apl.* **9** 178
[3] Rif'at M F, Wati M and Suyidno S 2020 *Berk. Ilm. Pendidik. Fis.* **8**
[4] Kurniawan S 2016 Pendidikan Karakter: Konsepsi dan Implementasinya Secara Terpadu di Lingkungan Keluarga, Sekolah, Perguruan Tinggi, dan Masyarakat (Yogyakarta: Ar-Ruzz Media)
[5] Çağırgan D and Soytürk İ 2021 *Elem. Educ. Online* **20** 456–67
[6] Erlisnawati, Sapriya and Budimansyah D 2020 *Int. J. Sci. Technol. Res.* **9** 4874–7
[7] Desmita D 2014 Psikologi perkembangan peserta didik (Bandung: PT. Remaja Rosdakarya)
[8] Toharudin U, Hendrawati S and Rustaman A 2011 Membangun literasi sains peserta didik (Bandung: Humaniora)
[9] Kurniawan W, Pathoni H, Muliawati L, Kurniawan D A, Romadona D D, Ningsi A P and Dari R W 2020 *Univers. J. Educ. Res.* **8** 5581–8
[10] Maison, Darmaji, Astalini, Kurniawan D A and Indrawati P S 2019 *Humanit. Soc. Sci. Rev.* **7** 48–56
[11] Mastuang M, Erliana E, Misbah M and Miriam S 2017 *J. Pendidik. Inform. dan Sains* **6** 132–143
[12] Kemendikbud P 2015 Hasil Trends in International Mathematics and Science Study (TIMSS)
[13] Handayani B T, Arifuddin M and Misbah M 2017 *J. Ilm. Pendidik. Fis.* **1** 143–154
[14] Khaeruddin K, Ibrahim M and Supardi Z A I 2011 *JPPS (Jurnal Penelit. Pendidik. Sains)* **1** 6–18
[15] Purnomo H Y, Mujasam M and Yusuf I 2016 Pancaran **5** 1–14
[16] Widoyoko E P 2016 Evaluasi program pembelajaran Evaluasi Program Pembelajaran (Yogyakarta: Pustaka Belajar)
[17] Suprihatiningrum J 2016 Strategi pembelajaran teori dan aplikasi (Jogjakarta: Ar-ruzz Media)
[18] Sumantri M S 2015 Strategi pembelajaran (Jakarta: PT. Rajagrafindo Persada)
[19] Sunarya N and Darmawan C 2018 *Advances in Soc. Scie. Educ. and Humanit. Res.* **251** 620–3
[20] Ariyana A, Enawar E, Ramdhani I S and Sulaeman A 2019 J. *English Educ. Teach. (JEET)* **3** 522–32
[21] Carin A A 1993 Teaching science through discovery (New York: Macmillan)
[22] Toharudin U, Hendrawan S and Rustaman A 2011 Membangun literasi sains peserta didik (Bandung: Humaniora)
[23] Rahmi K, Zainuddin and Suriasa 2013 *Berk. Ilm. Pendidik. Fis.* **1** 4–10
[24] Cahyo A N 2013 Panduan aplikasi teori-teori belajar mengajar teraktual dan terpopuler (Yogyakarta: DIVA Press)
[25] Arifuddin M, Aslamiah M, Misbah M and Dewantara D 2020 *J. Phys. Conf. Ser.* **1422** 012001
[26] Haryadi R and Pujiastuti H 2019 *J. Penelit. Pengemb. Pendidik. Fis.* **5** 183–92
[27] Hikmawati H, Kosim K, Doyan A, Gunawan G and Kurniawan E 2021 *J. Phys. Conf. Ser.* **1779**
[28] Bahtiar B and Dukomalamo N 2019 *Biosf. J. Pendidik. Biol.* **10** 57–65
[29] Mayub A, Suryani E and Farid M 2020 *J. Pendidik. IPA Indones.* **9** 287–93

[30] Inde K H, Kaleka M B U and Ilyas I 2020 *J. Sci. Educ. Res.* **4** 11–4
[31] Ashfahani A, Haryono H and Kustiono K 2020 *Innov. J. Curric. Educ. Technol.* **9** 72–7
[32] Fadlilah N, Sulisworo D and Maruto G 2020 *Univers. J. Educ. Res.* **8** 3648–54
[33] Dwiantoro G A da. K and C D 2013 P *J. Pendidik. Olahraga dan Kesehat.* **1** 463–468
[34] Hosnan M 2014 Pendekatan sintifik dan kontekstual dalam pembelajaran abad 21 (Bogor: Ghalia Indonesia)
[35] Sabriani S 2012 *J. Chem.* **13** 39–46
[36] Sulistyaningsih Y F da. C and H B 2011 *J. Spirits* **2** 1–13

Application of virtual reality assisted probing prompting model to improve critical thinking skills and student learning outcomes

Mahdian, A M Rahman, Leny, Rusmansyah and P Saadi

Chemical Education Study Program FKIP Lambung Mangkurat University,
Jl. Brig. Jend.. H. Hasan Basry, Banjarmasin 70123 South Kalimantan, Indonesia.

rusmansyah@ulm.ac.id

Abstract. This study aims to find out (1) the improvement of students' critical thinking skills to the application of probing prompting learning models assisted by virtual reality media in colloidal system materials, (2) improvement of students' learning outcomes against the application of probing learning models assisted by virtual reality media on colloidal system materials, (3) the response of students to the application of probing learning models assisted by virtual reality media on colloidal system materials. tudy pretest-posttest design. This study sample was a student of class XI MIPA 4 at SMA Negeri 8 Kota Banjarmasin. Data collection uses to test and non-test techniques. Data analysis techniques use descriptive and inferential analysis techniques. The results showed that: (1) the application of The Probing Prompting learning model assisted by Virtual Reality media could improve the critical thinking skills of students, (2) the application of the learning model probing prompting assisted by virtual reality media on colloidal system materials can improve students' learning outcomes, (3) the application of Probing Prompting learning model assisted by Virtual Reality media gets a good response from students.

1. Introduction

The world of education in the 21st century today requires students to solve problems in life-related to science. Therefore, learning patterns that are still teacher-centred must be changed to student-centred. The 2013 curriculum itself has the fundamental view that knowledge cannot simply move from educator to student. Instead, students are subjects who must have the ability to actively seek, cultivate, construct, and use knowledge [1].

Chemistry learning in schools requires students to explore chemical concepts in a structured and orderly manner. In addition, chemical learning also has a function and purpose, including fostering a scientific attitude that includes a critical perspective to scientific statements and is not easily believed without observation results [2]. The weak learning process is one of the problems in the world of education. Students are less encouraged to develop critical thinking skills. Students are only directed to memorize information without observation. Students must remember and accumulate various information without understanding and connecting with daily life [3].

Chemicals related to problem-solving and often found in everyday life are colloidal system materials. Colloidal system material is mainly in the form of concepts and is widely applied in everyday life. The characteristics of colloidal material containing theory and application require students to improve critical thinking skills to understand and solve problems thoroughly and see their meaning [4].

Critical thinking skills and learning outcomes can also be trained through the classroom learning process by acquiring new knowledge through problem solving and cooperation. Critical thinking skills involve analyzing arguments, solving problems scientifically, and generating insights into specific things and interpretations. This suggests that critical thinking leads to deep thinking about solving and specific issues [5], [6]. One way to realize improving students' critical thinking skills is by choosing a suitable model. Innovative learning models that can improve and train essential thinking skills are the Probing Prompting model. Probing Prompting learning model is learning by how the teacher presents a series of questions that are guiding and digging. A thought process relates each learner's knowledge and experience to new knowledge being learned [7]–[10].

The Probing Prompting model also complies with the demands of the 2013 curriculum. The probing prompting model is a model that directs students to learn independently, while teachers are only facilitators in the learning process so that learning is centred on the learner [11]. The delivery of a chemical concept will be easily understood and remembered by continuously repeating learning or assisted by a learning medium [12]. Virtual Reality media is a technology that allows users to interact easily in an environment or an event simulated by a computer in an application through a smartphone. Thus, it can be used in the learning process [13]. In pandemics like this, probing prompting models and virtual reality media can also be used in online learning. Because this model does not require face-to-face in its implementation, but by using platforms such as zoom and goggle meet, this model can be implemented so that teachers can give questions to students to answer a problem.

Based on the description above, this study was conducted to improve students' critical thinking skills and learning outcomes by applying the Probing Prompting model assisted by Virtual Reality media on colloidal system materials.

2. Method

The type of research used is pre-experimental using a one-group pretest-posttest design [14]. The population in this study is all students in SMAN 8 Banjarmasin school year 2020/2021. The study sample of class XI MIPA 4 students with purposive sampling techniques. The free variable is the Probing Prompting model assisted by Virtual Reality media. In contrast, the bound variable is critical thinking skills and research data retrieval learning outcomes from May to June 2021.

Research instruments in the form of test and non-test-intrusions. The test instrument used in this study is a matter of description (essay) given at the beginning and end of the lesson (pre-test and post-test) aims to find out the increase in critical thinking skills and learning outcomes and find out how students can answer the given questions. The test used to measure the skills of critical thinking and learning results in the form of a description test consists of 10 points. Non-test instruments used in this study are student response questionnaires that aim to determine the students' response to the learning. Model Probing Prompting assisted by Virtual Reality media of 10 points.

Aiken's V scale created the critical thinking skills problem sheet, learning outcomes, and students' responses. Aiken's V scale was organized in a statement followed by five respondents who showed the level and were given the final score. Descriptive data analysis to analyze critical thinking, learning, and students' response questionnaires while inferential analysis for rhythmic thinking skills aims to test hypotheses that have been proposed, there are differences or no differences in improvement. This analysis uses the t-test to find outH0 is accepted orH0 is rejected. Before performing another test, perform a normality test and a homogeneity test of the initial test*(pre-test)*and the final test*(post-test)*.

3. Result and Discussion

Probing Prompting learning is based on Virtual Reality media in learning that involves students guiding and digging into questions and associating knowledge with previous experiences with new knowledge to be learned. As for the Probing Prompting stage, according to Sudarti, there are seven stages in its implementation 1) confronting students to new situations through images or text that have problems, 2) teachers waiting for students' answers, 3) asking questions by learning goals, 4) teachers waiting for students' answers, 5) confirming answers, 6) teachers responding to answers, 7) teachers asking final questions. Thus the concept of material learned by students becomes more meaningful and can be applied in everyday life [15, 16].

In learning, in addition to models that support active students, there must be exciting media so that students' attention can be focused and facilitate in understanding the material provided by teachers [17]. *Virtual Reality* media has components capable of providing an engaging look and refers to the use of interactive simulations for users with the opportunity to engage in environments that may seem and are not similar to real-world events [18–21].

Data on critical thinking skills and learning outcomes obtained through pre-tests and post-tests are analyzed descriptively and inferentially. At the same time, the response questionnaire to the Probing Prompting model is based on Virtual Reality media.

3.1 Critical Thinking Skills
Critical thinking skills are a process where knowledge and abilities are applied to solve problems, make decisions, analyze emerging opinions, and conduct investigations or research based on facts, information, and data obtained to produce valid information and conclusions. Critical thinking skills defined that critical thinking is a disciplinary process that is intellectually active and skilled at conceptualizing, applying, analyzing, synthesizing, and evaluating the information gathered has six skill stages [22]. The criteria of critical thinking are FRISCO, i.e., focus, reason, inference, situation, and clarity [22] Overall the increase in the average achievement of critical thinking skills can be seen in Figure 1.

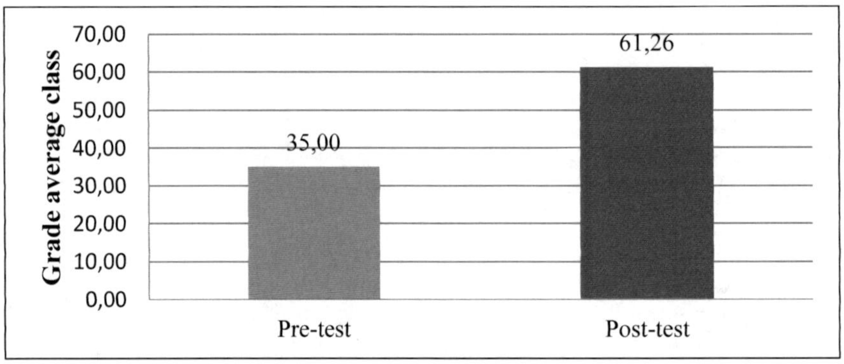

Figure 1 Average pre-test and post-test critical thinking skill values

The average value of students' critical thinking skills at the *post-test* was higher than at the *pre-test*, 61.26 at the *post-test* and 35.00 at the *pre-test.* Differences in the value of critical thinking skills before learning showed that the *Probing Prompting* model assisted by Virtual *Reality* media affected essential thinking skills in colloidal materials. This happens because students are more active in the learning process and more enthusiastic in solving the problems given.

The results of the pre-test and post-test critical thinking skills of each indicator can be seen in Figure 2.

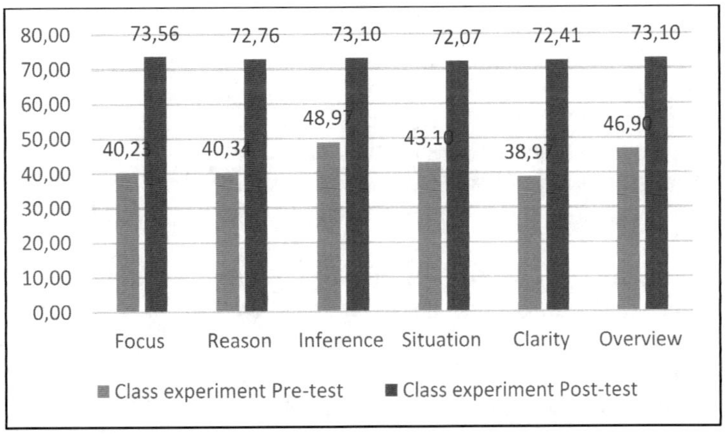

Figure 2 Pre-test results and post-test critical thinking skills of each indicator

After treatment, critical thinking skills at the time of post-test each indicator improved. This is inseparable from the influence of the probing prompting learning model assisted by virtual reality media that always trains students continuously at every meeting with teachers as mentors, especially when identifying and formulating problems. The learning process of the probing prompting model emphasizes problem-solving, discussion, and percentage and is supported by critical thinking tasks specifically designed to improve students' critical thinking skills. Learning that makes students active with teachers as mentors to direct students' thinking will enhance critical thinking skills.

This is in line with the research [23] which states that it is necessary to learn more systematic, structured, role transfer activities to students and oriented to the surrounding environment or daily life so that it will improve students' critical thinking skills. Differences in critical thinking skills improvement in indicators of focus, reason, inference, situation, and clarity increased significantly. The increased tests of students' critical thinking skills on focus indicators are continuously trained at each meeting with the teacher as a guide, especially when identifying and formulating problems. The findings are in line with Anisah & Carlian's (2020) findings, which states that the percentage of achievement results of focus indicators have increased after learning that trains in identifying and formulating problems [24, 25].

Indicators of critical thinking skills are trained when teachers provide opportunities for students to formulate answers to existing problems. In problem, learning is given, then students give reasons based on relevant facts/evidence and convey it when learning takes place. For example, on the critical thinking skills test, students can explain the nature and characteristics of colloids in food. Filsaime (2008) states that a person's critical thinking skills can be improved if he can provide arguments accompanied by supporting evidence to maintain his opinion. This causes critical thinking skills indicators of reason to increase significantly.

Indicators of critical thinking skills are trained when teachers test different hypotheses and students choose the correct hypothesis alternatives. The ability to select and identify problems to make reasoned conclusions, form hypotheses and estimates in considering relevant information and expand the consequences of data or evidence concludes students were increasing. On the test of critical thinking skills, students are already able to determine temporary hypotheses based on available discourse on how to make tofu and colloidal properties that exist in the manufacture This is in line [26] research that shows that using probing prompting models can improve critical thinking skills of inference indicators.

Situation indicators for critical thinking skills are trained when students are faced with the actual situation of a problem. Students must state reasoning, justify reasoning based on evidence and compare with the real situation. Some students are already able to compare and elaborate with colloidal properties in actual conditions, students in delivering their answers based on literature and data from various sources.

On indicators of the clarity of students in stating the results of reasoning, justifying reasoning must be based on evidence from various sources and present reasoning in the form of convincing arguments. According to Layn & Ruslan (2017), students' critical thinking skills are well stated if able to state reasoning following relevant information and facts [27]. Based on the critical thinking skills test, students can already present arguments or reasoning based on evidence on why detergents can remove fat and the event of smoothing kitchen spices before being mixed with food and its relationship with colloidal materials. Strong reasoning can be determined after students find assumptions that have good reason to trust the source.

Critical thinking skills are trained to review or evaluate when analyzing data and determining the most reasonable opinion resulting from data analysis. Based on the critical thinking skills test, students have been able to review or examine examples of events so that they can conclude in their entirety based on facts and evidence about the role of protective colloids.

The t-test was conducted on pre-test data and post-tested students' critical thinking skills in experimental classes tested for homogeneity and normality. Pre-test and post-test data should be normal and homogeneous. The pre-test data and post-test results of the learner's critical thinking skills can be seen in Table 1.

Table 1 Test results-t pre-test data and post-test CTS Students

Result	dk	x	Md	calculated	ttable5%	Conclusion
Pre-test	29	35	26.26	11.01	2.03	There are differences in critical thinking skills
Post-test	29	61.26				

Based on Table 1, the price t-calculated and t-table where t-calculated > t-table (11.01 > 2.03), which means Ho rejected and H1 accepted, so it can be said that there is a significant difference between the average pre-test value and post-test critical thinking skills of students in the moments before and after learning that applies the Probing Prompting model assisted by Virtual Reality media.

3.2 Learning outcomes

Learning outcomes in the field of knowledge of students are measured by learning using the Probing Prompting model assisted by Virtual Reality media. Test results are done before being given treatment (pre-test) and after given treatment(post-test). The Probing Prompting learning model assisted by Virtual Reality media shows the learning atmosphere becomes more directed and uplifting for students in the following learning.

This affects students' learning outcomes; 29 completed students, as many as three people, have not completed 26 students at the pre-test. While at the time of post-test, students are complete as many as 29 people, and nothing is not absolute. The pre-test is the initial test before the application of treatment to measure students' initial knowledge about the colloidal system. At the same time, the post-test is a test done at the end of learning after the application of treatment to measure students' level of understanding of colloidal system material that has been taught. The percentage of completion of learning outcomes can be seen in Figure 3.

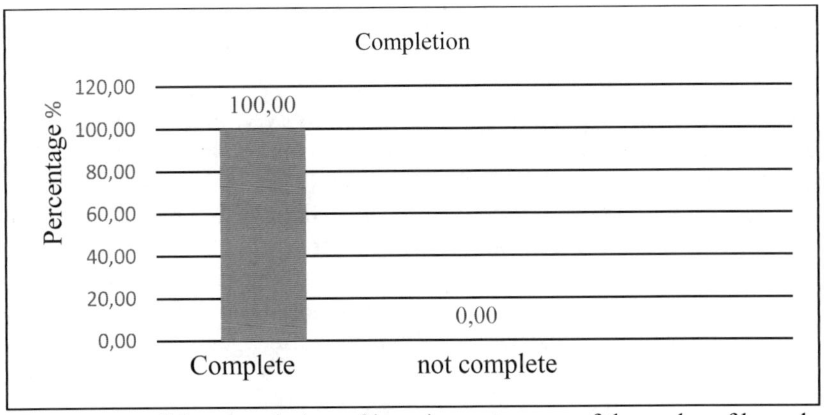

Figure 3 Percentage of the completion of learning outcomes of the realm of knowledge

Based on Figure 3 above, it is explained that the application of the probing prompting model assisted by Virtual Reality media makes the learning outcome of class XI MIA 4 students' knowledge reach completion of 100%, which means students have above the specified KKM value of 75. The improvement of students' learning outcomes can be influenced by the high critical thinking skills of students. With the critical thinking skills, students will use the ability to solve problems, formulate influential factors, find out various facts in the issue, and make the right decisions to improve students' results [28], [29].

This is in line with Wicaksono & Candra's research (2016) which states there is a significant influence of critical thinking skills with increasing learning outcomes, so students answer more critical problems, especially in deep thinking and straightforward and realistic reasons [30], [31]. Therefore, the Probing Prompting learning model assisted by Virtual Reality media can improve students' learning outcomes and achieve a minimum score of ≥ KKM based on the test of learning results provided. Based on the study results, students' completion is 100%, with an N-gain of 0.57 in the moderate category.

3.3 Student response

Students' response to using a probing prompting model assisted by virtual reality media is given at the end of learning, namely after the post-test. The response of students used a questionnaire containing seven positive questions and three negative questions. The results of the assessment of the students' response to learning in colloidal material are presented in Figure 4.

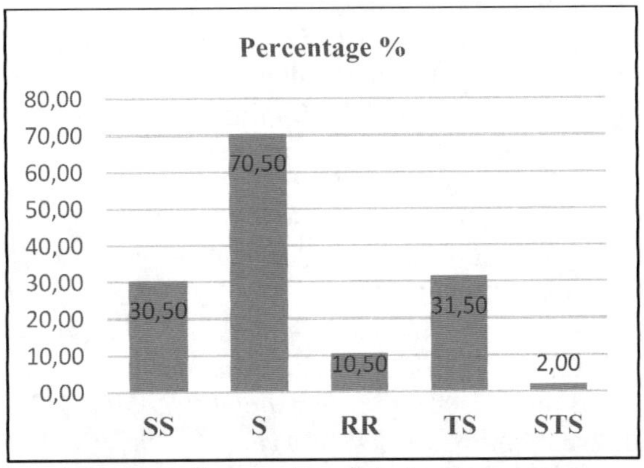

Figure 4 The results of student's response

Figure 4 shows the study for students' response to the probing prompting learning model assisted by virtual reality media in the colloidal system material of the student's experimental class responded well. This can be seen from 30.50% of students who responded strongly agree, 70.50% of students who responded agreed, 10.50 students responded hesitantly, and 2% of students responded disapprovingly. Thus, based on a table of 18 average scores, students in the experimental class meet the response level of the criteria range well. Therefore, it can be concluded that students in the experimental class responded positively to probing prompting models assisted by virtual reality media. This is supported by Anisah & Carlian's (2020) research, which states that the probing prompting model can train critical thinking skills [24].

The effectiveness of the probing prompting model is shown by increasing critical thinking skills in solving the given questions. The results showed that the activity of students in learning showed an active category with a percentage of 80.52%, an increase in critical thinking skills with an average N-gain of 0.89 in the high category. This supports the positive response to the probing prompting model assisted by virtual reality media, which is reasonably practical and effective.

4. Conclusion

Based on the data above, it can be concluded that there is an increase in critical thinking skills and student learning outcomes before learning and after learning using the Probing Prompting model assisted by Virtual Reality media. Furthermore, students' positive response to using the Probing Prompting model assisted by Virtual Reality media on the colloid system material of students makes learning more enjoyable. In addition, it makes it easier for teachers to convey material or information in learning. Therefore, the Probing Prompting model assisted by Virtual Reality media tested can be considered for chemistry subject teachers to practice critical thinking skills and learning outcomes through online learning.

Acknowledgement

The authors would like to thank the Dekan of FKIP ULM, the leadership of SMAN 8 Banjarmasin City, and all parties involved who have given permission to research and assisted in completing the study and to prepare the article.

References

[1] Heryadi D and Sundari R S, 2020 *Int. J. Educ. Res.* **8**, February p. 207–216.
[2] Fernanda A Haryani S and Prasetya A T, 2019 *J. Inov. Pendidik. Kim.* **13**, 1 p. 2326–2336.
[3] Kasih and Winarti, 2020 *J. Chem. Educ.* **3**, 4 p. 34–45.
[4] Agustiana J, 2019 *SPEKTRA J. Kaji. Pendidik. Sains* **5**, 1 p. 91.
[5] Muttaqiin A, 2015 *Edusentris* **2**, 2 p. 116.
[6] Intan P Imanda R and Alvina S, 2021 *Chim. Didact. Acta* **9**, 1 p. 1–7.
[7] Suherman, 2003 *Study and Learning* Bandung: Indonesian University.
[8] Alfian M Dwijayanto and Sunami, 2017 *Unnes J. Mathmathics Educ.* **6**, 2 p. 249–257.
[9] Fajar R F, 2020 *Int. J. Languange Educ. Cult.* **9**, 1 p. 86–95.
[10] Pahamzah J, 2021 *J. Southwest Jiaotong Univ.* **55**, 6 p. 1–25.
[11] Muthmainnah Hapizah and Yusuf M, 2019 *J. Math. Educ.* **1**, 1 p. 27–38.
[12] Khairunnisa Saadi P and Leny, 2017 *J. Chem. Educ.* **1**, 1 p. 151–157.
[13] Kusuma Wirawan and Arthana R, 2017 *Natl. J. Inform. Engenering* **6**, 3 p. 294–304.
[14] Sugiyono, 2015 *Research and Methods Kuantitatif and Kualitatif* Bandung: Alfabeta.
[15] Septarina E and Sodikin, 2019 *Indones. J. Sciene Math. Educ.* **02**, 1 p. 46–54.
[16] Tuerah P Regar and Javier, 2020 *J. Pshycosocial Rehabil.* **24**, 2.
[17] Rahmawati and Jayanti, 2019 *J. Chem. Educ.* **3**, 19–20.
[18] Abdussalam Sulthoni and Munzil, 2018 *Pendidikan* **3**, 9 p. 1160–1167.
[19] Chang G Chen and Jong, 2020 *J. Illum. Eng. Soc.* **28**, 7 p. 915–929.
[20] Ikhsan J Sugiyarto K H and Astuti T N, 2020 *Int. J. Interact. Mob. Technol.* **14**, 8 p. 183–195.
[21] Wang L Lei J Wang Q and Ren Y, 2020 *Br. J. Educ. Technol.* **51**, 6 p. 2034–2049.
[22] Ennis, 2011 *The Nature Of Critical Thinking : An Outline Of Critical Thinking Dispotitions*

and Abilities Urabana: University of Illions.
[23] Zaini M Masylni and Syahmani, 2018 *IOSR J. Res. Method Educ.* **8**, 2 p. 29–33.
[24] Anisah T and Carlian Y, 2020 *J. Islam. Prim. Educ.* **3**, 2 p. 43–51.
[25] Fajar M and Jayanti, 2019 *J. Chem. Educ.* **3**, 1 p. 50–58.
[26] Sahayu, Sherina S Jampel N and Jayanta I N L, 2018 *J. Elem. Sci.* **2**, 3 p. 321–328.
[27] Layn M R, 2017 *J. Educ. Math. Sains* **5**, 1 p. 77–80.
[28] Marceillina Novita D Wiryokusumo I and Walujo D A, 2019 *Int. J. Educ. Technol. Learn.* **5**, 1 p. 9–14.
[29] Lutfia A Asyhari A and Saidy, 2020 *J. Phys. Conf. Ser.* **1572**, 1 p. 1–6.
[30] Wicaksono C, 2016 *J. Pendidik. Sains* **2**, 2 p. 85–92.
[31] Sulastri Imran and Firmansyah A, 2019 *J. Creat. Tadaluko* **3**, 1 p. 90–103.

Feasibility of STEM-based basic chemistry teaching materials to improve students' science literature in wetland context

M Kusasi[1], F Fahmi[2]*, R E Sanjaya[1], M Riduan[1], N Anjani[1]

[1]Study Program of Chemistry Education, Departement of Mathematics and Natural Science Education, Faculty of Teacher Training and Education, University of Lambung Mangkurat, Banjarmasin City, South Kalimantan, Indonesia

[2]Master Program of Natural Science Education, Postgraduate Program, University of Lambung Mangkurat, Banjarmasin City, South Kalimantan, Indonesia

ibnusuwandy@gmail.com ; mkusasi_kimia@ulm.ac.id

Abstract. Development research has been carried out to produce teaching materials for basic chemistry learning materials based on STEM in a wetland context to improve scientific literacy that is valid, practical, and effective. This type of research is Research and Development (R & R&D) with a 4D development model modified into 3D (define, design, and develop). The trial of the developed teaching materials was carried out in a class in the first Basic Chemistry course consisting of 28 students. Data collection techniques used validation sheets, readability questionnaires, response questionnaires, observation sheets, and scientific literacy tests. Data analysis used descriptive analysis. The results showed that the developed teaching materials had met the following criteria: (1) Valid; judging from the value of the feasibility aspects of content, presentation, language, and media, each of which obtained very valid criteria. (2) Practical; reviewed from the results of the readability of teaching materials in individual and small group tests, the criteria are practical and very practical. Students' responses to teaching materials have very practical criteria, and observations of the ability of teachers to use teaching materials and manage their respective classes get very good criteria. (3) Effective, seen from the results of the N-gain scientific literacy showing a high increase. With the results of this development, it can be said that the STEM-based teaching materials for basic chemistry learning to improve scientific literacy in the context of wetlands are suitable as teaching materials.

1. Introduction

The country of Indonesia has entered the 21st century, marked by the increasing demand for high-quality human resources. Besides being required to master the knowledge to be taught, a prospective teacher is also required to have a set of knowledge and technical teaching skills.

Based on research results from the Program for International Student Assessment (PISA), it shows that the scientific literacy ability of Indonesian students is generally below the average compared to the international average score [1]. Therefore, for prospective teacher students, it is crucial to have high scientific literacy skills so that later they will improve students' scientific literacy skills.

The selection of appropriate models and approaches in learning can improve scientific literacy skills and student learning outcomes. The model and approach chosen is a project-based learning model with a STEM approach. The project-based learning model is innovative, encouraging students collaboratively to research and create projects that apply their knowledge of discovering new things, proficient in the use of technology, and able to solve real-world problems [2]. The STEM approach combines art with STEM subjects to increase student engagement, creativity, innovation, problem-solving skills, and other cognitive benefits [3].

The use of learning teaching materials used for direct student involvement will be interesting and increase the success of a learning process, especially the influence of technological developments. Teaching materials are often developed in the form of e-modules or electronic modules. One of the applications made is flipbook-based, where it looks like a digital book that looks interesting and more motivated.

Objects that are used as problems in each learning material can be sourced from the environment around students. One of them stems from the issue of the wetland context, which is a large area of land owned by South Kalimantan [4]. One part of the branch of chemistry that can directly contact the wetland environment is the sub-material of learning basic chemistry, namely acids and bases, and colloids. Based on this explanation, researchers are interested in developing teaching materials for basic chemistry learning based on a project-based learning model with a STEM approach to improving scientific literacy and student learning outcomes in the context of wetlands.

2. Method

This Research and Development (R&D) uses a model 4D development of 3D modification, namely define, design, and development [5]. The research procedure can be seen in Figure 1 below.

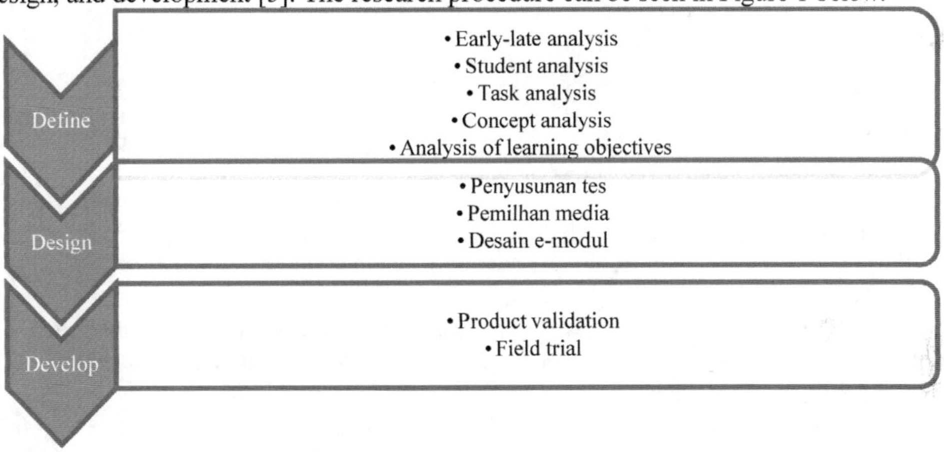

Figure 1 A Model 4D Development of 3D Modification

The dissemination stage was not carried out because the research time was close to the end of the even semester of the 2020/2021 academic year, so there was insufficient time. The research was carried out at the Chemical Education Study Program FKIP ULM Banjarmasin for second-semester students of class A2 for the academic year 2020/2021 for individual and small group trials and semester two students of class A1 for the academic year 2020/2021 for limited trials.

Data collection techniques are carried out through non-test instruments, including validation questionnaires, readability questionnaires, student response questionnaires, observation sheets on the ability of lecturers to use teaching materials and manage classes. Then, the test instruments in the form of objective questions and essays to measure students' scientific literacy skills were given before and after treatment.

Data analysis used descriptive analysis. Determination of the percentage value (P) in the validity and effectiveness test using the following formula.

$$Percentage\ value\ (P) = \frac{Total\ score\ given}{Overall\ score}\ x\ 100\%$$

The results obtained for the validity test are interpreted in Table 1 with the following criteria.

Table 1 Criteria for the validity of teaching materials [6]

No	Percentage value	Validity criteria	Information
1	85.01 – 100.00	Very valid	No need to revise
2	70.01 – 85.00	Valid	No need to revise
3	50.01 – 70.00	Less valid	Minor revision
4	1.00 – 50.00	Invalid	Major revision

The average score obtained for the practicality test is interpreted in Table 2 below.

Table 2 Criteria for the practicality of teaching materials [7]

No	Average score $\bar{x}$	Criteria
1	$\bar{x} > 3.25 – 4.00$	Very practical
2	$\bar{x} > 2.50 – 3.25$	Practical
3	$\bar{x} > 1.75 – 2.50$	Less practical
4	$\bar{x} > 1.00 – 1.75$	Not practical

The results obtained for the effectiveness test regarding scientific literacy skills are interpreted in Table 3 below.

Table 3 Criteria for scientific literacy ability [8]

No	Percentage value	Criteria
1	86.00 – 100.00	Very high
2	72.00 – 85.00	High
3	58.00 – 71.00	Medium
4	44.00 – 57.00	Low
5	0.00 – 43.00	Very low

3. Results and Discussion

Teaching materials developed using 3D model stages are presented as follows.

3.1 Define

3.1.1 Early-late analysis

The final preliminary analysis was carried out through field studies and literature studies aimed to discover the fundamental problems students face in learning related to acid-base and colloidal materials. It found out the difficulty of students understanding the material due to the lack of attractive teaching materials available.

3.1.2 Student analysis

Student analysis includes an analysis of students' academic ability and motivation, which aims to examine the characteristics of students in line with the design to be developed, such as students having the ability to do their projects and having adequate facilities to use smartphones devices.

3.1.3 Task analysis

Task analysis aims to thoroughly examine what tasks students will carry out through the developed teaching materials. Task analysis includes lessons that train students' scientific literacy skills in a wetland context.

3.1.4 Concept analysis

Concept analysis aims to identify concepts and determine the content of the teaching materials that are developed and then presented in a concept map.

3.1.5 Analysis of learning objectives

Analysis of learning objectives is made by adjusting existing tasks and concepts and then connecting them with the behavior of the subject to be achieved, namely increasing students' scientific literacy.

3.2 Design

3.2.1 Test preparation

The preparation of the test is the stage of preparing essay questions for scientific literacy instruments.

3.2.2 Media preparation
The media used as a medium for making teaching materials is a flipbook, where previously, the initial file form of teaching materials was in pdf form. Then converted into the form of a digital book with the addition of an attractive display and symbols.
3.2.3 Formatting
The choice of format includes selecting material content, strategies, learning methods, and learning models in the teaching materials.
3.2.4 Preliminary design
This stage is the first design carried out to design the developed e-module.

3.3 Develop
3.3.1 Product validation by experts
The validity test aims to determine the level of product validity, which is measured based on the assessment of the validator consisting of 3 lecturers of Chemistry Education FKIP ULM Banjarmasin as material experts and one education practitioner FKIP ULM Banjarmasin as media experts. The results of the assessment of the validators are shown in Figure 2 below.

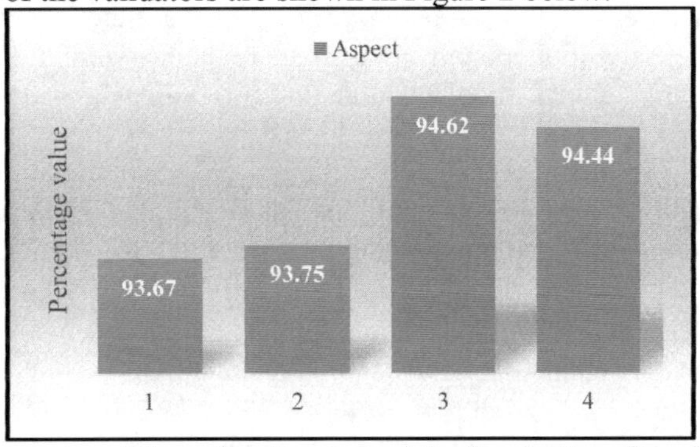

Figure 2 The results of the validation test of the feasibility of teaching materials

Based on Figure 2, it is known that the percentage scores for all aspects of the feasibility of teaching materials get high scores with very valid criteria. The content aspect gets a percentage value of 93.67% because the teaching materials have presented appropriate and accurate material content and have other supporters to clarify the material. The material content is proven to be up-to-date. The presentation aspect gets a percentage value of 93.75% because the teaching materials have been prepared with good presentation techniques and other supports. The form of learning encourages students to be interactive and participative, and teaching materials are arranged according to the rules, consisting of an introduction, content, and cover. The language aspect obtained a percentage value of 94.62% because the teaching materials had been presented in a language that was straightforward, communicative, dialogical, interactive, in accordance with the intellectual development and social, emotional level of students, coherent and integrated, as well as the consistent use of terms, symbols or icons. The media aspect gets a percentage value of 94.44% because the teaching materials have presented an attractive appearance and content and are in accordance with the characteristics of teaching material.

3.3.2 Product validation by experts
Field trials include individual, small group, and limited trials. Individual and small group trials were conducted to determine the practicality of teaching materials in terms of the readability questionnaire of teaching materials. The average score of the readability questionnaire for individual and small group trials is shown in Figure 3 below.

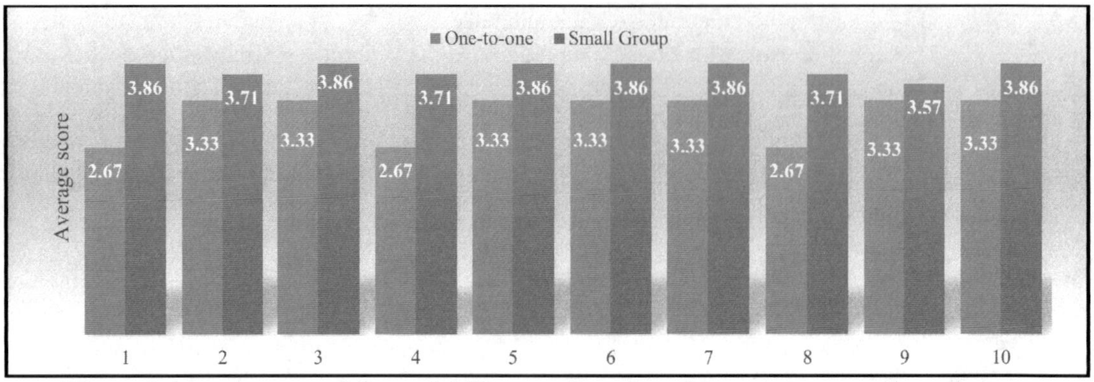

Figure 3 Average score of the readability questionnaire of the one-to-one and small group

Based on Figure 3, it is known that almost all readability statements obtained an average score in very practical criteria in individual trials although statement 1 (cover design), statement 4 (display color combinations), and statement 8 (ease of understanding the material) obtained practical criteria. Therefore, before being retested in the small group trial, revisions were made by changing the cover design, changing color combinations for display, and adding videos to clarify the material's content. Hence, in the small group trial, all the readability statements obtained an average score in very practical criteria. Choosing a harmonious and attractive color combination can provide a special attraction for students, and another purpose of this visual display is to facilitate the delivery of messages [9]. Then, the role of video in learning is a medium for demonstrating several material concepts, helping stimulate student curiosity about the chemical material to be studied and increasing student interest in learning.

The assessment of the readability of teaching materials is based on several things, such as the number of clear and not blurry images found to clarify the material's content. As revealed by Amir [10], that with the help of pictures helps facilitate the learning process and the achievement of learning objectives. Writing sentences in teaching materials are easy to understand because there is no misinterpretation due to wrong writing. Arizona [11] revealed that if there are spelling errors and improper use of punctuation in writing, then the meaning of the writing can be different from the intent or purpose, causing sentences to be challenging to understand. Teaching materials contain quiz time and formative tests that are easy for students to understand and answer to increase their understanding and motivation to learn. Puryati [12] stated that when students can answer practice questions, a sense of satisfaction will arise and make them more successful in their studies.

Limited trials were conducted to determine the practicality and effectiveness of teaching materials. The practicality of teaching materials is viewed from student response questionnaires and observation sheets on lecturers' ability to use teaching materials and manage classes. Meanwhile, the effectiveness of teaching materials is viewed from the scientific literacy ability of students. The average score for student responses is shown in Figure 4 below.

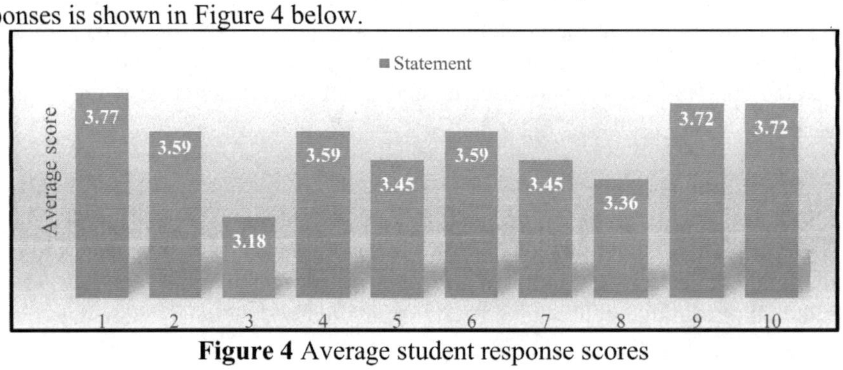

Figure 4 Average student response scores

Based on Figure 4, it is known that almost all statements obtained an average score in very practical criteria, except for statement 3 regarding the ease of solving acid-base and colloid problems which got practical criteria. The use of teaching materials can increase students' willingness to take part in learning. Zaharah & Susilowati [13] stated that learning using electronic modules could improve student learning motivation. The preparation of teaching materials using communicative language makes it easier for students to understand the material's content. The teaching materials themselves have implemented a project-based learning model syntax with a STEM approach not to cause students to feel bored during learning. As revealed by Rozikin., Lesmono., & Bachtiar [14], it is stated that through the application of the project-based learning model, it has a significant influence on interest in learning. One of which is marked by students' attitude being more concentrated and attentive when participating in learning because they feel they have a vital role in learning, especially related to doing projects with a STEM approach.

The use of teaching materials encourages students to learn independently. As stated by Setiyadi., Ismail., & Gani [15], teaching materials have self-instruction properties characterized by the presence of several command sentences that are easy to understand. There are practice questions, a summary of the material, and the contents of learning materials packaged in small units, making it easier to study thoroughly. The use of these teaching materials makes students happy to learn acid-base and colloid materials included in basic chemistry because they discuss the interrelationships of everyday life in the context of a wetland. As stated by Yildiz & Baltaci [16], contextual learning is more interesting for students because it has a connection to the experiences they have experienced.

The teacher's ability to use teaching materials is seen from the assessment of 3 observers based on three indicators: instructions for teaching materials, their contents, and ease of use. The average score for each indicator of the ability of teachers to use teaching materials is shown in Figure 5 below.

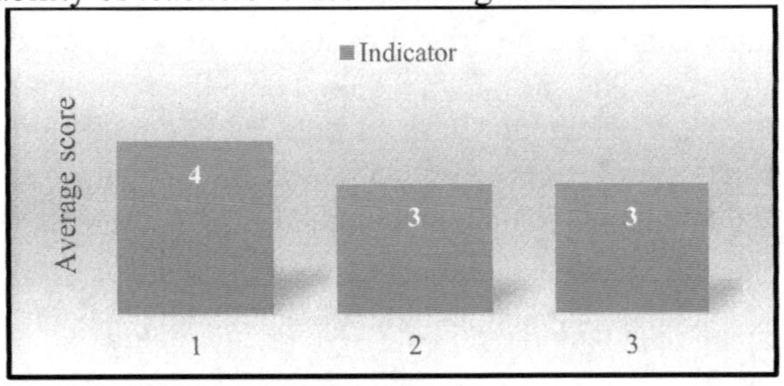

Figure 5 Average score indicator of the ability of lecturers to use teaching materials

Based on Figure 5, it is known that the indicator indicators obtain very practical criteria. Examples of use accompany instructions delivered by the teacher. Providing optimal direction and guidance from the teacher will make students use their potential in participating in the lecture process [17]. Content indicators and ease of use obtain practical criteria. The teacher shares the link for teaching materials through the WhatsApp group, then conveys the critical points of the subject matter in the teaching material so that later students can independently learn more about the material's content. As Johanes [18] stated, the lecturer is to act as a facilitator so that there is no need to intervene too deeply and does not take away the students' right to learn in the true sense.

The ability of the lecturer to manage the class is seen from the assessment of 3 observers based on four dimensions, namely introduction, core, closing, and time allocation. The average score obtained from the statement of the ability of the teacher to manage the class is shown in Figure 6 below.

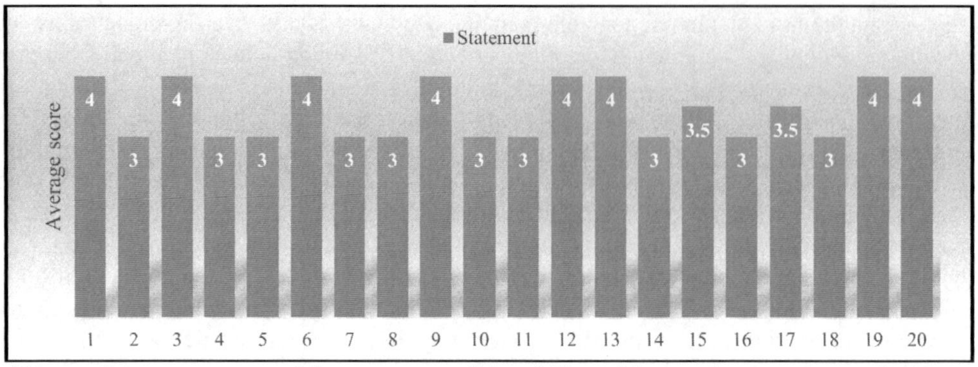

Figure 6 Average score of the statement of the ability of the teacher to manage the class

Based on Figure 6, it is known that the preliminary dimensions in statement 1 obtain very practical criteria. Before entering the core of learning, the teacher instructs the students to open the teaching materials that have been distributed. The core dimensions in statements 2 to 18 have very practical criteria. In this dimension, the stage of running the syntax of the project-based learning model with the STEM approach. According to Hendra., Arsa., & Krisnawati [19], the project-based learning model is meaningful learning that brings students to an impressive learning experience.

Moreover, the experience was obtained from problems in the context of wetlands. The closing dimension in statement 19 has very practical criteria. In this dimension, the learning process is closed with students making conclusions. According to Rahmawaty., Nurhayati., & Arsyad [20], learning with self-discovery of concepts and meaning can better understand concepts and student retention at the end of learning. The dimension of time allocation in statement 20 obtains very practical criteria. The time allocation is carried out in accordance with the RPP that has been made. Yatmini [21] revealed that RPP is an absolute requirement for the implementation of a conducive learning process.

The effectiveness of teaching materials in students' scientific literacy ability is seen from the N-gain obtained. The results of the N-gain of students' scientific literacy abilities are shown in Figure 7 below.

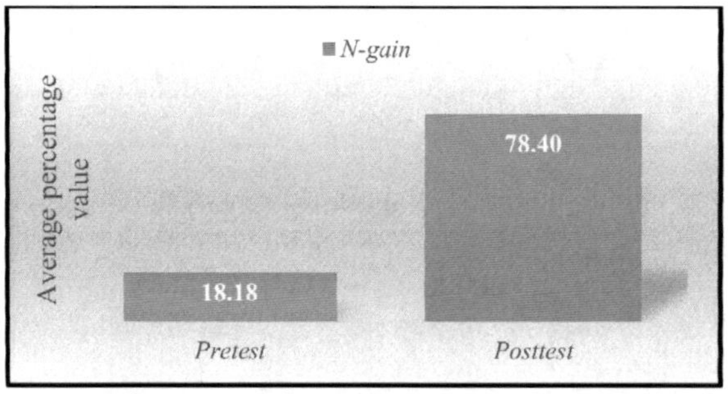

Figure 7 Average percentage of pretest and posttest scores of students' scientific literacy skills

Based on Figure 7, it is known that the average posttest result is higher than the pretest; this is because the posttest has used teaching materials based on a project-based learning model with a STEM approach during learning. As revealed by Khotimah., Suhirman., & Raehanah [22], using the project-based learning model can improve students' scientific literacy skills. It is also supported by the research results of Rusyati., Permanasari., & Ardianto [23] that through learning with the STEM approach, students can train students' scientific literacy skills. The application of the project-based learning model prepares students to solve problems through the stages of searching, analyzing, and synthesizing results that will be conveyed to other students. As revealed by Wulandari [24], through the application of the project-based learning model, students will explore, assess, interpret, and synthesize to produce various forms of learning outcomes. The application of the STEM approach trains students to build their knowledge in the scientific method. As revealed by Munawar., Roshayanti., & Sugiyanti [25], the STEM approach is seen as an approach that can encourage students to develop curiosity so that they can build knowledge by exploring, observing, discovering, and investigating how something happened. Through the implementation of this student-centered learning pattern, they can increase their learning activities [26].

4. Conclusion

Based on the results of discussion, it can be concluded that basic chemistry learning teaching materials for acid-base and colloid materials based on STEM integrated project-based learning models to improve scientific literacy in the context of developed wetlands are feasible because of: (a) Obtaining very valid criteria for use in learning with a score of 93.67% (very valid) in the aspect of content validity; 93.75% (very valid) on the aspect of the validity of the presentation; 94.62% (very valid) on the aspect of language validity; and 94.44% (very valid) on the media validity aspect; (b) Obtaining practical criteria because the average score on the individual test is 3.20 (practical), small group test is 3.68 (very practical), student response questionnaires are 3.54 (very practical), lecturers' ability observation sheets use e -module is 3.33 (very practical), and the observation sheet on the ability of e-module to manage class is 3.45 (very practical); (c) Obtaining effective criteria can contribute to a high increase in student scientific literacy results in the limited trial class, as seen from the N-gain value between pretest and posttest, namely 0.75 and 0.79, respectively.

References

[1] Toharudin, Hendrawati S and Rustaman A 2011 *Membangun Literasi Sains Peserta Didik* (Bandung: Humaniora)
[2] Suranti N M, Gunawan and Sahidu H 2016 *Jurnal Pendidikan Fisika dan Teknologi* **2** 73-9
[3] Keane L and Keane M A 2016 *An International Journal* **21** 61-82
[4] Iriani R, Herlina A, Irhasyuarna Y and Sanjaya R E 2019 *Jurnal Inovasi Pendidikan IPA* **5** 54-68
[5] Thiagrajan, Semmel D S and Semmel M I 1974 *Instructional Development for Training Teachers of Exceptional Children* (Washinton DC: National Center for Improvement Educational sySTEM)
[6] Akbar S 2015 *Instrumen Perangkat Pembelajaran* (Bandung: Remaja Rosda Karya)
[7] Widoyoko E P 2012 *Teknik Penyusunan Instrumen Penelitian* (Yogyakarta: Pustaka Pelajar)
[8] Djali and Muljono P 2008 *Pengukuran dalam Bidang Pendidikan* (Jakarta: PT Grasindo)
[9] Heinich, Molenda and Russel 1996 *Instructional Media and New Technologies of Instructional* (Englewood Cliffs N J: Prentice-Hall)
[10] Amir A 2016 *Jurnal Eksakta* **2** 34-40
[11] Arizona N 2016 *Jurnal Kata* **1** 93-8
[12] Puryati E 2017 *Jurnal PRISMA Universitas Suryakancana* **11** 192-201
[13] Zaharah and Susilowati A 2020 *Jurnal Ilmiah Pendidikan Biologi* **6** 145-58
[14] Rozikin A, Lesmono A and Bachtiar R 2018 *Jurnal Pembelajaran Fisika* **7** 108-15
[15] Setiyadi M W, Ismail and Gani H A 2017 *Journal of Educational Science and Technology* **3** 102-12
[16] Yildiz A and Baltaci S 2016 *Journal of New Horizons in Educational* **6** 155-66
[17] Suhaida, D and Azwar I 2018 *Jurnal Pendidikan Sosial* **5** 1-19
[18] Johanes 2018 *Forum Ilmiah* **15** 133-8
[19] Hendra B K, Arsa G R and Krisnawati L 2017 *Jurnal Pendidikan Teknik Elektro Undiksha* **6** 75-85
[20] Rahmawaty S, Nurhayati and Arsyad M 2020 *Prosiding Seminar Nasional Fisika PPs UNM 2 Universitas Makassar* 33-36
[21] Yatmini 2016 *Jurnal Ilmiah Mandala Education* **2** 172-84
[22] Khotimah H, Suhirman and Raehanah 2020 *SPIN* **2** 13-26
[23] Rusyati, Permanasari A and Ardianto D 2019 *Journal of Science Educational and Practice* **2** 10-22
[24] Wulandari F E 2016 *Jurnal Pedagogi* **5** 247-54
[25] Munawar M, Roshayanti F and Sugiyanti 2019 *Jurnal Ceria* **2** 276-85
[26] Novianti D, Rusmansyah and Kusasi M 2017 *Journal of Chemistry and Education* **1** 15-27

Visual learning style-based chemistry mental model representation through transformative online learning

A Winarti, A Almubarak and P Saadi

Chemistry Education Study Program, Teacher Training and Education,

Lambung Mangkurat University

almubarak_kimia@ulm.ac.id

Abstract. Analysis of learning styles is one of the main things teachers need to do before carrying out teaching. The study of learning styles can provide an overview of how a teacher designs a learning concept by students' learning styles. The learning process will show students' mental models with their respective learning styles so that these mental models become the primary material for how teachers develop students' cognitive. This research aimed to describe students' mental models in terms of students' visual learning styles. The method used is descriptive with qualitative and quantitative approaches with transformative-based learning concepts. The research results show that chemistry education students for chemistry learning innovation courses only have visual learning styles of 71.43% and audio by 28.55% and do not have kinaesthetic learning styles. This research focuses on visual learning styles to see students' mental models. The conclusion is that students still need cognitive strengthening, especially the ability to interpret phenomena at the sub microscopic level. With the visual learning style, students are expected to transform their cognitive so that they have mental structures and models relevant in theory and terminology.

1. Introduction

A mental model is a person's cognitive reflection on an idea or idea of a particular phenomenon [1]. In the context of learning chemistry, each student has initial knowledge called "preconceptions" that students will explain and show something based on basic reasoning and the results of student observations [2]. However, the preconceptions of these students need to be changed into "scientific concepts", meaning that each student needs to undergo a cognitive transformation so that students can understand a phenomenon in a concrete way [3,4]. Especially for prospective teachers, mental models are crucial because they are related to the teacher paradigm in delivering material [1]. Research [1] implements the TripleChem learning model in the ministry of students on organic chemistry in the context of mental models. Research results show that students experienced a significant increase from 1.36% to 62.73% [1]. Other research shows that improving students' mental models can be achieved by using the concept of representation [5]. It is in line with [6] that the idea of representation is a basic science of how to understand chemistry as a whole so that students do not experience misconceptions.

According to [7], students' mental models are dynamic, so teachers have to analyze students' character. In addition, the research results by [8] that students tend to construct their mental models in a non-scientific manner so that students' ideas are not in harmony and are not scientifically relevant to concepts. This condition is referred to as preconceptions by [2] where students will still maintain their perspective on a phenomenon so that this needs to be changed into a scientific concept. In addition, [3] added that students use everyday language to explain their knowledge and description of an event. In contrast, students need to use scientific terminology to understand the context of a phenomenon easily. However, students still have difficulty constructing their knowledge, which impacts growing mental models [9]. Learning needs analysis is one strategy to determine their understanding and progress, such as learning style analysis [10,11].

Each student has different learning styles, such as visual, audio, and kinesthetic [12]. Then [12] added that learning styles significantly affect managing the information obtained, including learning outcomes. If learning styles affect learning outcomes and students' mindsets [12], then the formation of mental models is influenced by how learning styles are owned [13]. Giving problems to students based on their learning styles makes it very easy for students to understand the context of the subject matter and construct their knowledge to reduce misconceptions that occur gradually [3,13]. That is, it is highly recommended that teachers analyze learning style assessments of students so that teachers know the abilities of students and how to improve students' understanding to become more scientific [2,9,14].

Visual learning style is essential in interpreting a phenomenon [13,15]. Visual learning style is a person's ability to explain something by looking at and paying attention to image media [13]. Cognitively, visual learning styles experience the ease of understanding when students are presented with something with pictures [13]. Visual learning style is certainly in line with the concept of chemical representation at the submicroscopic level; the submicroscopic and symbolic levels are the essential parts in understanding chemical matter [2,6,16,17]. According to [17], students have difficulty explaining what happens in chemical reactions at the particulate (submicroscopic) level. Then, using a medium such as a magnetic ball media (physically an image or a molecular engineering model) makes it easier for students to construct their understanding in learning [17]. Cognitive transformation is undoubtedly needed so that students experience changes in preconceptions into scientific concepts so that students do not experience misconceptions and form scientific mental structures and models [3,6,9,17–19]

Transformative learning is an alternative way of changing students' cognition to be more scientific. The concept of transformative learning is considered the best concept to use, especially in higher education [20]. The role of transformative learning is very influential in changing students' perspectives and how students describe a problem [21]. Transformative learning will broaden students' horizons in learning, and they can see solutions from various perspectives so that students grow into mature learners [3,4,22,23]. On the other hand, Covid-19 is a substantial case affecting all countries globally and affecting all sectors such as the economy, politics, industry, and education. Especially in the field of education, it is necessary to have innovations in learning so that the quality of learning and teaching is needed.

During the pandemic, virtual or online learning is considered an appropriate solution where students continue to carry out the learning process and interact with teachers through various platforms [24,25]. Communication and sharing of multiple things through online learning are considered digital learning concept that brings many benefits [26]. Other studies also suggest computer-based learning with e-learning integration where teachers and students use platform technologies such as Edmodo, Social Media, Blogs, Coursera, Youtube, etc. [25]. The use of this technology is constructive for students in online activities with easy and affordable accessibility. This technology is highly recommended for students who find it difficult to obtain references, especially during the current pandemic [25].

In connection with the explanation above, learning needs to transform from face-to-face concepts in the classroom to virtual areas so that the learning process continues during the pandemic without leaving the essence and context of the knowledge being learned [27–29]. The concept of online transformative learning is an innovative solution in constructing students' knowledge. In addition, participants will be cognitively facilitated because, before learning, the teacher analyzes students learning styles so that the learning process, interaction, and the concept of the assignment are oriented to the suitability of students' learning styles. The idea of this research will facilitate the interaction pattern between the teacher and the participants even though it is done online. Psychological experts [30] say that in schools, the student environment is referred to as a microsystem where a classroom is a place for interaction between teachers and students. This interaction significantly affects the internals of students, including the development of mental models based on their learning styles [3,31–33]. Based on the explanation, this research aims to analyze the mental models of participants (students) who have a visual learning style through transformative learning.

2. Method

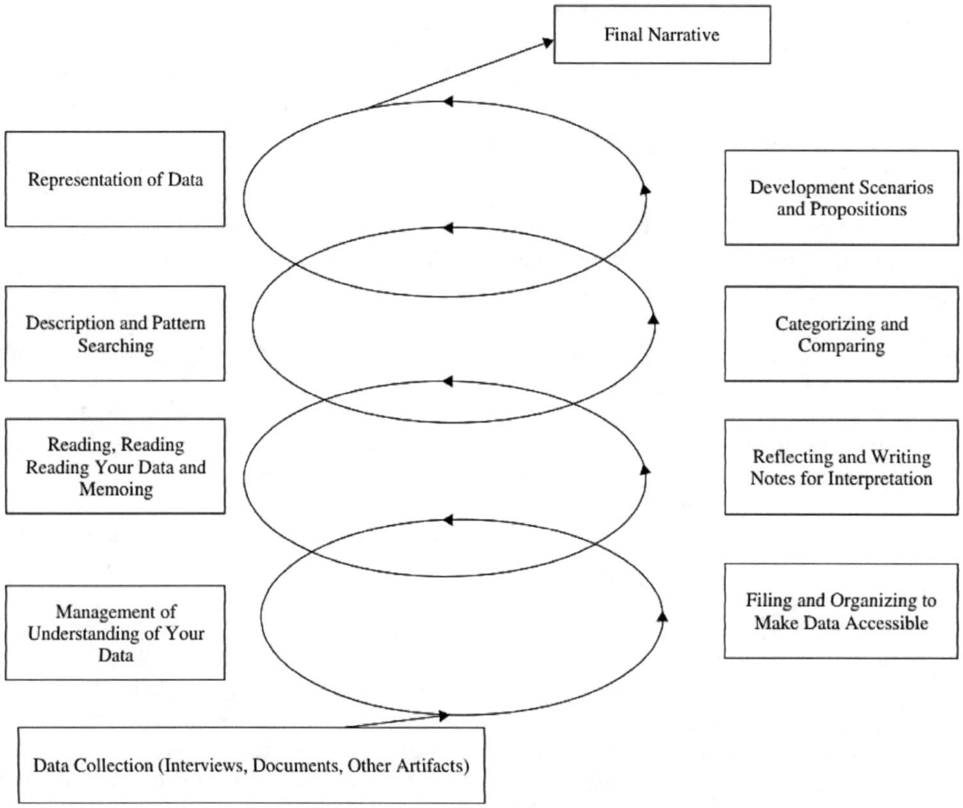

Figure 1. Cresswells' analysis spiral [34]

The method used is descriptive with qualitative and quantitative approaches [34]. The sample used in chemistry education students, FKIP ULM, with chemistry learning innovation courses. Qualitatively, the data obtained are data from student analysis results from assignments and student interviews. Quantitatively, the data were obtained from the analysis of learning styles conducted before the learning process occurred. Transformative learning is a learning model used, especially in the current pandemic era. One of the sample indicators used is students who have experienced and have felt the impact of the pandemic situation being discussed. The sampling technique used is Intense case sampling. According to [34], "intense case sampling occurs when the researcher selects people who have had an intense experience or intense feeling related to the topic of interest". The process of analyzing research results uses the concept of study validity by Cresswells' spiral analysis, which then explains the topics brought into the research [34].

3. Discussion

The primary orientation of this research is to find out how the representation of students' mental models is based on students' learning styles. Learning style analysis is an initial analysis that is carried out before starting the teaching process by distributing a learning style questionnaire instrument. This learning style analysis is included in learning assessment, where teachers are highly recommended to determine students' learning needs so that it is easy for teachers to design a teaching and learning process. After that, the teacher makes teaching materials by adopting the concept of transformative learning and problem-based materials. The teaching materials presented consist of theoretical explanations, resumes to materials and things that are not yet understood, and problems based on representation. Then, the learning process is carried out online with a transformative learning model due to the ongoing pandemic situation. The sample used was students of Chemistry Education, Faculty of Teacher Training and Education, Lambung Mangkurat University, considering they were part of those affected by the pandemic. The following is a sketch of the transformational learning carried out.

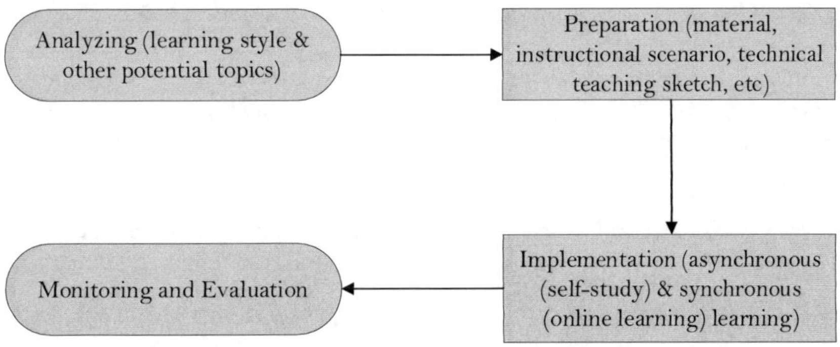

Figure 2. Transformative learning implementation

A preliminary review of the research data results shows that chemistry education students only have a visual learning style of 71.43% and an audio learning style of 28.55%. That is, kinesthetic learning styles are not owned by them (students). The implementation of this learning style results is then integrated into the material, the brainstorming process during learning, summarizing the material and things that are difficult to understand, and the process of working on the task. The understanding carried out is adjusted to their learning style so that when they do something, they carry it out optimally. Research shows that teaching based on learning style analysis dramatically affects the quality of the teaching process carried out [32]. In addition, the use of technology is not the only strategy in improving students' cognitive abilities in learning, but how the learning presentation provided includes the process of solving problems [32]. Assessment of learning styles is part of the cognitive, affective, and even psychological aspects because each student has a unique mental model pattern and differs from one and another [2,3,9,32]. If the learning presentation includes assignments according to their learning style, participants will be too early in understanding and interpreting the phenomenon under study [2,9,13,32].

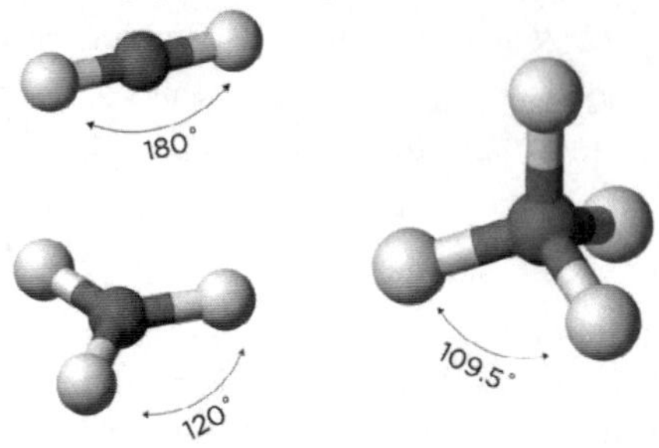

Figure 3. Visualization of Carbon atoms with other atoms [35]

Explanation of the material. *In the earth's crust, the element carbon is a minor component. However, carbon compounds (i.e., magnesium and calcium carbonates), form common minerals (e.g., magnesite, dolomite, marble, or limestone). The corals and shells of oysters and clams are calcium carbonate. Carbon is widely distributed as coal and in the organic compounds that make up petroleum, natural gas, and plant and animal tissues. The raw sequence of chemical reactions called the carbon cycle involves converting carbon dioxide in the atmosphere to carbohydrates through photosynthesis in plants. The consumption of these carbohydrates by animals, oxidation through metabolism to produce carbon dioxide and other products, and the return of carbon dioxide to the atmosphere are among the most important biological processes* [35].

The presentation of the material is taken from [35], but this material is only as supplementary material. That is, students need to train their reasoning power and critical thinking in solving a given problem. The following are questions and visualization of students' mental models, which are the basis for analyzing their understanding.

Figure 3 shows the visualization of the C molecule with other atoms. Write and explain scientifically why the bond angles of the three molecules are different and draw orbitals to strengthen your scientific perspective?. Figure 4 shows how the students' mental models of the cases presented. According to the analysis process, it is found that 1) there are students who use quotes or citations on the results of their analysis even though they are encouraged to interpret the phenomenon according to their perspective. The second finding is that the orbital images presented are standard images that can be traced through *websites* and various technological media. The orbital designation in question is quite representative. Still, in this context, they are directed to be completely honest in developing their ideas so that they have the potential to build their analysis results in any visual form, not having to quote directly and in whole from websites, etc. The third reason is the answer given is related explicitly to the "molecular bond angle," where the student's response is less concrete in explanation. Although students use scientific terminology in their descriptions, they are considered less than optimal in executing the issues given. In representation, visual learning styles have adequate sketching abilities, so that their reasoning power should be explored through their analysis. However, this will take time as optimization of spatial capabilities is difficult due to weaknesses in interpreting the particulate level [2,9]. On average, students have the same tendency in interpreting their explanations, where when online learning through zoom meetings, their participation increases considerably from each session. However, when they are given time to explain their opinion directly, they are pretty passive. Then, on weekly assignments, every time students start the resume process and work on the problems presented, they tend to explain based on references and various works of literature. Even though they can define from their perspective so that it can be seen how their mental model is with the given study, this is in line with the idea [2]. Those students tend to stick with their opinions and the observations they experience, so this becomes difficult when they are cognitively transformed. This finding is in line with research [32] that students with a "visual" learning style manage information based on what they observe and direct observation. This experience can strengthen them in defending their point of view. They need to understand the concept of representation in a concrete way so that the results of their observations (visual learning styles) are scientific. In addition, cognitive transformation can make students observe a phenomenon and interpret the acquired knowledge themselves [3,6,32]

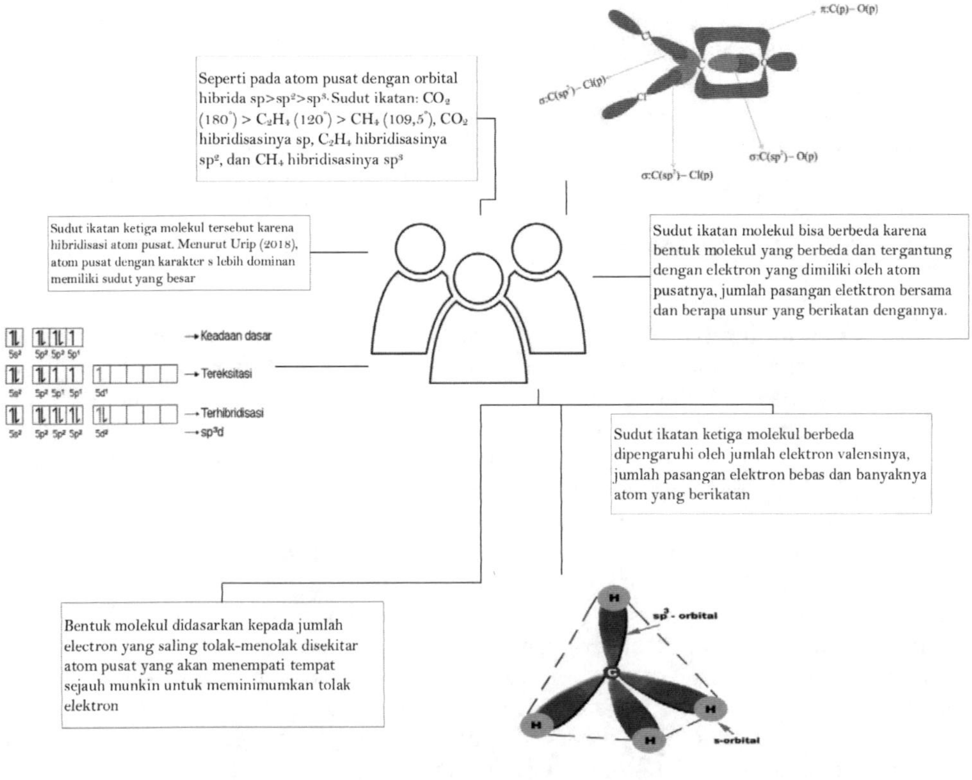

Figure 4. Description of Student Mental Model for visual learning style

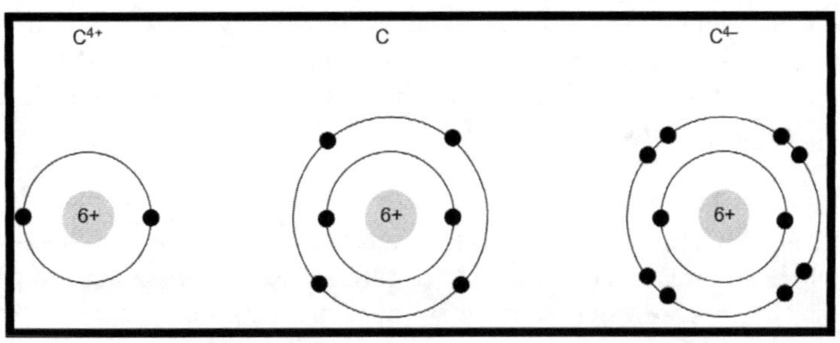

Figure 5. Visualization of the C atom: interpreting the stability of the carbon atom in various forms [36]

Figure 3 shows the visualization of the C molecule with other atoms. Write and explain scientifically why the bond angles of the three molecules are different and draw orbitals to strengthen your scientific perspective?. Figure 4 shows how Figure 5 is a presentation of the issues given to students for analysis related to the carbon atom with the question: which is more stable C^{4-}vs $C^{4+}$and C vs C^{4+}?. This presentation is a submicroscopic visualization of the C atom in which students are invited to interpret how their mental model of the carbine structure is shown in Figure 5. The following describes the student's mental model for the questions related to Figure 5 given.

- Between C^{4-}vs C^{4+}, the more stable is C^{4+}, because the C atom requires 4 electrons to achieve a stable octet. In addition, C^{4-} also has a complete electron configuration, namely by attracting four free electrons into the outer shell so that it is durable;
- Between C vs C^{4+}, the more stable is C^{4+}, because the C^{4+}atom has an electron configuration that follows the duplet rule of 2, C^{4+} loses 4 electrons, and their structure becomes 2, so it is stable;
- Non-metal elements tend to form anions rather than cations, so C^{4-} is more durable than C^{4+};
- The C^{4-} is more stable than C^{4+}, because to achieve stability of C (carbon) atoms by adding 4 electrons to complete the octet rule;
- The C is more stable than C^{4+} because the C atom requires a lot of energy to give or lose 4 electrons to become C^{4+}.

The mental model is shown (linked to Figure 5) shows that students understand enough in theory and terminology how to interpret carbon atoms in various forms. The basic concept of stability has been fulfilled cognitively as they have understood that the C atom requires 4 electrons to achieve octet stability, the C^{4+} atom has an electron configuration that follows the duplet rule of 2, and the C atom requires a lot of energy to be given or lose 4 electrons to become C^{4+}. Although the description of the mental model of students is relatively weak in Figure 4, in Figure 5 they are quite mastered in theory and terminology. The trigger factor for the formation of this mental model is that they are given time to work on the problems given and look for a variety of literature to strengthen their perspective in explaining. However, students still need to train their reasoning power and transform cognitively in interpreting particulates to explain in-depth and concretely. In this context, [17] suggest using tools or media to stimulate their reasoning so that they are easy to understand chemical molecules in submicroscopic and symbolic level.

Overall, online transformative learning based on learning styles is the right solution in developing students' mental structures and models. Analysis of learning styles is an important thing to do at the beginning of learning so that teachers know the level of understanding of students before learning. In addition, assignments and problems based on their learning style make it easier for students to develop cognitively. The presentations given by adopting the concept of correct representation of chemistry train their reasoning power so that they train them in understanding a phenomenon. Although, achieving submicroscopic understanding is still considered difficult for students, but with gradual treatment by considering learning styles such as visuals, integration of representative chemistry concepts, transformative learning concepts, and other components, it certainly makes it easier for them (students) to interpret so that they do not experience misconceptions.

4. Conclusion

Visual learning style is the main point that needs to be analyzed by the teacher. This is related to the analysis of learning needs where this analysis (learning style) makes it easier for teachers to provide learning in the classroom. The results of the research show that chemistry education students for chemistry learning innovation courses only have visual learning styles with % of 71.43% and audio of 28.55% and do not have kinesthetic learning styles. This research focuses on visual learning styles to see how their mental models are. The conclusion is that students still need cognitive strengthening, especially how to interpret phenomena at the submicroscopic level. With visual learning styles, they are expected to be able to transform cognitively so that they have mental structures and models that are relevant in theory and terminology.

In relation to the student's mental model of the presented case, if the mental model is analyzed, it is found that 1) there are students who use quotes or citations on the results of their analysis even though they are encouraged to interpret the phenomenon according to their perspective. The second finding is that the orbital images presented are standard images that can be traced through *websites* and various technological media. The orbital designation in question is actually quite representative, but in this context they are directed to be completely honest in developing their ideas so that they have the potential to develop their analysis results in any visual form, not having to quote directly and in full from websites, etc. The third reason is the answer given specifically related to the "molecular bond angle" where the student's answer is less concrete in explanation. Although, students simply use scientific terminology in their explanations, they are considered less than optimal in executing the issues given. In representation, visual learning styles have adequate sketching abilities so that their reasoning power should be explored through their analysis. However, this will take time as optimization of spatial capabilities is difficult to achieve due to weaknesses in the interpretation of the particulate level [2,9].

On average, students have the same tendency in interpreting their explanations in which as online learning through zoom meetings their participation increases considerably from each meeting. However, when they are given time to explain their opinion directly they are quite passive. Then, on weekly assignments, every time they start the resume process and work on the problems presented, they tend to explain based on references and various literature, even though they can explain from their own perspective so that it can be seen how their mental model is with the given study. This is in line with the idea [2] that students tend to stick with their opinions and with the observations they experience so this becomes difficult when they are cognitively transformed. This finding is in line with research [32] that students with a "visual" learning style manage information based on what they observe and direct observation so that this experience is able to strengthen them in defending their point of view. In fact, they need to understand the concept of representation in a concrete way so that the results of their observations (visual learning styles) are scientific. In addition, cognitive transformation can make students not only observe a phenomenon, but they are able to interpret the acquired knowledge themselves [3,6,32].

References

[1] Suja I W, Redhana I W and Sudria I B N 2020 *J. Phys. Conf. Ser.* **1503**
[2] Barke H-D, Hazari A and Yitbarek S 2009 *Misconceptions in Chemistry (Addresing Perceptions in Chemical Education)* (Berlin, Heidelberg: Sense Publisher)
[3] Barke H-D, Harsch G and Schmid S 2012 *Essentials of Chemical Education* (Verlag Berlin Heidelberg: springer)
[4] Mezirow J 1991 *Tranformative Dimensions of Adult Learning* (San Francisco: Jossey-Bass)
[5] Yuanita L and Ibrahim M 2015 *Online J. New Horiz. Educ.* **5** 30–45
[6] Gilbert J K and Treagust D F 2009 *Multiple Representations in Chemical Education, MOdels and Modeling in Science Education* (Berlin, Heidelberg: Springer Science + Business) pp 1–8
[7] Majid A and Prahani B K 2017 *J. Res. Method Educ.* **7** 120–4
[8] Halim N D A, Ali M B, Yahaya N and Said M N H M 2013 *Procedia - Social and Behavioral Sciences* vol 97 (Elsevier B.V.) pp 224–8
[9] Darmiyanti W, Rahmawati Y, Kurniadewi F and Ridwan A 2017 *JRPK J. Ris. Pendidik. Kim.* **7** 38–51
[10] Özgür S D 2018 *Cypriot J. Educ. Sci.* **13** 521–8
[11] Taçgin Z, Uluçay N and Özüağ E 2016 *Turkiye Kim. Dern. Derg. Kisim C Kim. Egit.* **1** 147–64
[12] Rijal S and Bachtiar S 2015 *J. Bioedukatika* **3** 15

[13] Sulistyawati A H, Parubak A S and Suparman A R 2018 *QUANTUM J. Inov. Pendidik. Sains* **9** 100–6
[14] Apipah S, Kartono and Isnarto 2018 *J. Phys. Conf. Ser.* **983**
[15] Leopold L 2012 *TESL Canada J.* **29** 96
[16] Chen X, de Goes L F, Treagust D F and Eilks I 2019 *Educ. Sci.* **9**
[17] Ryan S and Herrington D G 2014 *J. Chem. Educ.* **91** 860–3
[18] Hill B, Winmar G and Woods J 2020 *Aust. J. Indig. Educ.* **49** 2–13
[19] Mezirow J 1997 *New Dir. Adult Contin. Educ.* **1997** 5–12
[20] Fraser S P 2015 *J. Transform. Educ.* **13** 140–60
[21] Quennerstedt M 2019 *Sport. Educ. Soc.* **24** 611–23
[22] Supatmi S, Setiawan A, Rahmawati Y, Education C, Program S and Jakarta U N 2019 *African J. Chem. Educ.* **9** 18–37
[23] Mahardika E, Ridwan A and Rahmawati Y 2018 *Educhemia (Jurnal Kim. Dan Pendidikan)* **3** 51–65
[24] Ray S and Srivastava S 2020 *J. Proteins Proteomics* **11** 77–80
[25] Sukendro S, Habibi A, Khaeruddin K, Indrayana B, Syahruddin S, Makadada F A and Hakim H 2020 *Heliyon* **6** e05410
[26] Wargadinata W, Maimunah I, Dewi E and Rofiq Z 2020 *Tadris J. Kegur. dan Ilmu Tarb.* **5** 141–53
[27] Dickson-Karn N M 2020 *J. Chem. Educ.* **97** 2955–9
[28] Tsao S-F, Chen H, Tisseverasinghe T, Yang Y, Li L and Butt Z A 2021 *Lancet Digit. Heal.* **7500** 1–20
[29] Yuliati Y and Saputra D S 2020 *J. Elem. Edukasia* **3** 142–9
[30] Ireson J 2008 *Learners , Learning and Education Activity* (London & New York: Routledge)
[31] Rasmawan R 2020 *Int. J. Instr.* **13** 747–62
[32] Syofyan R and Siwi M K 2018 The Impact of Visual, Auditory, and Kinesthetic Learning Styles on Economics Education Teaching **57** 642–9
[33] Yazicilar Ö and Güven B 2009 *Procedia - Soc. Behav. Sci.* **1** 2578–83
[34] Schreiber J and Asner-Self K 2011 *Educational Research* (New Jersey: John Wiley & Sons, Inc)
[35] Britannica E of E 2021 Carbon: Chemical Element *Br. Encycl.*
[36] Taber K S 2002 *Chemical Misconceptions: Prevention, Diagnosis and Cure. Vol. 2: Classroom Resources* (Royal Society of Chemistry)

Development of flipbook e-module problem-based learning (PBL) learning model to increase students' learning outcomes in oxidation-reduction reaction material

Leny[1], K Husna[1], Rusmansyah[1], M Kusasi[1], Syahmani[1] and H Zuwida[2]

[1]Chemical Education Study Program FKIP Lambung Mangkurat University, Indonesia
[2]SMAN 12 Banjarmasin, Jl Alalak Utara RT 02 Banjarmasin 70125 South Kalimantan, Indonesia.

rusmansyah@ulm.ac.id

Abstract. Research and Development has been carried out to produce a valid, practical and effective e-module *flipbook* model of *Problem-Based Learning* (PBL) to improve learning outcomes on redox material. This type of research is *Research and Development* (R&D) adopting the ADDIE development model. The e-module trial was carried out in Class X MIPA at SMA Negeri 12 Banjarmasin for the 2020/2021 Academic Year. Validation and observation sheets, readability questionnaires, response questionnaires, and learning outcomes tests were occupied to collect the data. In analyzing the data, descriptive analysis was used. The results showed that the developed e-module has met the following criteria: (1) Validity; the value of the content feasibility aspect of 91% (very valid), presentation of 96% (very valid), language of 95% (very valid), and media of 94.5% (very valid). (2) Practicality, seen from the readability results of e-module on individual tests of 3.5 (very good) and small groups of 3.4 (very good), student responses to e-modules of 3.3 (very good), and responses to teacher teaching activities using e-module of 3.7 (very good). (3) Effectiveness, views and learning outcomes of knowledge with an N-gain value of 0.80 (high) and the result of learning attitudes of 83.26 (very good).

1. Introduction

Education is a deliberate and planned effort to help develop the potential and ability of learners to benefit the interests of their lives as human beings and as citizens for the future. Changing the world requires knowledge and ability. Such knowledge and skills are acquired through education [1]. So, teachers who play an essential role as facilitators in learning must integrate their ability to utilize technological advances as an innovation in the learning process.

In teaching and learning in schools, it is essential that there is media as a learning process used by learners independently and maintains the spirit of learning learners. In addition, the teaching materials that are not varied and not attractive will make learners feel bored and difficult to follow the learning process. Further, the chemistry lessons are striking on concepts and calculations, which learners still consider difficult. Therefore, a material oxidation-reduction reaction that deals with the idea of daily life so that the low learning outcomes are obtained by learners [2].

Improving quality in the learning process involves finding the suitable model, method, or medium to carry out the teaching and learning process [3]. As previously described, it can be minimized with the help of learning media, namely e-modules with problem-based learning models that learners can study independently to understand the material thoroughly.

The selection of the suitable model to improve the learning outcomes of learners to solve problems in everyday life is model *problem-based learning.* The problem-based *learning* model is a learning model that uses problems as a basis for learning materials for students. In line with this, the role of teachers in this learning model is more instrumental as a guide and facilitator so that students learn to think and solve their problems [4].

The use of teaching materials in the form of exciting modules can increase the success of a learning process, especially the influence of technological developments. Modules are often developed in the form of e-modules (electronic modules). Furthermore, E-modules can reduce paper use because e-modules can be made using various *software* without reducing its function as a source of information. One of which is *flipbook software* was in its application using *smartphone* technology, laptops, internet, etc.

This *flipbook-based* e-module aligns with the 4.0 industrial revolution that applies the digital world in various aspects of life. Currently, the development of the world of technology also impacts the habits of students in their daily lives who use *smartphones* as the use of digital technology in the learning process.

Based on the exposure, researchers are interested in researching the development of interactive learning multimedia, namely *flipbook-based* modules with *Problem Based Learning* (PBL) learning models, in improving learning outcomes in the oxidation-reduction reaction material of class X of the 2020/2021 school year at SMA Negeri 12 Banjarmasin.

2. Method

Research and Development (R&D) uses the development of ADDIE. Research Conducted at SMA Negeri 12 Banjarmasin class X MIPA school year 2020/2021. In individual trials conducted in class X MIPA 1, small group trials are conducted in class X MIPA 2, and limited trials are conducted in class X MIPA 3.

Data collection techniques are carried out through non-test instruments that include validation questionnaires, legibility questionnaires, response questionnaire learners, the implementation of teacher learning using e-modules, and the learning outcomes of learners' attitudes. Then, the test instrument in the form of multiple-choice questions to measure the learning outcomes of learners' knowledge given before and after treatment.

3. Research and Discussion

E-modules developed using the stages of model ADDIE *(Analysis, Design, Development, Implementation, Evaluation)* is presented as follows:

3.1 Stage Analysis

At the analysis stage, the activities carried out are observing problems in schools, determining the object of the experiment, making a concept analysis, and analyzing learning objectives.

3.2 Stage Design

The design stage aims to design the contents and outline of a product to be developed. At the design stage, the activities carried out are preparation of the test, media election and initial design.

3.3 Stage Development

At the stage of development of the activities carried out are as follows:

3.3.1 Validation of products by experts

The validity test aims to find out the validity level of e-modules measured based on the assessment of validators consisting of 3 lecturers of Chemical Education FKIP ULM Banjarmasin, one chemistry teacher of State High School 12 Banjarmaisn as material experts, and one lecturer of Educational Technology FKIP ULM Banjarmasin as a media expert. The results of the validator's assessment are shown in Figure 1.

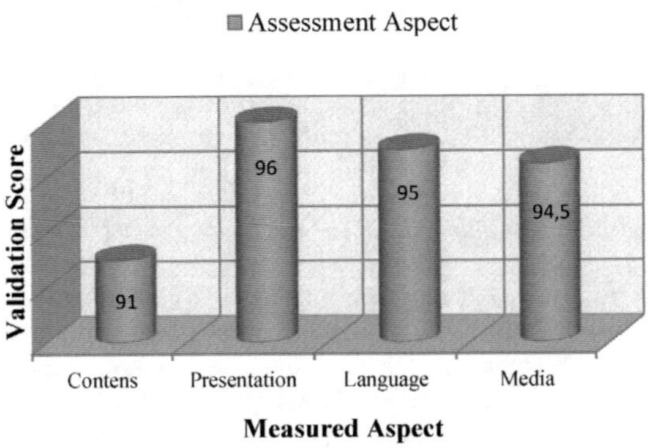

Figure 1. E-module eligibility validation results

Based on Figure 1 known percentage of value for the aspect of content obtained a percentage value of 91% by meeting several criteria, namely the conformity of the material loaded with basic competencies, learning indicators, and learning goals to be achieved, in addition to the support of learning materials and the updating of the material.

The presentation obtained 96% since it shows good presentation techniques equipped with various other supporters such as compelling images and videos in carrying out the learning process to encourage learners to be more interactive and participative. Moreover, the presentation is equipped with introductions, contents, and covers.

The language aspect got 95% by meeting several criteria, namely presented with a straightforward, communicative, dialogical language according to learners' intellectual development and emotional level, direct and integrated, and the consistent use of terms, symbols or icons.

The media aspect obtains a percentage value of 94.5% by meeting several criteria, namely presenting attractive displays and content that corresponds to the composition of colors, images, letter layout, sound or music, videos, and usage instructions, and has been in accordance with the characteristics of media use, attractiveness and the presence of flipbook media *elements.* The validity test aims to find out the validity level of e-modules measured based on the assessment of validators consisting of 3 lecturers of Chemical Education FKIP ULM Banjarmasin, one chemistry teacher of State High School 12 Banjarmaisn as material experts, and one lecturer of Educational Technology FKIP ULM Banjarmasin as a media expert. The results of the validator's assessment are shown in Figure 3.1.

3.3.2 Field trials

Field trials include individual, small-group, and limited trials. Individual and small group trials are conducted to determine the practicality of e-modules reviewed from *flipbook-based* e-modules readability. The average score of individual and small group trial readability questionnaires is shown in Figure 2.

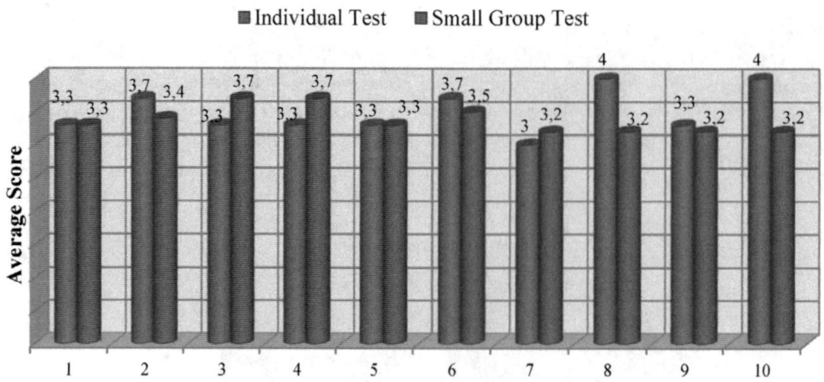

Figure 2. Results of individual and small group readability tests

Based on Figure 2, it is known that statements 2, 6, 8, and 10 obtain an average score in the very well category on individual trials. Whereas, statement 2 which is about the cover design; statement 3, which is about clear images; statement 4, which is about colors and letters and matching images; statement 5, which is about sentences; statement 7, which is about the content of the material; and statement 8 which is about the material studied obtain an average score of practical categories. So, revisions are carried out first before being tested again in small group trials.

The assessment of the readability of the e-module is based on several things. The e-module can trigger the involvement of students in learning because it is designed attractively and simply [5]. The delivery of the material relies on aspects of the text. It is carried by multimedia components such as images, graphics, videos, and animations to deliver more interactive materials and increase interaction between students and teachers. Attractiveness display includes color combination selection, typeface selection, audio selection, video, and animation in designing e-modules [6].

Presentation of exciting material as challenging as any material provided if packaged more enjoyable, easy to understand, and read will make students more motivated to learn it and not feel bored following the learning [7].

The use of video in the learning process also allows learners to witness an event that cannot be detected directly. In addition, learning with video media fosters interest and motivates always to pay attention to lessons [8], making learning more effective and students responding well in video learning media.

Furthermore, you tried to be limited to knowing the practicality and effectiveness of *flipbook-based modules*—the practicality of e-modules reviewed from the student responses, the teacher's teaching activity in using e-modules. At the same time, the effectiveness of e-modules is examined from the learning outcomes of knowledge and attitudes of learners. Here is the average score of the learner's response is shown in Figure 3.

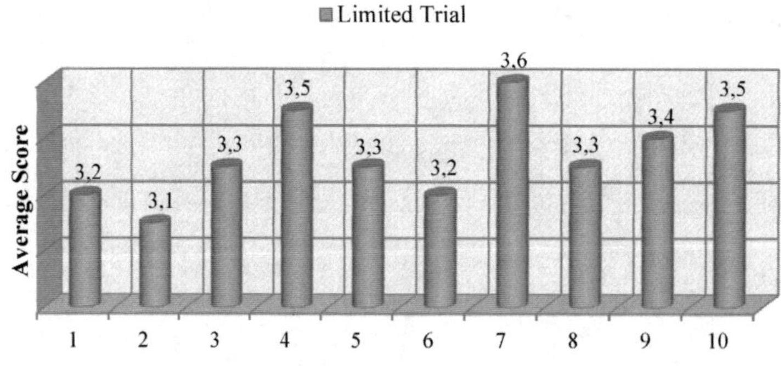

Figure 3. Limited trial student response results

Based on Figure 3, statements 3, 4, 5, 7, 8, 9, and 10 obtained an average score of category very well, except statement 1 regarding high willingness in the following learning, statement 2 about understanding learning materials, and statement 6 regarding the curiosity of learners gaining good categories. The use of e-modules by using *flipbook software* can provide positive things for learners in assisting the learning process by facilitating it with clear and easy-to-understand material. Some videos are beneficial so that it is interesting for them to learn independently both in the classroom and outside school [9].

Learning oxidation-reduction reaction material with flipbook-based e-modules using the PBL model makes it easier for learners to understand. The results show that the e-modules used can make it easier for learners to understand the material because the e-module presents concise material and communicative language to help learners understand the material more easily.

Using e-modules as teaching materials can facilitate and motivate students in the learning process because it is equipped with material descriptions, animations, videos, answer sheets, and final formative test questions [10].

The results of the student response show that the interest of learners is outstanding to learn during the learning process because in the learning process using the stages or synths *of problem-based learning* models to make learners have more high interest during learning. Learning using suitable models and media will improve students' good skills and high motivation [11].

Flipbook-based e-modules using the PBL model will add to learners' curiosity to learn the material presented. In addition, after using *flipbook-based* e-modules, student's critical attitude can be fostered.

Flipbook-based e-modules using the PBL model foster awareness of learners' thought processes in solving problems related to the material. They present learning steps that can make learners construct for themselves and consciously solve the given problem. Hence, using the PBL learning model, the student can achieve the right indicators to solve problems [12].

Contextual learning is more interesting for learners due to the facts used in the learning process to relate the experience possessed by learners; this can be seen from the real problems of the student's daily life contained in teaching materials. Therefore learners can connect the experience that has been learned before, and it is easy for learners to produce learning to be more meaningful [13].

Implementing the PBL model can improve students' activeness and learning achievement [14] and positively and significantly influence students' learning outcomes [15]. Thus e-modules with PBL learning models are otherwise effective in improving problem-solving skills [16].

The ability of teaching activities of teachers using e-modules is seen from the assessment of 3 observers. Here is the average score of the acquisition of teacher teaching activities in using e-modules shown in Figure 4.

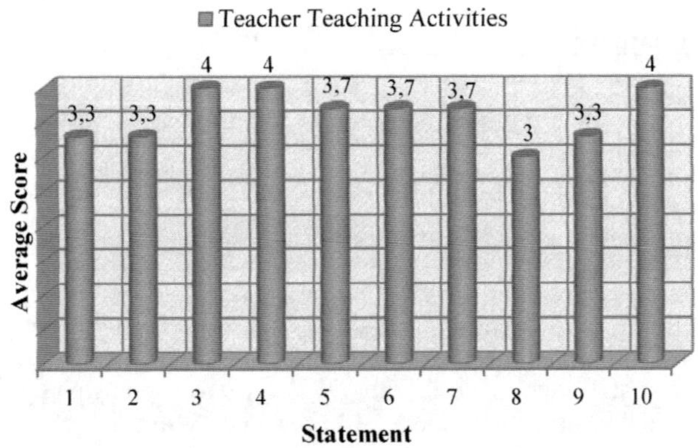

Figure 4. Average teacher teaching activity score using e-modules

Based on Figure 4, responses to teacher teaching activities using e-modules applying the average score criteria are very practical. At the time, the teacher used an e-module based electronic media *flipbook* on the oxidation-reduction reaction material using the steps of the PBL model that can make learners learn independently so that the achievement of learning indicators and goals can be achieved. It follows the characteristics of user-friendly e-modules, which means that e-modules should be friendly with the user.

PBL learning there are parts of the problem that learners must solve to train the ability of learners in reasoning, solving problems, finding and declared effective in improving the ability to solve problems [17]. In addition, it can provide opportunities for learners to interact with teachers and between learners with their groups [18]. Although the learning process is only through online meetings, learners look active during the learning process and work on problems that prove learners are responsible for the tasks assigned.

3.4 Stage Implementation

The effectiveness of an e-module development can be known through the learning results of knowledge N-*gain* in limited trials conducted after individual and small group trials. The analysis of this study was conducted the first test of sample prerequisites on data measurement results of learning knowledge before and after using *flipbook-based* e-modules. Unfortunately, the results obtained from learning knowledge before and after being treated are distributed abnormally. It suggests that the prerequisite test was not met; therefore, nonparametric analysis was used for hypothesis testing to determine differences in knowledge learning outcomes before and after using *flipbook-based* e-modules. Based on the hypothesis test results using nonparametric analysis with the Wilcoxon Signed Rank Test showing the rejection ofH0 and acceptance ofH1, there is a difference in the average learning outcome of knowledge before and after using *a flipbook-based* e-module. The following are the results of learning knowledge with the Wilcoxon Signed Rank Test, which can be seen as follows:

Table 1. Wilcoxon Signed-Rank Test

		N	Mean Rank	Sum of Ranks
posttest – pretest	Negative Ranks	0[a]	,00	,00
	Positive Ranks	23[b]	12,00	276,00
	Ties	0[c]		
	Total	23		

a. postest < pretest

b. postest > pretest

c. postest = pretest

Test Statistics[a]

	posttest – pretest
Z	-4,242[b]
Asymp. Sig. (2-tailed)	,000

a. Wilcoxon Signed Ranks Test

b. Based on negative ranks.

Furthermore, at this stage, learning activities are carried out based on the Learning Implementation Plan (P) using the PBL model. E-modules are effective if there is a change in learners' learning outcomes before*(pretest)*and after*(posttest)* using *flipbook-based* e-modules. The average learning score of knowledge from a limited trial conducted in class X MIPA 3 at SMAN 12 Banjarmasin is presented in Figure 5.

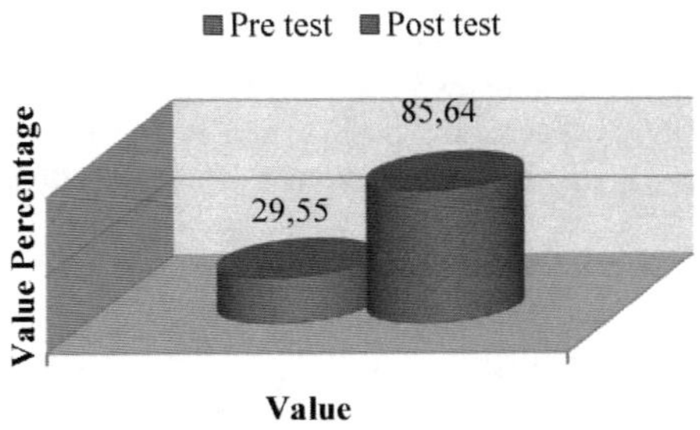

Figure 5 Limited trial *pre-test and post-test* values

Figure 5 shows that *obtained a pretest* value of 29.55 and a *posttest* average of 85.64; through these values, an increase from the learning outcomes of learners with an average N-*gain* of 0.80. It is because of the use of *flipbook-based* e-modules using the previously revised PBL model that limited trial classes experience an increase in learning outcomes.

PBL-based e-modules provide innovative learning resources for learners. Learning is more effective and engaging and facilitates learners to actively achieve learning goals, [19] thus increasing learners' scores from pretest to posttest on oxidation-reduction reaction material.

The advantages of developing flipbook-based e-modules applied to oxidation-reduction reaction materials with the ADDIE development model appeal to learners [20], making it easier for learners to understand learning materials and motivate and improve learning outcomes. Flipbook Maker-based e-modules effectively improve learning outcomes or achievement indexes and strengthen the students' character [21]. In addition to being attractive, learners can also increase motivation and interest in learning because it is easier to use in making electronic teaching materials [22].

The PBL model applied is also influential for the learning process because the PBL learning model can increase teacher activity, student activity, student affectivity, student thinking, critical thinking, knowledge learning outcomes, and students can respond well[23]. In addition, applying the Problem Based Learning model contributes significantly to learning outcomes so that it affects learners' learning outcomes on oxidation-reduction reaction material [24].

Effectiveness to the learning outcomes of learners' attitudes was measured during the learning process at two meetings consisting of curiosity and cooperation. The percentage of the value of learning outcomes of the learner's attitude obtained is shown in Figure 6.

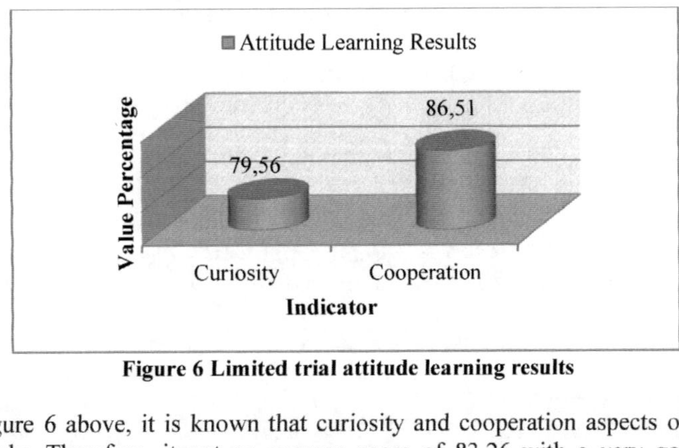

Figure 6 Limited trial attitude learning results

Based on Figure 6 above, it is known that curiosity and cooperation aspects obtained 79.56 and 86.51, respectively. Therefore, it got an average score of 83.26 with a very good category. It is undoubtedly affected by the use of *flipbook-based* e-modules using the PBL model with the role of teachers as facilitators who guide the learning process so that learners have good curiosity and cooperation.

Learners have an attitude of curiosity because the videos provided in the e-module can be studied further by learners; the questions that arise in the material and questions also add to learners' curiosity. Curiosity will arise when learners or students ask questions and seek answers from statements [25].

Furthermore, the attitude of cooperation of learners is evidenced by the attitude of students actively discussing following their respective groups. Because with the attitude of collaboration owned by learners, they will be easy to socialize with their friends. The attitude of cooperation is essential to be instilled in students to be easy to socialize wherever they are and foster an open mindset, and easy to accept differences [26].

3.5 Evaluation Stage

This stage activity is carried out to analyze the use of e-module-based learning devices at the implementation stage whether there are still shortcomings. If there are no deficiencies and revisions, then the learning device is suitable for use. Finally, the evaluation stage is evaluated, which includes formative evaluation in the form of individuals, small group evaluations, and field trials.

4. Conclusion

Based on the results of research and discussion, can be made conclusions as follows: (a) The e-module *flipbook* model of *Problem Based Learning* (PBL) on the redox material developed is very valid. In terms of content it is 91% (very feasible), presentation is 96% (very feasible), language is 95% (very feasible) and media is 94.5% (very feasible; (b) the e-module *flipbook* model *Problem-Based Learning* (PBL) on the redox material has met the practical category with an average score on the individual test of 3.5 (excellent), the small group test of 3.4 (excellent), student questionnaire responses of 3.3 (excellent) and teacher teaching activities using e-modules of 3.6 (excellent); (c) The e-module *flipbook* model *Problem-Based Learning* (PBL) on the redox material has met the effective criteria. In the limited trial class, the learning outcomes of knowledge with the N-gain value between *pretest* and *posttest* are 0.80 (height), and attitude learning outcomes are 83.26.

References

[1] Sari A N I and Suharto B 2017 *Journal of Chemistry And Education* **1** 1-14
[2] Yakina, Kurniati T and Fadhilah R 2017 *Ar-Razi Jurnal Ilmiah* **5** 287–97
[3] Raihana E, Iriani R and Leny 2017 *Journal of Chemistry And Education* **15** 8-64
[4] Mentari, M U 2014 *Comparative Study of Student Chemistry Learning Outcomes Using the PBL (Problem Based Learning) Learning Model and the TPS (Think Pair Share) Learning Model* (Universitas Bengkulu)
[5] Shilpa S and Sunita M 2016 *International Journal of Home Science* **2** 214-15
[6] Winatha K R, Suhartono N and Agustini K 2018 *Technology and Vocational Education Journal* **15** 188
[7] Dewi M S A and Lestari M A P 2020 *Scientific Journal of Education and Learning* **4** 433-41
[8] Prastowo A 2016 *Creative Guide to Making Innovative Teaching Materials* (Yogyakarta: Diva Press)
[9] Romayanti C, Sundaryono A and Handayani D 2020 *J. Educ. Chem.* **4** 51-8
[10] Sari W, Jufrida and Pathoni H 2017 *Journal of Physical Education* 1-10
[11] Sardadevi N, Winarti A and Leny 2017 *Journal of Chemistry And Education* **1** 195-203
[12] Yulinar and Suherman 2019 *Journal of Mathematics Education and Education* **8** 233-9
[13] Yildiz A and Baltaci S 2016 *The Online Journal of New Horizons in Education* 155-160
[14] Ariyanti P, Martini K S and Agustina W 2015 *Journal of Chemical Education (JPK)* **4** 1-9
[15] Ningras A S K 2017 *Influence of Problem Based Learning (PBL) Model Using Multimedia On Student Learning Outcomes Class X SMAN 7 Mataram Year of Study 2015/2016* (Universitas Mataram)
[16] Zhafirah T, Erna, M and Rery R U 2021 *Proceedings of the National Seminar on Research and Devotion*
[17] Zhafirah T, Erna, M and Rery R U 2021 *Proceedings of The National Research and Service Seminar*
[18] Elinawati W, Duda H J and Julung H 2018 *Jurnal Sainsmat* 13-24
[19] Jaenudin A, Baedhowi and Murwaningsih T 2017 *Advances in Social Science, Education and*

Humanities Research (ASSEHR) 30-36
[20] Anandari Q S *et* al 2019 *Pedagogical Journal* **6** 416-36
[21] Asmi A R, Surbakti A N D and Hudaidah 2018 *Social Science Education Journal* **27** 1-10
[22] Silitonga F S and Khoirunnisa F 2018 *Journal of Zarah* **6** 63-7
[23] Lathifah N H, Kusasi M and Rusmansyah 2019 *Journal of Chemistry and Education* **3** 1-9
[24] Janah M C, Widodo A T and Kasmui 2018 *Journal of Chemical Education Innovation* **12** 2097-107
[25] Puspitasari M T, Santoso S and Muchsini 2015 *UNS art Journal* **1** 31-9
[26] Rahayu D, Puspita A M I and Puspitaningsih F (2020). *Educational Research Journal* **7** 111-22

The practicality of natural science learning devices on the concept of environmental pollution with problem-solving learning models

Fahmi[1], **H Fajeriadi**[2], **Y Irhasyuarna**[3], **Suryajaya,**[4] **Abdullah**[5]

[1]Master Program of Natural Science Education, Postgraduate Program, University of Lambung Mangkurat, Banjarmasin City, South Kalimantan, Indonesia

[2]Master Program of Biology Education, Postgraduate Program, University of Lambung Mangkurat, Banjarmasin City, South Kalimantan, Indonesia

[3]Study Program of Natural Science Education, Departement of Mathematics and Natural Science Education, Faculty of Teacher Training and Education, University of Lambung Mangkurat, Banjarmasin City, South Kalimantan, Indonesia

[4]Study Program of Physics, Faculty of Mathematics and Natural Science, University of Lambung Mangkurat, Banjarbaru, South Kalimantan, Indonesia

[5]Study Program of Chemistry, Faculty of Mathematics and Natural Science, University of Lambung Mangkurat, Banjarbaru, South Kalimantan, Indonesia

heryfaje@gmail.com

Abstract. Less varied learning methods make learning activities boring. Emphasis on mastery of concepts and teacher dominance make students' activities passive. The correct methods are needed to achieve the learning objectives. This study aims to evaluate the practicality of natural science learning devices for junior high school on the concept of environmental pollution using a problem-solving learning model. The research method uses Design Development Research by Tessmer. The data was taken at the small group evaluation and field test. Data were collected through observation of teacher and student activities with a scale of 1-4; self-assessment with a score of 1 (positive) and 0 (negative); and student response questionnaires on a scale of 1-5. The data were categorized and analysed descriptively. The results of the small group evaluation are: teacher activities get a good category; student activities with good categories in the preliminary and closing activities, but main activities are still quite good; self-assessment aspects show eight positive and two negatives, and student responses show nine positive and one negative. The field test results are: teacher and student activities get a "very good" category; self-assessment aspects show eight positive and two negative, and student responses are positive for all aspects.

1. Introduction

Learning is a process of interaction between students, educators, and learning resources. The general purpose of learning is to improve students' ability and increase the learning experience [1]. In addition, the learning environment also affects the success of learning, including the *IPA Terpadu* or Integrated

Science Learning. Integrated Science Learning based on Permendikbud Number 58 of 2014 is integrated science learning from the subjects of Biology, Physics, and Chemistry.

Science learning so far has not fully met expectations. Teachers carry out integrated science learning with more emphasis on mastering facts and concepts [2], using less varied learning methods [3][4], and teacher dominance in learning activities [5][6]. Limited time, facilities, a learning environment, and too many students per class are why science learning is carried out using memorizing concepts, theories, and laws. This reason indirectly causes students not to be accustomed to critical thinking [7]. As a result, students' interest in Integrated Science learning is low because it is considered that this lesson is boring.

In the PISA (Programme for International Student Assessment) study, Indonesian students' scientific ability in 2000, 2003, 2006, 2009, and 2012 has an average score of 393, 395, 395, 383, and 382 [8]. This result is below the international average score of 500, showing that the scientific ability of Indonesian students is still shallow and has only just arrived at the ability to recognize some basic facts and have not been able to communicate and relate these abilities in abstracts. Students are not trained to think critically and only memorize formulas compared to understanding concepts [9].

Critical thinking involves sorting out valuable ones from many ideas or making judgments before making decisions [10] [11]. There are five critical thinking indicators: identifying the problem, defining the problem clearly, exploring the problem and possible solutions, evaluating its application, and integrating understanding with existing knowledge [12]. In order to fulfill these critical thinking indicators, strategic learning planning is needed.

Experts in the World have developed and produced many learning strategies. This learning strategy aims to provide an alternative for educators to develop or improve students' problem-solving and thinking skills. One of the learning strategies in question is the Problem-solving Learning Model [13]. This learning model supports higher-order thinking activities, such as asking questions, finding problems, making hypotheses, planning investigations, expecting the desired results, conducting experiments, collecting and evaluating data, concluding, suggesting ideas, evaluating other ideas, and looking for alternative interpretations [14]. Learning with the problem-solving model occurs when students try to solve problems [15]. Skills in solving problems are done by proposing alternative solutions and testing their feasibility [11]. The ability to solve this problem includes understanding the problem, finding solutions, and communicating the results of the problem-solving experiment [16]. Students will always think critically and suggest ideas and solutions throughout the problem-solving process in learning activities with this model [17]. Learning activities using the Problem-solving Learning Model can be the right strategy applied to science subjects.

Not only critical thinking skills, but science learning also requires students to develop science process skills. This skill is essential for students to practice being honest, thorough, and process the information they have [18]. Science process skills consist of several activities, including observing, predicting, concluding, classifying, communicating, and measuring [19]. Science process skills include formulating problems, identifying variables, presenting data, analyzing data, formulating hypotheses, and conducting investigations or experiments [20][21].

Students need science to have critical thinking and problem-solving skills, become lifelong learners, and increase curiosity about the environment [17]. Environmental pollution is one of the main subjects in science subjects that are often discussed. This material is directly related to the current environmental issues. As in SMP Negeri 19 Banjarmasin, the lesson plans have not optimally implemented a problem-solving approach related to environmental issues. Analysis of the impact of learning activities on students' critical thinking skills is often overlooked.

The lesson plans need to be tested before being implemented, and the implementation results also need to be evaluated. In order to solve this problem, the researcher intends to develop learning devices that can improve critical thinking skills and scientific processes. The implication is that students can get used to suggesting ideas and solutions to environmental pollution problems.

2. Method

This study emphasizes the prototype phase of the development research design model. The research method uses a formative evaluation model from Tessmer. Tessmer's formative evaluation model consists of five steps: Self Evaluation, Expert Review, One-to-one Evaluation, Small Group Evaluation, and Field Test. The research is limited to the practicality test of the developed learning device [22].

The research was conducted at SMP Negeri 19 Banjarmasin. The students involved in the One-to-one Evaluation and Small Group Evaluation stages came from class VII C. Furthermore, the class used in the Field Test stage was Class VII A. The research subjects are described in table 1.

Table 1. Place of practicality in formative methods of evaluation

		Written Test-Based		Final Version	
		Expert Review (n=3)	One-to-one Evaluation (n=3)	Small Group Evaluation (n=5)	Field Test (n=30)
Practicality	Content		√wt	√to	√ft
	Interface		√wt	√to	√ft

Information:
*) = Content refers to the content of the support system
√ = primary attention of prototype and of formative evaluation Methods of formative evaluation: wt = walkthrough; ea = expert appraisal; ft = field trial; to = try-out

Practicality data were obtained from several instruments at several stages of formative evaluation, as shown in Table 1 above. The content practicality data was obtained at the One-to-one Evaluation stage using the students' readability instrument. The practicality of expectations is obtained at the Small Group Evaluation stage using the instrument of observing the implementation of learning devices, observing teacher activities, observing student activities, self-assessment, and student responses. Then, the actual practicality was obtained at the Field Test stage using the same instrument as the Small Group Evaluation stage.

Data analysis was carried out in several ways. The implementation data were analyzed using a Likert scale with a score of 1-4. The score for observing teacher and student activities uses an assessment rubric. The practicality analysis on the self-assessment questionnaire uses the Guttmann scale with a score of 1 if positive and 0 if negative. The analysis of practicality on students' questionnaire responses to learning activities was carried out using a Likert scale with a score of 1-5.

3. Results and Discussion

The product developed was an Integrated Science Learning device for the main subject of environmental pollution. The learning devices consisted of Syllabus, Material Analysis, Lesson Plans, Teaching Materials, Student Worksheets, and Assessment Sheets. The learning devices mentioned have gone through the Expert Review stage with valid results and can be used in the next stage, namely the practicality test.

The practicality of learning devices is divided into three: the content practicality, the expectations practicality, and the actual practicality, which are detailed as follows.

3.1 Content Practicality

The content practicality aspect seen from the user can easily understand the material presented [22]. The content practicality of the learning devices developed can be seen in table 2 and table 3. The content practicality includes the results of the student's readability test on teaching materials and Student Worksheets. The content practicality aims to review whether the content is easy to understand so that the implementation stage focuses on the use of the developed product.

Table 2. Assessment of teaching materials readability

No	Observed Aspects	Number of Students Assessing		Rating result
		Yes	Not	
1	Is the content of the teaching material interesting?	3	0	Interesting
2	Is the appearance of the teaching material attractive?	3	0	Interesting
3	The type of font used by comic sans	3	0	Interesting
4	Is the font size 12pt?	3	0	Interesting
5	The legibility and language used are appropriate for the age of the students	3	0	Already well
6	The terms used are precise and understandable	3	0	Already well
7	Use symbols and terms consistently	3	0	Already well
8	Use communicative language	3	0	Already well
9	Use effective language	3	0	Already well
10	In your opinion, is the description or explanation in this teaching material too difficult?	1	2	No

Table 3. Assessment of student worksheets readability

No	Observed Aspects	Number of Students Assessing		Rating Result
		Yes	Not	
1	Is the content of the Student Worksheets interesting?	3	0	Interesting
2	Is the appearance of the Student Worksheets attractive?	3	0	Interesting
3	The font type used by Book Antiqua	3	0	Already well
4	Is the font size 13 pt?	3	0	Already well
5	The legibility and language used are appropriate for the age of the students	3	0	Already well
6	The terms used are precise and understandable	3	0	Already well
7	Use symbols and terms consistently	3	0	Already well
8	Use communicative language	3	0	Already well
9	Use effective language	3	0	Already well
10	In your opinion, is the description or explanation in this Student Worksheets too difficult?	1	2	No

One-to-one evaluation of the developed learning devices, including teaching materials' readability and student worksheets, obtained a very good response. Sari has conducted a similar study about the content practicality of the science learning device for liquid materials to practice science process skills. The results of this study indicate that the learning devices developed are easy to use [23]. A similar study by Arsyad on the readability test of student worksheets shows that there are still tricky terms to understand, and it is necessary to choose more attractive images. The suggestions received at this stage have been improved [24]. Research on the development of high-level skills-based learning devices by Randa obtained an average readability response of students, namely practical [25].

3.2 Expectations Practicality

The expectations practicality includes the results of observing teacher activities, observing student activities, self-assessment by students, and student responses at the small group evaluation stage. This expectations practicality becomes the basis for determining whether the developed product can be continued at the field test stage. The expectation's practicality can be seen in table 4 to table 5.

Table 4. Observation of student activities at the small group evaluation stage

No	Activity types	Average Score Each Meeting				Average	Max Score	%	Category
		1	2	3	4				
1.	Preliminary	2	3	3	4	3	4	75	Good
2.	Main	2	2,3	2,5	2,8	2,4	4	60	Good Enough
3.	Closing	2	3	3	4	3	4	75	Good

The activities of students in small group evaluation in the preliminary and closing stages are in a good category, while the main stage for student activities are in the fairly good category. Meanwhile, student activities in main activities are still in a good enough category because students are still not used to using the problem-solving learning model. They must adjust first even though the teacher has been a good facilitator. These results need to be tested further through field tests.

Table 5. Student self-assessment at the small group evaluation stage

No	Observed Aspects	Score	%
1	I always work with friends in a group when doing group assignments	(+)	100
2	I record data carefully and according to facts	(+)	80
3	I perform tasks according to a predetermined schedule	(+)	80
4	I complete assignments based on various literature related to the subject matter being studied	(+)	100
5	I really like science lessons	(+)	100
6	I like science practicum activities	(+)	100
7	I always do my homework at home	(-)	40
8	I'm lazy to do my science homework	(+)	80
9	I have the ability to think critically	(+)	80
10	I have science process skills in learning science	(+)	80

Self-assessment at the Small Group Evaluation stage of ten aspects observed, nine aspects stated positive, and one aspect in number seven stated negative, namely the aspect of doing science homework at school. This aspect number seven obtained a negative value because only a small number of students did their homework at home according to the learning procedure. These results still need to be tested further through the field test stage.

Student responses at the Small Group Evaluation stage of the ten aspects observed: nine aspects state positive and one aspect in number four state negative. Although science learning devices can motivate students to think critically, students do not feel challenged to think critically in solving problems on the subject matter being taught. These results need to be tested further at the Field Test stage to determine their actual practicality.

3.3 Actual Practicality

Actual practicality can be seen in table 6 to table 7. Actual practicality includes the results of self-assessment by students and student responses at the Field Test stage.

Teachers and students have been able to carry out their roles very well so that learning goes as expected. In line with Sulistyanngkarti's research, students' involvement and mastery of concepts in the process of learning activities can be an indicator of learning success [26]. The results of observations by Tampubolon on the activities of teachers and students in implementing problem-based learning devices to improve problem-solving abilities of high school students obtained practical criteria. The results of this observation are considered a reference that the developed learning device can be continued on the effectiveness test [27].

Tabel 6. Student self-assessment at the field test stage

No	Observed Aspects	Score	%
1	I always work with friends in a group when doing group assignments	(+)	100
2	I record data carefully and according to facts	(+)	93
3	I perform tasks according to a predetermined schedule	(+)	100
4	I complete assignments based on various literature related to the subject matter being studied	(+)	90
5	I really like science lessons	(+)	100
6	I like science practicum activities	(+)	97
7	I always do my homework at home	(-)	30
8	I'm lazy to do my science homework	(-)	23
9	I have the ability to think critically	(+)	77
10	I have science process skills in learning science	(+)	90

Self-assessment at the Field Test stage of the ten aspects observed, eight aspects stated positive, and two aspects in numbers seven and eight stated negative, namely the aspects of doing homework at school and being lazy to do science homework. Suarta and Gusti agreed that students, through self-assessment, could see their strengths and weaknesses, and then the assessment results became an improvement goal [28][29].

Table 7. Student responses at the field test stage

No	Observed Aspects	%
1	Learning devices with the Problem-solving model can improve critical thinking skills, so I am very interested in learning about the subject matter of environmental pollution.	92
2	Learning devices with the Problem-solving model can improve science process skills, so that I enjoy learning the subject matter of environmental pollution.	91
3	Learning devices with Problem-solving models motivate me to improve critical thinking skills in studying the subject matter of environmental pollution.	89
4	Learning devices with the Problem-solving model challenged me to think critically in solving problems on the subject matter of environmental pollution.	94
5	Learning devices with the Problem-solving model can improve my scientific process skills to solve problems on the subject matter of environmental pollution.	92
6	Learning devices with the Problem-solving model, which involves practical activities and discussions, allows me to think critically in solving problems related to the subject matter of environmental pollution.	96
7	Learning devices with the Problem-solving model that involves practical activities and discussions allow me to improve science process skills in solving problems related to the subject matter of environmental pollution.	92
8	Learning activities in small groups on the Problem-solving model allow me to work together with friends in groups. They can improve my ability to think critically to solve problems related to the subject matter of environmental pollution.	92
9	Learning activities in small groups on the Problem-solving model allow me to work together with friends in groups. They can improve science process skills in solving problems related to the subject matter of environmental pollution.	92
10	Learning devices with the Problem-solving model can make me express ideas without any limitations so that it further enhances the critical thinking skills that I previously had.	94

Observation of student responses at the Field Test stage obtained positive results for all aspects. Problem-solving models influence positive student responses to improve critical thinking skills and science process skills. The feeling of being challenged to think critically initially received a negative response at the small group evaluation stage, but it received a positive response at this stage. In line with Hidjrawan's research, the response given by students to the problem-solving learning model is positive with good criteria [30]. Students' positive responses arise because they are given the freedom to be more active in asking questions, giving opinions, finding problems, analyzing, discussing, and solving problems independently [31][32].

4. Conclusion

Based on the results and discussion, the conclusions from the implementation of science learning devices, with the concept of environmental pollution, use the problem-solving learning model. The practicality of the contents of the one-to-one test results obtained very good criteria. The expectation practicality includes observing teacher and student activities as a whole obtained good criteria, although the main activities of student activities were still good enough. Nine out of ten aspects of self-assessment scored positive. Nine out of ten aspects of student responses received positive scores. The actual practicality includes the observation of teacher and student activities as a whole obtained very good criteria. Eight of the ten aspects of self-assessment scored positive. All aspects of student responses received positive scores. The results of each test show that the science learning devices developed are practical and can be continued in the effectiveness test.

References

[1] Fahmi and Irhasyuarna Y 2019 *Pengantar Pendidikan: Manusia, Pendidikan, dan Perkembangan Zaman* (Banjarmasin: Program Studi Magister Keguruan IPA PPs ULM)
[2] Af'idayani N, Setiadi I and Fahmi 2018 *Eur. J. Educ. Stud.* **4** 177–82
[3] Rahayu A B, Hadi S, Istyadji, Zaini M, Sholahuddin A and Fahmi 2018 *Eur. J. Altern. Educ. Stud.* 107–17
[4] Suryani E, Ws R and Nugraha A 2018 *PEDADIDAKTIKA J. Ilm. Pendidik. Guru Sekol. Dasar* **5** 100–8
[5] Adilah D N and Budiharti R 2015 *Pros. Semin. Nas. Fis. dan Pendidik. Fis. Ke-6* vol 6 212–7
[6] Yasiro L R, Wulandari F E and Fahmi 2021 A *J. Banua Sci. Educ.* **1** 69–72
[7] Ningsih N W 2013 *Pendidik. Sains* **1** 54–9
[8] Setiadi D 2014 *J. Pijar Mipa* **9** 1–8
[9] Putri M H, Fahmi and Wahyuningsih E 2021 *J. Banua Sci. Educ.* **1** 79–84
[10] Fahmi, Setiadi I, Elmawati D and Sunardi 2019 *Eur. J. Educ. Stud.* **6** 342–51
[11] Tawil M and Liliasari 2013 *Berpikir Kompleks dan Implementasinya dalam Pembelajaran IPA* (Makassar: Badan Penerbit UNM)
[12] Fahmi 2018 *Pengembangan Perangkat Pembelajaran untuk Melatihkan Keterampilan Berpikir Kritis Peserta Didik SMP pada Materi Klasifikasi Benda* (Universitas Lambung Mangkurat)
[13] Mukhopadhyay R 2013 *IOSR J. Humanit. Soc. Sci.* **8** 21–5
[14] Alebous T and Ayasra A H 2016 *Int. J. Educ.* **8** 1
[15] Huda M 2019 *Model-Model Pengajaran Dan Pembelajaran Isu-Isu Metodis Dan Paradigmatis* (Yogyakarta: Pustaka Pelajar)
[16] Greiff S, Holt D V. and Funke J 2013 *J. Probl. Solving* **5** 71–91
[17] Seyhan H G 2014 *Eurasian J. Phys. Chem.* **6** 142–61
[18] Prasasti P A T 2017 *Profesi Pendidik. Dasar* **4** 19–26
[19] Delismar D, Asyhar R and Hariyadi B 2013 *Edu-Sains J. Pendidik. Mat. dan Ilmu Pengetah. Alam* **2**
[20] Hidayat M L, Rufaida A D and Abadi R 2013 *Ilmu Pengetahuan Alam SMP/MTs Kelas VII Semester 1* (Jakarta: Intan Pariwara)
[21] Suryaningsih Y 2017 *Bio Educ.* **2** 279492
[22] Akker J van den, Bannan B, Kelly A E, Nieveen N and Plomp T 2007 A *Proc. Semin. Conducted at the East China Normal University, Shanghai University, P.R.Chinae East China Normal University, Shanghai University, P.R.China* (Shanghai (PR China): Enshede the Nether-lands: Netzodruk, Enschede) pp 23–6
[23] Sari R N, Ibrahim M and W. W 2016 *Pendidik. Sains Pascasarj. Univ. Negeri Surabaya* **5** 658–71
[24] Arsyad M, Zaini M and Aziati K 2019 *Pros. Semin. Nas. Pendidik. Fis. MOTOGPE*
[25] Randa M, Maimunah M and Yuanita P 2020 *J. Cendekia J. Pendidik. Mat.* **4** 1197–206
[26] Sulistyaningkarti L, Utami B and Haryono H 2016 *J. Pendidik. Kim.* **5** 1–9
[27] Tampubolon R, Sinaga B and Mulyono M 2021 *Parad. J. Pendidik. Mat.* **14** 16–28
[28] Gusti D A and Ratnawulan R 2021 *J. Penelit. Pembelajaran Fis.* **7** 77–84
[29] Suarta I M, Hardika N S, Sanjaya I G N and Arjana I W B 2015 *J. Penelit. dan Eval. Pendidik.* **19** 46–57
[30] Hidjrawan Y and Khaldun I 2016 *J. Pendidik. Sains Indones.* **4** 121645
[31] Irwandi I and Fajeriadi H 2020 *BIO-INOVED J. Biol. Pendidik.* **1** 66
[32] Komisia F, Tukan M B and Uron Leba M A 2021 *Indones. J. Educ. Sci.* **3** 98–104

Development of e-modules coloid materials based on PjBL-STEM to improve scientific literature and student learning outcomes of wetlands context

Rusmansyah, N Anjaini and M Kusasi

Chemical Education Study Program, FKIP, Lambung Mangkurat University.
Jl. Brigjend. H. Hasan Basry, Banjarmasin 70123, South Kalimantan, Indonesia.

rusmansyah@ulm.ac.id

Abstract. Development research has been carried out that aims to produce an e-module, a valid, practical, and effective STEAM-PJBL approach on colloidal material. This type of research is Research and Development (R&D) with a 4D development model modified into 3D (define, design, development). The e-module trial was carried out in the School Chemistry 2 class, which consisted of 22 students. Data collection techniques used validation sheets, readability questionnaires, response questionnaires, observation sheets, scientific literacy tests, and learning outcomes tests. Data analysis used descriptive analysis. The study results indicate that the developed e-module has met the following criteria: (1) Validity, judging from the value of the module feasibility aspects, has very valid criteria. (2) Practicality, seen from the readability results of e-modules on individual and small group tests obtaining practical and very practical criteria, student responses to e-modules obtained very practical criteria, as well as observations of lecturers' ability to use e-modules and manage their respective classes. -each obtained very practical criteria. (3) Effectiveness, seen from the results of the N-gain scientific literacy and knowledge learning outcomes, each of which shows an increase in height and the criteria for learning outcomes of attitudes and skills to obtain good and very skilled criteria.

1. Introduction

The state of Indonesia has entered the 21st century, characterized by the increasing demand for human resources (HR) that have high quality. Therefore, for a prospective teacher, in addition to being required to master the scientific to be taught, it is also necessary to have a set of knowledge and technical skills.

Based on the results of research from the *Programme for International Student Assessment* (PISA) show that the scientific literacy skills of Indonesian learners, in general, are below average when compared to the average international score [1]. Therefore, for prospective teacher students, it is essential to have high scientific literacy skills to be later able to improve the scientific literacy skills of learners.

The selection of models and approaches to informal learning can train scientific literacy skills and student learning outcomes. The model and approach chosen is the *project-based learning* model with the STEAM approach (STEAM-PjBL Approach). The *project-based learning* model is innovative, encouraging students collaboratively to research and create projects that apply their knowledge from discovering new things, proficient in the use of technology, and able to solve a real-world problem [2]. In addition, combining art with STEM subjects with the STEAM approach can enhance student engagement, creativity, innovation, problem-solving skills, and other cognitive benefits [3].

The use of teaching materials in the form of exciting modules can increase the success of a learning process, especially the influence of technological developments. Modules are often developed in the form of e-modules (electronic modules). One application of e-module creation is *flipbook-based,* where it looks like a digital book that looks more attractive.

Objects that are used as problems in every learning material can be sourced from the environment around students. One of them is sourced from the context of wetlands, which is a large land owned by South Kalimantan [4]. For example, one part of the branch of chemistry that can directly contact the wetland environment is the sub-material of the Chemistry School 2 course, colloidal. Based on these exposures, researchers are interested in researching the development of the e-module STEAM-PJBL approach in a wetland context on colloid materials to improve scientific literacy and student learning outcomes.

2. Methods

Research *and Development* (R&D) uses a 4D development model modified into 3D, namely *define, design,* and *development* [5]. The *dissemination* stage is not carried out because the research is done close to the end of the even semester of the 2020/2021 school year, so it does not have enough time. The research was conducted at the Chemical Education Study Program of FKIP ULM Banjarmasin in the 2nd semester of class A2 of the 2020/2021 school year for individual trials and small group trials, as well as students of semester 2 of class A1 of the 2020/2021 school year for limited trials.

Data collection techniques used non-test instruments that include validation questionnaires, readability questionnaires, student response questionnaires, observation sheets of lecturers' abilities using e-modules and managing classes, and observation sheets to measure student attitude learning outcomes and skills. Then, test instruments in the form of essay problems to measure the ability of scientific literacy and multiple-choice questions to measure the learning outcomes of student knowledge given before and after treatment. Data analysis used descriptive analysis—determination of percentage value (P) in the validity and effectiveness test using the following formula.

$$Percentage\ value\ (P) = \frac{The\ total\ score\ given}{Total\ overall\ score}\ x\ 100\%$$

The percentage of values obtained for the validated test is interpreted with Table 1. the following criteria [6].

Table 1. E-module validity criteria

Percentage value (P)	Validated criteria	Information
85,01 - 100,00	Very valid	No need to revise
70,01 - 85,00	Valid	No need to revise
50,01 - 70,00	Less valid	Minor revisions
01,00 - 50,00	Invalid	Major revisions

The average score obtained for the practicality test is interpreted in Table 2. The following criteria [7].

Table 2. E-module practicality criteria

Average score ($\bar{x}$)	Criteria
$\bar{x} > 3{,}25 - 4{,}00$	Very practical
$\bar{x} > 2{,}50 - 3{,}25$	Practical
$\bar{x} > 1{,}75 - 2{,}50$	Less practical
1,00 – 1,75	Impractical

The percentage of scores obtained for the effectiveness test is reviewed from the ability of scientific literacy to be interpreted with Table 3. the following criteria [8].

Table 3. Criteria for scientific literacy skills

Percentage value (P)	Criteria
86,00 - 100,00	Very high
72,00 - 85,00	Tall
58,00 - 71,00	Keep
44,00 - 57,00	Low
0,00 - 43,00	Very low

The percentage of values obtained for the effectiveness test is reviewed from the learning outcomes of knowledge interpreted with Table 4. the following criteria [9].

Table 4. Criteria for learning knowledge results

Percentage value (P)	Criteria
P ≥ 80.00	A
77.00 - P < 80.00	A-
75.00 - P < 77.00	B+
70.00 - P < 75.00	B
67.00 - P < 70.00	B-
67.00 - P < 67.00	C+
60.00 - P < 64.00	C
50.00 - P < 60.00	C-
40.00 – P < 50.00	D
0.00 – P < 40.00	E

The percentage of scores obtained for the effectiveness test is reviewed from the results of attitude learning interpreted with Table 5 [10].

Table 5. Criteria for attitude learning outcomes

Percentage value (P)	Criteria
P > 81.25 – 100.00	Very good
P > 62.50 - 81.25	Good
P > 43.75 - 62.50	Pretty good
25,00 – 43,75	Less good

The percentage of scores obtained for the effectiveness test is reviewed from the learning outcomes of the skill interpreted with Table 6. the following criteria [10].

Table 6. Skills learning outcome criteria

Percentage value (P)	Criteria
P > 81.25 – 100.00	Very skilled
P > 62.50 - 81.25	Skilled
P > 43.75 - 62.50	Quite skilled
25,00 – 43,75	Less skilled

3. Research and Discussion Results

Colloidal material e-modules that have been developed using 3D model stages are presented as follows:

3.1. Defining Stage (Define)

Activities at the stage of definition that is done there are several, namely can be described as follows: (a) *Early analysis*. The final initial research is done through field studies and literature studies that aim to find out the fundamental problems students face in learning related to colloidal materials, such as the difficulty of understanding the material because of the lack of exciting teaching materials. (b) *Student analysis*. Student analysis includes the analysis of academic ability and motivation of students who aim to study the characteristics of students by design to be developed such as students can create their projects and have adequate facilities to use smartphone devices. (c) *Task analysis.* Task analysis aims to thoroughly review what tasks students will undertake through developed e-modules. Task analysis includes tasks that train scientific literacy skills and student learning outcomes in the context of wetlands. (d) *Concept analysis*. Concept analysis aims to identify concepts and determine the material's content in e-modules developed to be presented in the concept map. (e) *Analysis of learning objectives*. Analysis of learning objectives is made by adjusting existing tasks and concepts then linking to the behaviour of the subject to be achieved, namely improving scientific literacy and student learning outcomes.

3.2. *Planning Stage (Design)*
Activities at the stage of design that is done there are several, namely can be described as follows: (a) Preparation of tests. The test preparation is the stage for preparing the essay problem for instrument scientific literacy and the multiple-choice question for learning instruments; (b) Media drafting. The media used to make e-modules is a flipbook, where the initial file form of the e-module was pdf. Then converted into the form of a digital book with additional interesting views and symbols; (c) Formatting. Format selection includes selecting material content, selection of strategies, learning methods, and learning models contained in e-modules; (d) Initial design. This stage is the first design to be done to design e-modules developed.

3.3. Development Stage (Develop)
Activities at the stage of development are carried out there are several, namely can be described as follows:

3.3.1 Validation of the product by an expert
The validity test aims to determine the level of product validity measured based on validator assessment consisting of 4 lecturers of Chemical Education FKIP ULM Banjarmasin as material experts and one lecturer of Educational Technology FKIP ULM Banjarmasin as a media expert. The results of the validator's assessment are shown in Figure 1.

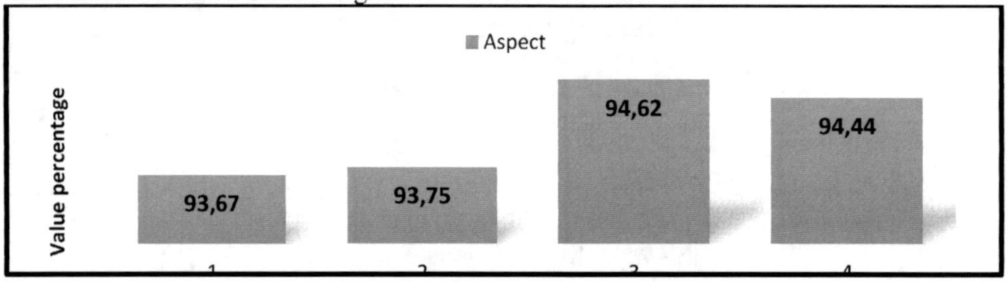

Figure 1. E-module eligibility validation results

Based on Figure 1, it is known that the e-module obtained a category worth using without any revision from the validators. Aspects of the content of the e-module have presented the content of the material appropriately and accurately, have other supporters to clarify the material, and the content of the material proved to be final. In the aspect of the presentation that has been arranged with good presentation techniques and equipped with various other supporters, the form of learning encourages students to be interactive and participative, and e-modules are arranged according to the rules that consist of preliminary content and closing sections.

Aspects of language have been presented with straightforward, communicative, dialogical, interactive vocabulary, following the student's intellectual development and social-emotional level and the consistent use of terms, symbols, or icons. The media aspect has presented an attractive look and content and has been by the characteristics of an e-module.

Field trials. Field trials include individual, small group, and limited trials. Individual and small group trials are conducted to determine the practicality of e-modules reviewed from the e-module readability questionnaire. The average score of individual and small group trial readability scores is shown in Figure 2.

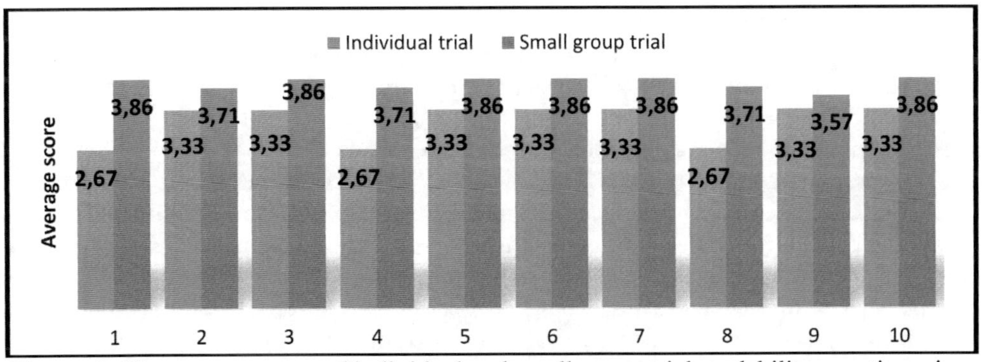

Figure 2. The average score of individual and small group trial readability questionnaires

Based on Figure 2, it is known that the individual test obtained an average overall score of 3.20 in practical criteria, and the small group test obtained an overall score average of 3.68 in efficient criteria. In individual trials, its suggestions with the cover's design, the colour combination of look, and the material content. Therefore, before being retested in small group trials, revisions were made by changing the cover's design, color combinations for the display, and adding videos to clarify the material's content. Choosing a harmonious and exciting color combination can give students its appeal, as well as another purpose of this visual display is to facilitate the delivery of messages [11]. Then, video in learning is a medium to demonstrate some material concepts, help stimulate students' curiosity about the chemicals to be studied and increase students' interest in learning [12].

The assessment readability of the e-module is based on several things, such as the number of clear and non-blurry images to clarify the content. As revealed by Amir [13], the help of images helps facilitate the learning process while achieving learning goals. Writing sentences in e-modules is easy to understand because there is no discovery of misinterpretation due to miswriting. Arizona [14] states that if there are spelling errors and improper use of punctuation in writing, then the meaning of the writing can be different from its purpose or purpose, causing the sentence to be challenging to understand. The e-module contains *quiz time* and formative tests that are easy for students to understand and answer to improve their understanding and learning motivation. Puryati [15] expressed that when students can answer the problem, there will be a sense of satisfaction and make them more successful in learning.

Limited trials are conducted to determine the practicality and effectiveness of e-modules. The practicality of e-modules is reviewed from student response questionnaires, observation sheets of lecturers' abilities using e-modules, and managing classes. Meanwhile, the effectiveness of e-modules is examined from the ability of scientific literacy and student learning outcomes. The average student response score is shown in Figure 3.

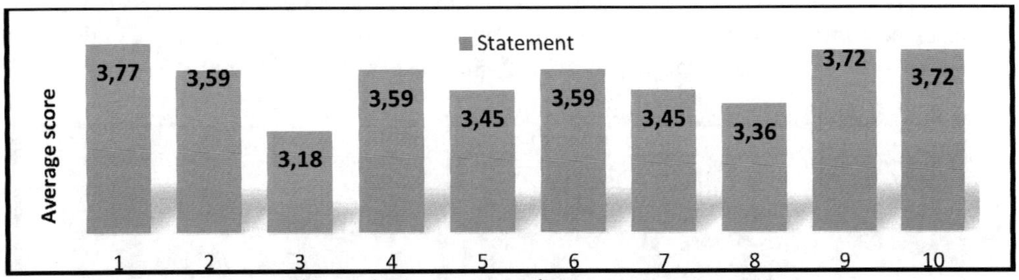

Figure 3. Average student response score

Based on Figure 3, students showed a response that the product developed was convenient to use. The use of e-modules can increase the willingness of students to follow the learning. Zaharah & Susilowati [16] stated that learning using electronic modules could improve student learning motivation. The preparation of e-modules using communicative language makes it easier for students to understand the material's content. The e-module itself has run a syntactic STEAM-PJBL approach not to cause students to feel bored during learning. As expressed by Rozikin, Lesmono, & Bachtiar [17], that through the application of project-based learning models has a significant influence on learning interests, one of which is characterized by the attitude of students more concentration and attention when the following learning because they feel they have a significant role in learning, primarily related to the task of doing projects with the STEAM approach.

The use of e-modules encourages students to study independently. As revealed by Setiyadi, Ismail, & Gani [18], that e-modules have *self-instruction* properties characterized by the inclusion of several command sentences that are easy to understand, and there are training questions, there are summaries of materials and contain the contents of learning materials packaged in small units, making it easier to learn thoroughly. These e-modules make students happy to study colloidal materials because it discusses the interrelationships of wetland-textured daily life. Yildiz & Baltaci [19] revealed that contextual learning is more attractive to students because it has a connection from the experiences they have experienced.

The ability of lecturers to use e-modules is seen from the assessment of 3 observers based on three indicators, namely e-module instructions, e-module content, and ease of use of e-modules. The average score of each indicator of a lecturer's ability to use e-modules is indicated in Figure 4.

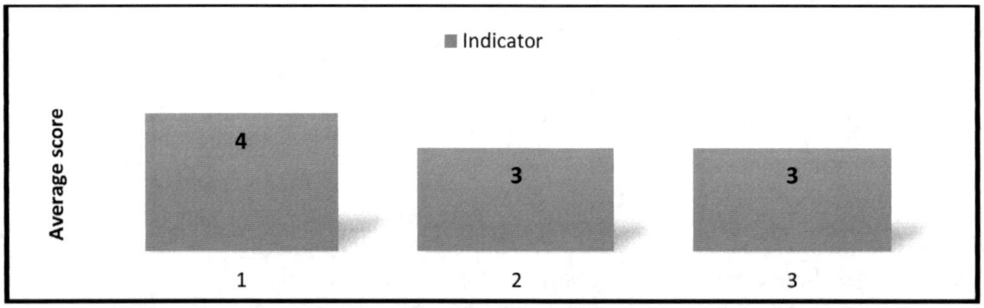

Figure 4. Average indicator score of lecturer's ability using e-modules

Based on Figure 4, it is known that the ability of lecturers using e-modules obtains very practical criteria that show these abilities are done very well. Examples of the use of e-modules accompany instructions submitted by lecturers. Providing optimal direction and guidance from lecturers will make students make the best use of their potential in following the lecture process [20].

The content indicator and ease of use of the e-module acquire practical criteria. Lecturers share e-module links through the Whataspp group, then convey the essential points of the subject matter in the e-module so that students can independently learn more about the content of the material. As Johannes [21] expressed, lecturers act as facilitators, so there is no need to intervene too deeply and do not take away students' right to study in the true sense.

The ability of lecturers to manage classes is seen from the assessment of 3 observers based on four dimensions, namely introduction, core, closing, and time allocation. The average score of the acquisition from the statement of the lecturer's ability to manage the class is shown in Figure 5.

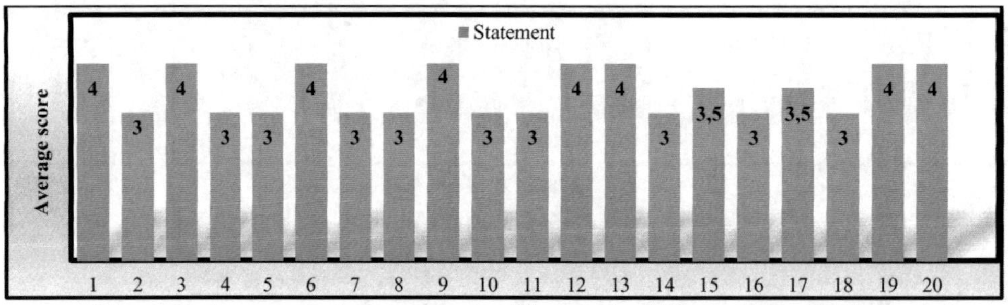

Figure 5. Average statement scores of lecturer's ability to manage classes

Based on Figure 5, it is known that the ability of lecturers to manage classes obtains very practical criteria and shows that ability is done very well. In the preliminary dimension, before entering the core of learning, lecturers instruct students to open e-modules that have been shared. The core dimension is the stage of running the syntactic of the project-based learning model with the STEAM approach. According to Hendra, Arsa, & Krisnawati [22], the project-based learning model is meaningful learning that brings students an impressive learning experience. Moreover, the experience is obtained from the problem of wetland texture. In the closing dimension, the student concludes. According to Rahmawaty, Nurhayati, & Arsyad [23], learning by discovering itself concepts and meaningful can result in a good understanding of concepts and retention of students at the end of learning. Finally, the allocation of time is carried out following the RPP that has been made. Yatmini [24] expressed that RPP is an absolute requirement for implementing a conducive learning process.

The effectiveness of e-modules is reviewed from the scientific literacy skills of students judging from the N-gain obtained. The results of N-gain of students' scientific literacy skills are shown in Figure 6.

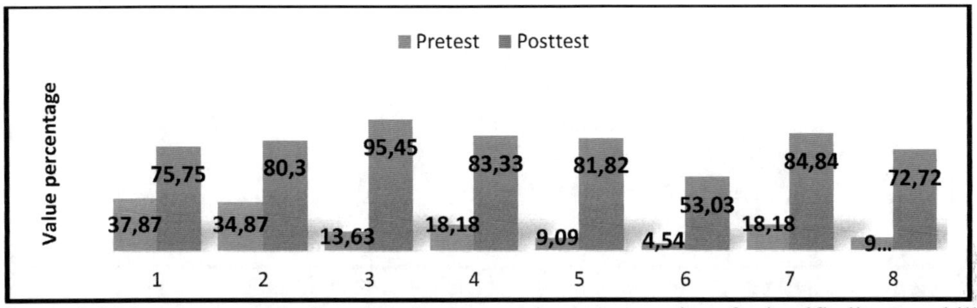

Figure 6. Value percentage of students' pretest and posttest grades of scientific literacy skills

Based on Figure 6, N-gain is obtained by 0.75 and shows an increase in the percentage of the value obtained from pretest to high posttest. Thus, according to Khotimah, Suhirman, & Raehanah [25], that through the application of project-based learning models can improve students' scientific literacy skills and also supported by the results of Rusyati, Permanasari, & Ardianto [26], that through learning with STEM approach can train students' scientific literacy skills.

The application of the project-based learning model trains students to solve problems through the stages of the search, analysis, and synthesis of results that will be delivered to other students. Wulandari [27] expressed that through the application of project-based learning models, students will explore, assess, interpret, and synthesize to produce various forms of learning outcomes. The application of the steam approach trains students to build their *knowledge scientifically*. As revealed by Munawar, Roshayanti, & Sugiyanti [28], the steam approach is seen as an approach that encourages students to develop curiosity and ask questions so that students can build knowledge around their world by exploring, observing, discovering, and investigating how things happen. Through the implementation of student-centred learning patterns, they can increase their learning activities [29].

Effectiveness on the learning outcomes of student knowledge is seen from the N-gain obtained. The results of N-gain student knowledge learning results are shown in Figure 7.

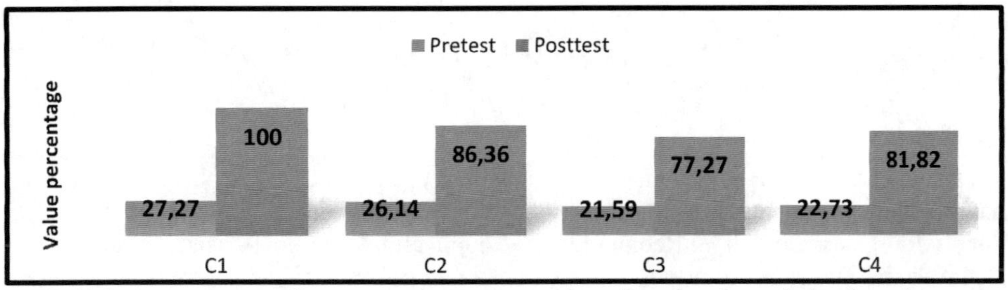

Figure 7. Value percentage of pretest and posttest grades of student knowledge learning outcomes

Based on Figure 6, N-gain obtained by 0.79 shows an increase in the percentage of the value obtained from pretest to high posttest. Thus, based on the results of Ma'sumah & Mitarlis's [30] research, that through learning that applies the project-based learning model with STEM approaches successfully affects student learning outcomes by showing an increase from pretest to posttest.

Project assignments presented from the project-based learning model provide opportunities for students to learn from hands-on experience. Through project-based learning, students have a better mastery of concepts and make students more active as learning centres. In contrast, lecturers as facilitators, so that with the functional role of students cause the learning results obtained also increase [31][32].

Effectiveness to student attitude learning outcomes is measured during the learning process at two meetings consisting of curiosity, responsibility, and cooperation. The percentage of student attitude learning outcomes is shown in Figure 8.

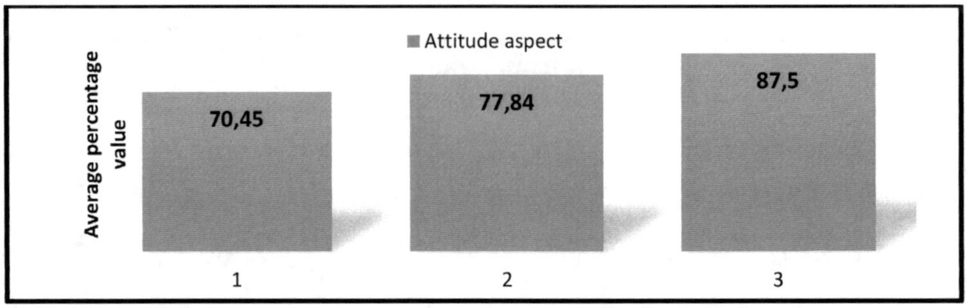

Figure 8. The average percentage value of each aspect of attitude learning outcomes

Based on Figure 8, it is known that students show a good attitude during learning using e-modules. Furthermore, students are given group tasks to create projects related to the wetland environment with lecturers as facilitators who, without directly telling students how to create the project, so that students' attitude of curiosity, responsibility, and cooperation is trained independently.

According to the results of Setiono's research, Dadi &Yuliantini [33], the application of project-based learning models successfully increases students' curiosity in learning because this attitude of interest is a character that students must master to be able to learn independently. The project-based learning model is also able to improve student responsibility attitudes [34]. According to Rahayu, Puspita, & Puspitaningsih [35], the students' emergence of a cooperative attitude makes it easier for them to socialize and foster an open and accessible perspective to accept differences.

Effectiveness to the learning outcome of student skills was measured at the 2nd meeting related to the results of the water purification project. The percentage of student skills learning outcomes is shown in Figure 9.

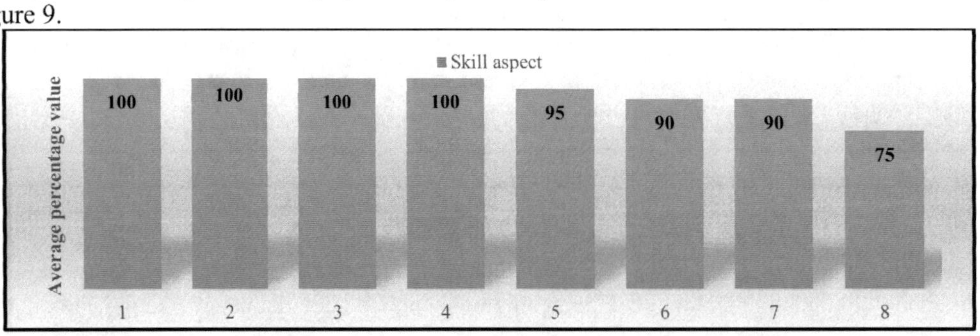

Figure 9. The average percentage of grades from each aspect of learning skills results

Based on Figure 9, students show highly skilled abilities in project creation and percentage of results. According to Rati, Kusmaryanti, & Rediani [36], the application of project-based learning models gives students the flexibility to find solutions and problem solving from various sources. From the reference sources established to make students carry out a targeted and integrated project work process.

4. Conclusion

Based on the results of research and discussion can be concluded that the e-module colloidal material based on STEAM-PJBL approach to improving scientific literacy and learning outcomes in the context of wetlands developed: (a) Obtaining a very valid criterion for use in learning with a score of 93.67% (very valid) on the validity aspect of the content; 93.75% (very valid) on the validity aspect of presentation; 94.62% (very valid) on the validity aspect of language; and 94.44% (very valid) on the aspect of media validity; (b) Obtaining practical criteria because of the average score on individual tests of 3.20 (practical), small group tests of 3.68 (very practical), student response questionnaires of 3.54 (very practical), observation sheets of lecturers' ability to use e-modules of 3.33 (very practical), and observation sheets of e-module ability to manage classes by 3.45 (very practical); (c) Obtaining effective criteria can contribute to significant improvements from the results of scientific literacy and student knowledge learning outcomes in limited trial classes seen from the N-gain value between pretest and posttest is 0.75 and 0.79, consecutively. Then, the results of learning attitudes obtained a score of 78.59 with good criteria, and the learning skills result obtained a value of 93.18 with very skilled criteria.

References

[1] Toharudin, Hendrawati S and Rustaman A 2011 *Building Student Scientific Literacy* (Bandung: Humanities)
[2] Suranti N M, Gunawan and Sahidu, H 2016 *Journal of Physics and Technology Education* **2** 73-79
[3] Keane L C and Keane M A 2016 *Engineering Design and Technology Education: An International Journal* **21** 61-82
[4] Iriani R, Herlina, A and Sanjaya R E 2019 *Journal of Educational Innovation IPA* **5** 54-68
[5] Thiagrajan, Semmel D S and Semmel M I 1974 *Instructional Development for Training Teachers of Exceptional Children* (Washinton DC: National Center for Educational Improvement system)
[6] Akbar S 2015 *Instruments of Learning Devices* (Bandung: Teenager Rosda Karya)
[7] Widoyoko E P 2018 *Assessment of Learning Outcomes in School* (Yogyakarta: Student Library)
[8] Djali and Muljono P 2008 *Measurement in the field of education* (Jakarta: PT. Grasindo)
[9] ULM Building Team 2017 *Academic Guidelines of Lambung Mangkurat University 2017* (Banjarmasin: Lambung Mangkurat University)
[10] Widoyoko E P 2012 *Engineering of Preparation of Research Instruments* (Yogyakarta: Student Library)
[11] Heinich, Molenda and Russell 1996 *Instructional Media and New Technologies of Instructional* (Englewood Cliffs N. J.: Prentice-Hall)
[12] Atika D, Nusmowati M and Nurhayati S 2018 *Journal of Chemical Education Innovation* **12** 2149-58
[13] Amir A 2016 *Journal of Exact* **2** 34-40
[14] Arizona N 2016 *Journal of Words* **1** 93-8
[15] Puryati E 2017*Prisma University Suryakancana Journal* **11** 192-201
[16] Zaharah, Susilowati A 2020*Scientific Journal of Biological Education* **6** 145-58
[17] Rozikin A, Lesmono A and Bachtiar R 2018 *Journal of Physical Learning* **7** 108-15
[18] Setiyadi M W, Ismail and Gani H A 2017 *Journal of Educational Scientific and Technology* **3** 102-12
[19] Yildiz A and Baltaci S 2016 *Journal of New Horizons in Educational* **6** 155-66
[20] Suhaida D and Azwar I 2018 *Journal of Social Education* **5** 1-19
[21] Johanes 2018 *Scientific Forum* **15** 133-138
[22] Hendra B K, Arsa G R and Krisnawati L 2017 *Journal of Electrical Engineering Education Undiksha* 75-85
[23] Rahmawaty S, Nurhayati and Arsyad M 2020 *Proceedings of the National Seminar on Physics PPs UNM 2 University of Makassar* 33-6
[24] Yatmini 2016 *Scientific Journal of Mandala Education* **2** 172-84

[25] Khotimah H, Suhirman and Raehanah 2020 *SPIN* **2** 13-26
[26] Rusyati, Permanasari A andArdianto, D 2019 *Journal of Scientific Educational and Practice* **2** 10-22
[27] Wulandari F E 2016 *Journal of Pedagogy* **5** 247-54
[28] Munawar M, Roshayanti F and Sugiyanti 2019 *Cheerful Journal* **2** 276-85
[29] Novianti D, Rusmansyah and Kusasi M 2017 *Journal of Chemistry and Education* **1** 15-27
[30] Ma'sumah A and Mitarlis 2021 *Journal of Chemical Education Innovation* **3** 22-34
[31] Astuti I D, Toto and Yulisma L 2019 *Journal of Education and Biology* **11** 93-8
[32] Wahyuni H T 2020 *VARIABLE* **3** 1-4
[33] Setiono P, Dadi S and Yuliantini, N 2021 *Journal of Basic Education* **5** 1-8
[34] Rifai S S, Uswatun D A and Nurasiah I 2019 *Journal of Veteran IPA Education* **3** 127-37
[35] Rahayu D, Puspita A M and Puspitaningsih F 2020 *Journal of Educational Research* **7** 111-22
[36] Rati N W, Kusmaryanti N and Rediani. (2017 *Indonesian Journal of Education* **6** 60-71

The effect of the group investigation-guided inquiry (GI-GI) learning model to improve students' collaboration and science process skills

Indrawati, I K Mahardika, J Prihatin, Supeno, S Astutik, Sudarti, and I Wicaksono*

Faculty of Teacher Training and Education, University of Jember, Jember, Indonesia

iwanwicaksono.fkip@unej.ac.id

Abstract. The GI-GI learning model is a combination of the Group Investigation and the Guided Inquiry model. The research aimed to examine the effect of the GI-GI learning model on collaboration and students' science process skills. The type of research carried out was Quasi-Experimental and used a post-test only control group design and purposive sampling covering four high schools in the Besuki Residency, two classes as the experimental class and the control class. The research instrument used tests, observations, and documentation. The results showed a significant effect of the GI-GI learning model on collaboration and students' science process skills. This research implies that the GI-GI learning model can condition students to research in groups to interact actively with friends and educators to exchange opinions, knowledge, or experiences, find and solve problems, and hypothesize through investigation, exploration, and discussion outside or inside in class.

1. Introduction

Science learning plays a significant role in the educational process because science can generate interest in developing science. Science is a study about the universe and its contents, as well as the events that occur in it, and is produced by experts through a series of scientific processes that are carried out carefully [1]. Science is related to finding out about nature systematically so that it is not only mastery of a collection of knowledge in the form of facts, concepts, or principles but also a discovery process [2]. Therefore, learning science teaches students to be active in finding concepts, principles, and theories to develop knowledge.

The results of the Research Group research based on road maps in 2018, 2019, 2020 provide an overview of follow-up that is relevant to needs, including (1) the model used tends to be teacher-centered, this results in learning that does not involve students actively through scientific activities; (2) students are not familiar with the nuances of collaboration in understanding science subject concepts; and (3) the need for a learning model that trains science process skills that are used to find a concept or theoretical principle, to develop a pre-existing concept that is an overall directed scientific skill. The results of other studies show that the learning process in the classroom still uses the lecture and discussion method, which does not emphasize the process aspect [3]. Collaboration and science process skills are essential for every student to develop science and develop their knowledge.

The learning model that refers to a student-centered approach is included in the Constructivism Theory. The constructivism approach is learning that makes students the center of the learning process.

One learning model that refers to Constructivism Theory is the GI-GI (Group Investigation-Guided Inquiry) learning model. The GI-GI learning model is a combination of the Group Investigation model with the Guided Inquiry model, which has a syntax consisting of four phases, namely building a concept (Constructing of Concept), proposing/asking for guidance from an instructor or teacher (Guiding), formulating and testing hypotheses (Hypothesizing and Testing), communicating and assessing results (Communicating and Assessing).

Collaboration has been accepted as an essential skill for achieving effective and meaningful learning and work outcomes. In today's time, collaboration is not only crucial; everyone needs it. Students must collaborate as a learning process to plan and work together, weigh different views/perspectives, and participate in discussions by brainstorming, listening, and supporting others [4]. Collaborative indicators include: contributing actively, working productively, demonstrating flexibility & compromising, managing projects well, demonstrating responsibility, and showing respect.

Science process skills are essential for every student as a provision to develop science. They are expected to gain new knowledge or develop the understanding they already have [5]. Fundamental science process skills indicators consist of observing, classifying, measuring, communicating, drawing conclusions, and predicting. The indicators for integrated science process skills consist of formulating hypotheses, naming variables, controlling variables, making operational definitions, conducting experiments, interpreting, designing investigations, applying concepts [6]. Science process skills emphasize the growth and development of several science process skills in students to process information so that new things are found that are useful in the form of facts, concepts, and the development of attitudes and values.

Therefore, the Research Group needs to look at the effect of the GI-GI learning model on students' collaboration and science process skills.

2. Method

The type of research conducted is Quasi-Experimental and uses a post-test-only control group design [7]. In the experimental pattern, the post-test-only control group design uses two groups, namely the group that is treated using the GI-GI (Group Investigation-Guided Inquiry) learning model is called the experimental group. In contrast, the group that is not treated is the control group that uses ordinary learning applied in school. Determination of the sample in this research using purposive sampling, the example in this research was four high schools in the Ex-Residency of Besuki, each with two classes as the experimental and control classes. The research implementation time is in the Odd Semester of 2021/2022. The research instrument uses tests, observations, and documentation. The technique of analyzing student collaboration data is by giving a score for each collaboration task for each student as follows [8].

$$\%\, score\, for\, each\, task = \frac{Total\ scores\ of\ all\ students}{Total\ maximum\ score} \times 100\%$$

Table 1. Guidelines for converting percentage intervals into categories

Percentage (%)	Category
$80 < X \geq 100$	Very good
$60 < X \leq 80$	Good
$40 < X \leq 60$	Enough
$20 < X \leq 40$	Less
$0 < X \leq 20$	Very Less

Data analysis technique by giving the percentage of science process skills as follows [9].

$$P_{Kps} = \frac{P}{N} \times 100\%$$

Explanation:

P_{Kps} = Percentage of students' science process skills
P = Total score obtained from each indicator
N = Total maximum score

Further analysis for the criteria of science process skills is as follows [8].

Table 2. Criteria for students' science process skills

Percentage of Science Process Skills	Science Process Skills Category
≥ 85%	Very good
71% – 84%	Good
56 – 70	Enough
41 –55	Less
≤ 40%	Very Less

After obtaining the collaboration scores and students' science process skills, the data normality test was conducted using SPSS software with a 5% confidence level. If the data obtained are typically distributed, then parametric statistical analysis is used, namely, the independent sample t-test. If the data obtained are not normally distributed, non-parametric statistical analysis, namely the Wilcoxon test, is used.

H_0 = there is no significant difference between collaboration scores or science process skills in the experimental and control classes

H_a = there is a significant difference between collaboration scores or science process skills in the experimental and control classes

3. Result and Discussion

The research was conducted for three weeks in both classes. Learning in the experimental and control classes is carried out online. The experiment class uses the GI-GI learning model, observing collaboration indicators and students' science process skills. However, the control class uses the usual learning done by the teacher.

3.1. Collaboration

Table 3. The results of the collaboration between the experimental and control classes

No	Indicator	Experiment Class		Control Class	
		Percentage (%)	**Category**	**Percentage (%)**	**Category**
1	Contribute actively	86.13	Very good	70.42	Good
2	Work productively	85.12	Very good	63.65	Good
3	Demonstrate Flexibility & compromise	78.38	Good	49.13	Enough
4	Manage projects well	71.52	Good	53.52	Enough
5	Show responsibility	74.63	Good	59.63	Enough
6	Show respect	76.68	Good	62.68	Good
	Average	78.74	Good	59.83	Enough

Based on Table 3 above, the experimental class collaboration has an average of 78.74, categorized as good. In contrast, the control class has an average of 59.83, which is classified as sufficient. The results of statistical tests using independent sample t-test obtained the value of Sig. > 0.05 so that H_a is accepted; there is a significant difference between the experimental and control classes' collaboration scores. Collaboration is a philosophy about relating to others (how to learn and work), which is a way to deal with others by respecting differences, sharing power, and gathering knowledge from others. In collaborative classrooms, students will share goals, learn together, engage in meaningful assignments, and build prior knowledge to generate ideas and various products [9]. The results showed that collaboration skills provide students with opportunities to interact in activities working together to achieve goals by respecting differences, participating in discussions, brainstorming, listening, and supporting others [10].

The GI-GI learning model makes students more active in the learning process and better understand science concepts and principles. In contrast, the control class students are more dominant passive, and the teacher plays an active role in the learning process. When carrying out the process of actively contributing, students can conclude and present experimental data that has been obtained with the direction and guidance of the teacher. Thus students can interpret or interpret and analyze a problem given by the teacher to conclude. The advantages in the learning process are that students can play an active role in the learning process, can solve problems in the discussion, and play a direct role in the learning process so that it is easier for them to understand the material [11].

3.2. Science process skills

Table 4. The results of the value of science process skills in the experimental and control classes

No	Indicator	Experiment Class		Control Class	
		Percentage (%)	Category	Percentage (%)	Category
1	Observe	86.98	Very good	77.27	Good
2	Grouping	78.21	Good	70.42	Good
3	Interpret	74.44	Good	71.44	Good
4	Predict	74.51	Good	60.51	Enough
5	Formulating Hypotheses	68.39	Enough	53.53	Enough
6	Planning an Experiment	72.62	Good	57.62	Enough
7	Communicating	77.01	Good	72.34	Good
	Average	76.02	Good	Good	Enough

Based on the data in Table 4 above, the science process skills of the experimental class have an average of 76.02, which is categorized as good. In contrast, the control class has an average of 66.16, which is classified as sufficient. The results of statistical tests using independent sample t-test obtained the value of Sig. > 0.05 so that H_a is accepted; there is a significant difference between the scores of science process skills in the experimental class and control classes. Science process skills are essential for every student as a provision to develop scientific knowledge. They are expected to gain a new understanding or be able to create the knowledge they already have. Science process skills can be used to find a concept or theoretical principle, to develop a pre-existing idea that is a whole directed scientific skill (both cognitive and psychomotor) [12]. The results showed that students' intellectual abilities were because they used their minds or cognitive skills to carry out scientific process skills [13].

The GI-GI learning model involves physical movement and intellectual activity of students to find problem-solving strategies and strengthen the knowledge gained by students through discovery learning in groups to acquire knowledge independently through scientific methods in experimental activities. The GI-GI learning model can foster students' scientific attitudes. By researching in groups, students can interact actively with friends or teachers to learn new knowledge or information, find and analyze problems, exchange opinions, knowledge, and even experiences [14]. Students can also develop skills to respect ideas, dare to have thoughts, be responsible, and work together. Another advantage of the GI-GI learning model is that it can allow students to interact with their groups to discuss observations. Besides, students are directly involved in teaching and learning activities to be motivated to learn [15].

4. Conclusion

Based on the results of data analysis and discussion, it can be concluded that there is a significant effect of the GI-GI learning model on collaboration and students' science process skills. This research implies that the GI-GI learning model can condition students to research in groups to interact actively with friends and educators to exchange opinions, knowledge, or experiences, find and solve problems, and hypothesize through investigation, exploration, and discussion outside or inside. in class.

Acknowledgments
The successful implementation of this research cannot be separated from the assistance of various parties. Therefore the research team of the 21st Century Science Teaching Model Research Group (21st Century STM) expresses its gratitude to LP2M Jember University through Research Group Grants and Research Partner Schools with Assignment Agreement Number 2700 /UN25.3.1/LT/2021.

References

[1] Budiastra A A K, Wicaksono I and Erlina N 2020 *J. Educ. Gift. Young Sci.* **8** 1291–304
[2] Siahaan K W A, Lumbangaol S T P, Marbun J, Nainggolan A D, Ritonga J M and Barus D P 2020 *J. Basicedu* **5** 195–205
[3] Budiastra A A K, Wicaksono I and Sanjaya I G M 2020 *Int. J. Instr.* **13** 763–80
[4] Der Linden J, Erkens G, Schmidt H and Renshaw P 2000 *Springer* **3** 37–54
[5] Budiastra A A K, Erlina N and Wicaksono I 2019 *New Educ. Rev.* **57** 187–99
[6] Erlina N, Susantini E, Wasis, Wicaksono I and Pandiangan P 2018 *J. Balt. Sci. Educ.* **17** 972–85
[7] Rogers J and Révész A 2020 *Experimental and Quasi-Experimental Designs The Routledge Handbook of Research Methods in Applied Linguistics* (New York: Routledge)
[8] Yuliati Y 2017 *J. Cakrawala Pendas* **3** 21–8
[9] Juhji J and Nuangchalerm P 2020 I *J. Educ. Gift. Young Sci.* **8** 1–16
[10] Wicaksono I, Wasis and Madlazim 2017 *J. Balt. Sci. Educ.* **16** 549–61
[11] Joice B and Weil M 1980 *Model of Teaching* (London: Printice-Hall Inc)
[12] Listyawati M 2012 *J. Innov. Sci. Educ.* **1** 91–9
[13] Wicaksono I, Supeno and Budiarso A S 2020 *Int. J. Instr.* **13** 157–70
[14] Le H, Janssen J and Wubbels T 2018 *Cambridge J. Educ.* **48** 103–22
[15] Tyas E H and Naibaho L 2021 *Int. J. Res.* **9** 176–82

Accelerating teacher career through improving competence in scientific publications: Physics teacher perspectives

S Admoko[1]*, Z A I Supardi[1], Wasis[1], N Suprapto[1], A Realita[1], B K Prahani[1], S Muhayatin[2], M Irfa'i[3], Dakri[4], Suyanto[5], H S Hati[6], M Misbah[7]

[1] Physics Department, Faculty of Mathematics and Natural Sciences, State University of Surabaya, Indonesia
[2]State High School 1 Mantup, Lamongan, Indonesia
[3]State Junior High School 2 Laren, Lamongan Indonesia
[4]State High School 5 Tuban, Tuban, Indonesia
5State Senior High School 4 Semarang, Semarang, Indonesia
[6]State High School 2 Selong, East Lombok, Indonesia
[7]Physics Education Study Program, Universitas Lambung Mangkurat, Indonesia

*setyoadmoko@unesa.ac.id

Abstract. In promotion of teachers career to a higher level are required to write scientific papers. This study aims to describe the physics teacher's perspective on the promotion process they have experienced so far and the obstacles they face. This research is a descriptive research, data collection through questionnaire and information deepening is done through interviews. The sample of this research is 73 physics teachers. Based on the results of data analysis, it can be seen that more than 65% of physics teachers have difficulty in fulfilling scientific publication requirements for promotion, 85% of respondents have done classroom action research (CAR), but only about 44% have published in the form of scientific articles. In general, teachers have experience in conducting CAR, but they still lack experience in writing scientific articles. Most physics teachers expect local schools and education offices to cooperate with universities to organize training and mentoring of scientific publications. This study can be concluded that the main obstacle to the promotion of physics teachers is caused by the low ability to carry out scientific publications. The results of this study are expected to provide consideration to stakeholders in making decisions and actions related to teacher promotion, especially regarding the mandatory requirements for scientific publications.

1. Introduction

The government has enacted Law Number 20 of 2003 concerning the National Education System [1] Law Number 14 of 2005 concerning Teachers and Lecturers [2], Government Regulation Number 74 of 2008 concerning teachers mandates that teachers must have academic qualifications, competence, and educator certification. In connection with that, the Minister of National Education Regulation (Permendiknas) Number 10 of 2009 was issued concerning Certification for Teachers in Position. The existence of the above regulations has given new hope in the world of education, where the increased teacher income associated with educator certification should be balanced with predetermined academic competencies.

The government has also implemented the Regulation of the Minister for Empowerment of State Apparatus and Bureaucratic Reform No. 16 of 2009 concerning Teacher's Functional Positions and Credit Scores [3]. This regulation contains changes in which promotions and classes of teachers to move up to a higher class must make scientific papers or scientific articles. This requires all teachers, including junior and senior high school teachers, to be able to compile a scientific paper or article. Regarding scientific articles, many junior and senior high school teachers are not accustomed to making scientific articles which are then published in scientific journals [4,5]. The low writing ability of middle and high school teachers is also evident from the rare number of junior and senior high school teachers who participate in seminar activities by becoming accompanying speakers in national or regional seminars. The low ability of physics teachers in writing scientific papers or scientific articles has an impact on delaying the promotion of teachers for years, because one of the requirements is not fulfilled in the element of publication scientific article.

Writing skill for a teacher is a very important competency. Writing ability is closely related to the ability to communicate scientific ideas both theoretically and practically. Writing scientific publications aims to enrich the repertoire of knowledge or get a response from a professional audience to be involved in a scientific phenomenon at hand. In the context of language, writing a scientific work requires its own skills, through mastering the rules required in writing [6]. Writing skill is not a natural skill but a skill that can be trained and learned. This means that a person's skills to express his thoughts through writing can be honed and improved through exercises. The more often someone does write exercises, the better their writing skills will be [4]

One of the main elements assessed in the promotion/position of teachers to a higher level is continuous professional development, namely the development of teacher competencies that is carried out according to needs, gradually, continuously to improve their professionalism. Forms of sustainable development activities include scientific publications [3]. Types of scientific publications/innovative works for each level of position that must be fulfilled by teachers who will apply for promotion/teaching positions are presentations in scientific forums, publication of research results or innovative ideas in the field of formal education, and publication of textbooks, enrichment books and/or or teacher guidelines [7].

According to Subarkah, there are three barriers to promotion: first: Cultural Barriers, Second: Administrative Barriers, and Third: Structural Barriers [8]. Cultural barriers include: Teachers are more speech-oriented than writing, Teachers prefer teaching rather than writing, Low teacher involvement in seminars, workshops, training workshops, and so on. Administrative barriers include Weaknesses in filing letters, SK, Letters of Assignment, or other physical evidence, Weaknesses in presenting supporting evidence for promotion, Weak understanding of regulations regarding teacher positions, Inaccuracy of proposals for promotion files. At the same time, structural barriers include the limited staff of the teacher credit score assessment team at both the regional and central levels [8]. To find solutions to various obstacles to the promotion of teachers, a comprehensive understanding of the root cause of the problem from the perspectives of the various parties involved and especially the teacher as the subject of the perpetrator is needed [9].

This study focuses on the obstacles to a promotion stemming from the low knowledge and skills of physics teachers in scientific publications, with five research questions: (1) How do physics teachers respond to the current promotion system? (2) In which promotions and classes do most physics teachers experience obstacles and difficulties? (3) In fulfilling the credit score, do physics teachers experience difficulties when they are promoted to rank and class? (4) What is the general description of the physics teacher's understanding of the knowledge and skills of writing scientific papers or scientific articles? (5) What suggestions do physics teachers give to facilitate them in solving problems in promotion (scientific publications)?

2. Method

This research is a descriptive research, data collection through questionnaire [10-12]. The sample of this research is 73 physics teachers. The questionnaires are distributed to physics teachers by asking questions about the promotions they experienced and the obstacles and further about the teacher's obligation to conduct scientific publications for promotion requirements. Questionnaires were distributed online via Google Form with the target respondents being physics teachers who teach at public schools at random from various provinces. The collected data was then analyzed using descriptive statistics by presenting in the form of percentages and inferential statistics. The deepening of information on the data obtained is done by conducting interviews with certain respondents [13-15].

3. Result and Discussion

3.1. Characteristics of Respondents

Based on the data entered, there are 73 respondents of physics teachers with various educational backgrounds, 41 Bachelor's degree (S1) teachers (56%) and 32 Master's degree (S2) teachers (44%), meanwhile based on the place of assignment where teachers teach there are 45 teachers teaching in Senior High School (62%), 22 teachers teach at Junior high school (30%), 4 teachers teach at Vocational School (5%) and 2 teachers teach at Madrasa aliyah (3%) (see Figures 1a and 1b)

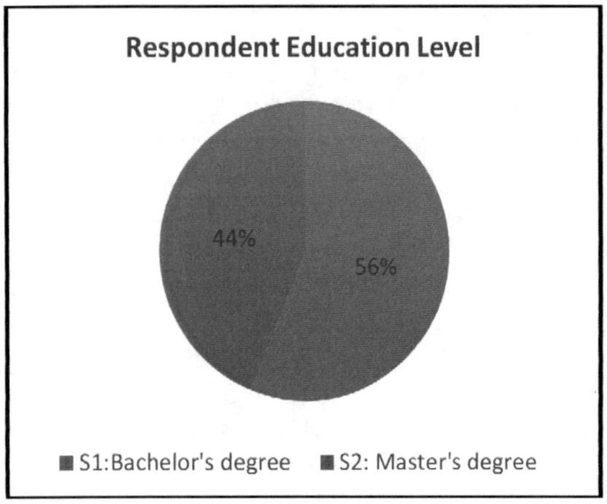

a.

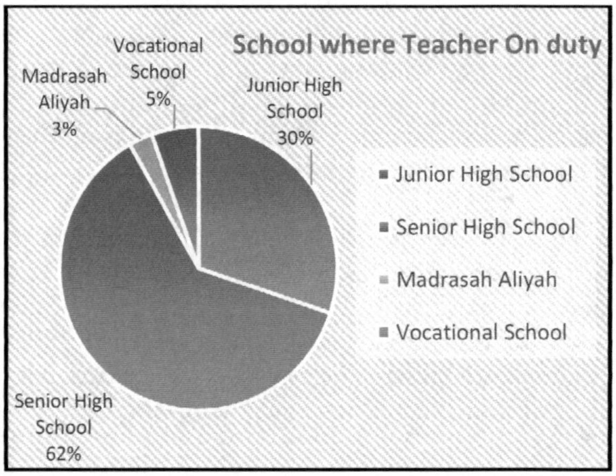

b.

Figure 1 a. Respondent's education level and b. Level of a schoolteacher on duty

3.2. The Perspective of the Physics Teacher of the Promotion System and the obstacles faced

The results of the physics teacher's assessment of the promotion system that is currently being implemented can be seen in Figure 2. where 52% of teachers gave an assessment of "It is good" while 48% of teachers gave an assessment of "Not good". The physics teachers who gave the assessment "Not good" gave several reasons, including (1) the use of information technology was not yet optimal (2) the submission file for rank was not simple and paperless (3) the standard for implementing the rules for promotion was the same between regions (districts and provinces) (4) the validity, objectivity, and transparency of the assessment of promotions are not yet optimal, and (5) the school and the education office have not been optimal in facilitating the resolution of obstacles faced by teachers.

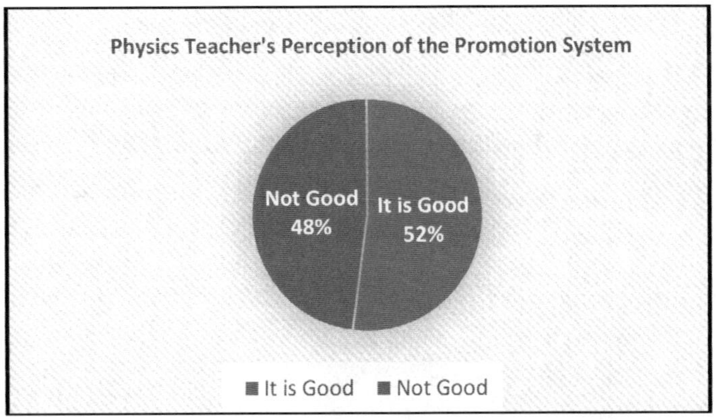

Figure 2. The physics teacher's perspective on the current promotion system.

Based on Figure 3a. and 3b. below, the obstacles faced in the promotion of the largest percentage that exceeds the percentage of the rank of the respondent group occur at the level of groups III B, IIIC and IVB. Based on this data, respondents from Group IIIB are only 6%, but the number of respondents who find it difficult to increase to IIIB is 12%; this is none other than because the increase at this level is the first increase made by the teacher so that they have difficulty because there is no previous experience. The number of respondents in group IIIC is only 12%, but the number of respondents who find it difficult to advance to IIIC is greater, namely 27%, this is because at this level increase, scientific publications that previously did not have these conditions began. Meanwhile, at the rise in the IVB level, the respondents were 30%, but those who had difficulty at this level reached 33%. This was due to the rise in the percentage of credit score requirements in the scientific publication aspect, which increased from 8 credit numbers at the IVA level to 12 credit numbers at the IVB level.

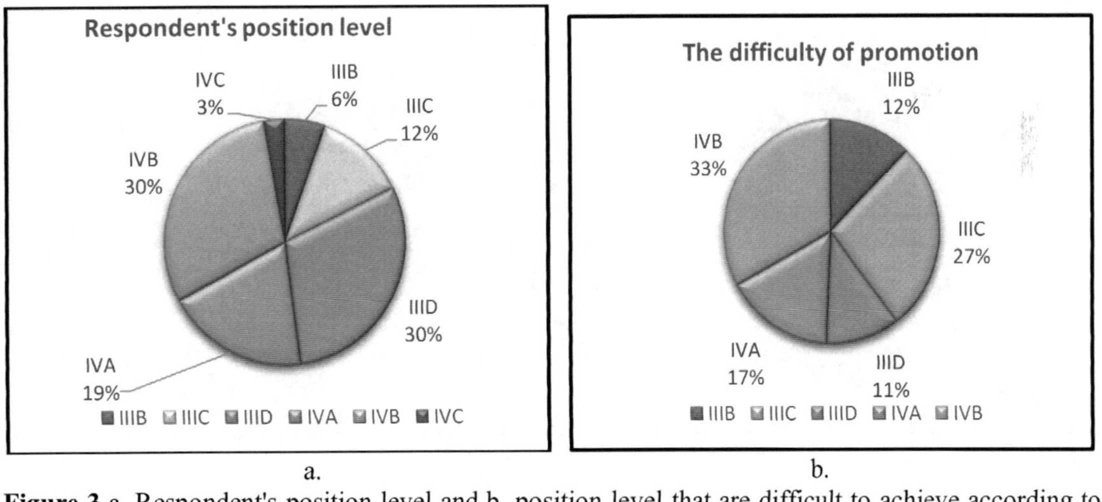

Figure 3 a. Respondent's position level and b. position level that are difficult to achieve according to the physics teacher

The level of difficulty in meeting the requirements for the credit score for promotion based on the physics teacher's perspective on the four aspects of the assessment is presented in Figure 4 below. The level of difficulty in fulfilling the credit score from the highest to the lowest, respectively, is 66% scientific publications, 23% innovative work, 8% self-development, and 3% learning or mentoring. This data is in line with previous data, which shows that with the application of scientific publication requirements and increasing credit scores on aspects of scientific publications, many physics teachers have difficulty getting promoted.

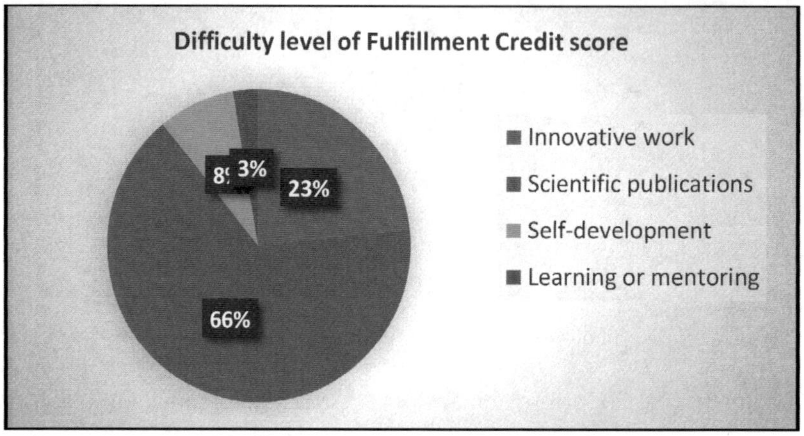

Figure 4. The level of difficulty in fulfilling credit scores in four aspects of assessment according to the physics teacher's perspective

The experience of physics teachers in carrying out classroom action research (CAR) and writing articles to be published in seminars or scientific journals is shown in Figures 5a and 5b. Based on the data presented in the figure, almost all respondents (85%) have experience doing CAR, but only a small percentage (44%) have published scientific articles. Of course, this needs further investigation to find out the cause of the low experience of scientific publications by physics teachers

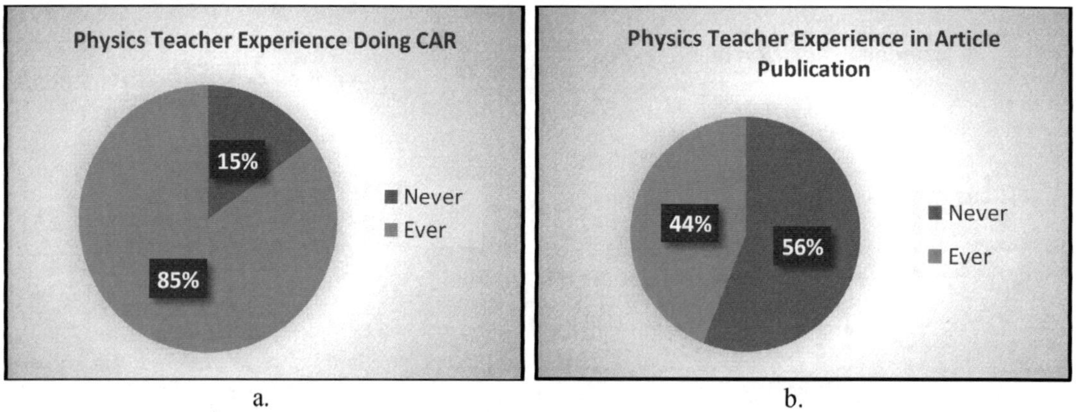

Figure 5 a. Physics teacher experience doing CAR and b. Physics teacher's experience in publishing articles

Based on the results of further research on the low scientific publications of physics teachers, four main reasons were obtained including they did not understand the systematics of writing scientific articles (16%), did not know the steps in publishing articles in scientific journals or seminars (26%), did not have enough time to write scientific articles. compiling scientific articles (44%), and difficulties in analyzing data in compiling scientific articles (14%) (see Figure 6)

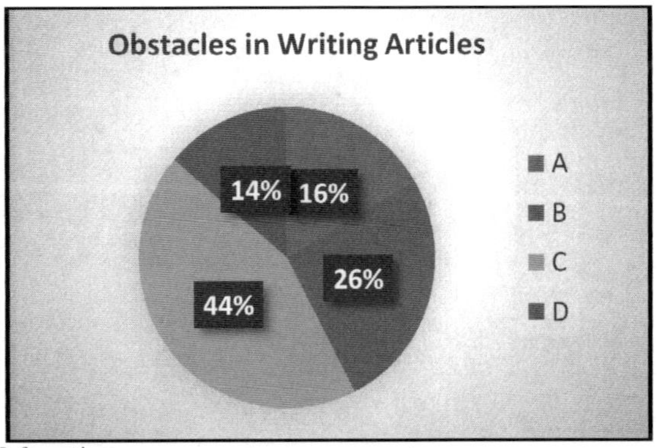

Information:
A. Do not understand the systematics of writing scientific articles
B. Do not know the steps in publishing articles in scientific journals or seminars
C. Not having enough time to compile scientific articles
D. Difficulty in analyzing data in compiling scientific articles

Figure 6. Sources of difficulties in writing scientific articles according to the physics teacher's perspective

Based on the perspective of self-assessment conducted by physics teachers on scientific publication material, the results can be seen in Table 1. below. Overall, the physics teacher's understanding of scientific publication material is still very low, with an average score of 2.12 which is included in the "Knowing little" category. Materials that are known in general tend to get higher ratings than materials that are relatively new among physics teachers, such as software for library management and the preparation of a scientific article that has novelty.

Table 1. Physics teacher's perspective on understanding scientific publication material

No.	Scientific Publication Material	Physics Teacher's Level of Understanding (%)				Average
		1	2	3	4	
1	Important aspects that need to be considered in writing scientific articles	6,85	54,79	36,99	1,37	2,33
2	The structure of writing in scientific articles	5,48	52,05	41,10	1,37	2,38
3	Code of ethics for publishing scientific articles	20,55	52,05	27,40	0,00	2,07
4	Software for citation management in the preparation of scientific articles	31,51	47,95	20,55	0,00	1,89
5	Compiling literature review in writing scientific articles	12,33	54,79	31,51	1,37	2,22
6	Writing a scientific article that has novelty	35,62	47,95	16,44	0,00	1,81
	Average	**18,72**	**51,60**	**29,00**	**0,68**	**2,12**

Description of Assessment Criteria:
1. Very knowing
2. Knowing enough
3. Little to know
4. Don't know

Figure 7. below shows some scientific publication materials that physics teachers really need in training. There are three materials that are highly expected by physics teachers, namely the writing of scientific journal manuscripts (82.2%), important aspects of journal scientific publications (72.6%), and best practice journal scientific publications (60.3%). Based on these three selected scientific publications material, physics teachers tend to choose the material that can be directly practiced in the preparation of scientific articles.

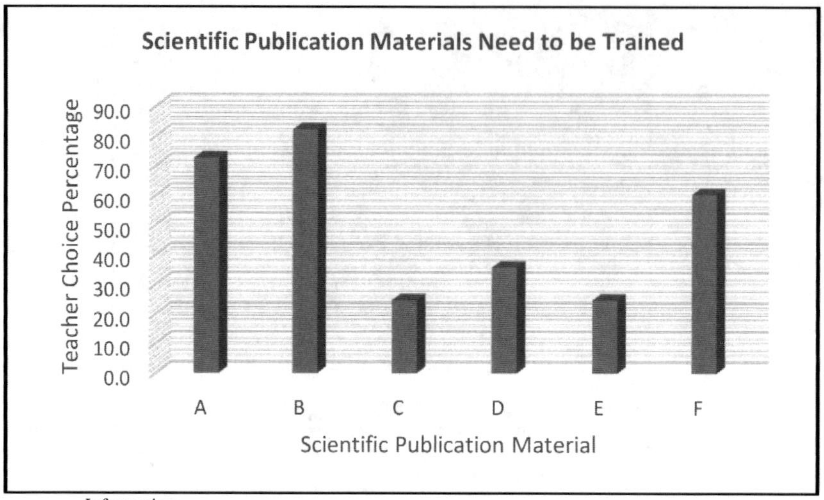

Information:
A. Important Aspects of Journal Scientific Publication
B. Scientific Journal Manuscript Writing
C. Code of Ethics for Journal Scientific Publications
D. Software for Library Management
E. Review of literature and the up-to-date content of scientific articles
F. Best Practice of Journal Scientific Publication

Figure 7. Expectations of physics teachers for scientific publications in training

3.3. Suggestions from physics teachers to facilitate them in solving scientific publications problems

Some inputs from teachers aimed at schools and the education office to resolve the obstacles faced in the publication of scientific articles are summarized in Figure 8 below. Based on these inputs, there are three highest inputs given by physics teachers: (1) Schools or local education offices collaborate with universities, (2) MGMP activities to improve knowledge and skills in writing scientific articles, and (3) Schools or local education offices hold training on writing scientific articles that are held regularly

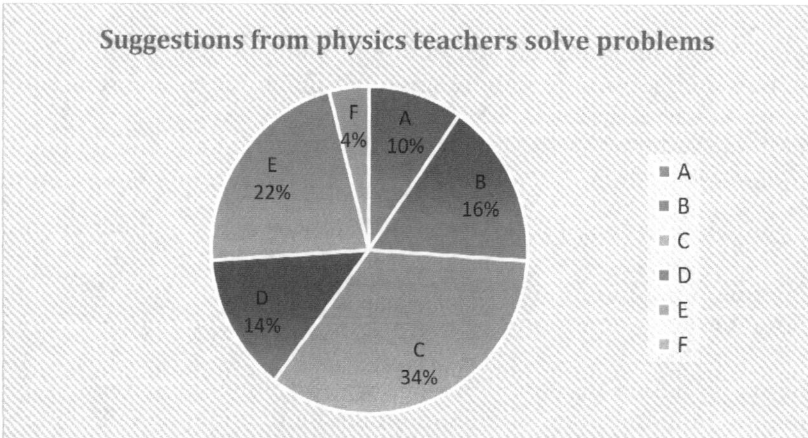

Information:
A. Schools and local education offices hold a scientific paper writing consultation center.
B. Schools or local education offices hold training on writing scientific articles which are held regularly
C. Schools or local education offices cooperate with universities in producing scientific works or scientific articles for teachers
D. Schools or local education offices need to conduct training to turn CAR into scientific articles
E. MGMP activates activities to improve knowledge and skills in writing scientific articles.
F. The local education office needs to hold periodic and routine scientific article writing competitions

Figure 8. Suggestions from physics teachers to solve scientific publications problems

Meanwhile, suggestions from physics teachers to improve the promotion system are:

1) Need to simplify the promotion process by utilizing information technology whose process can be monitored online. This needs to be done because not all regions have great attention to the application of this information technology. Some areas that have optimized information technology include Surabaya, Bengkulu, Tanah Datar, and Tapin [16-19].
2) It is necessary to simplify the files that will be used as a promotion submission.
3) Application of credit score assessment standards so that they can be carried out in a valid, objective, and transparent manner.
4) For promotion at a certain level, it is necessary to wait for a longer process of submission and assessment.
5) It is necessary to standardize the implementation of the same promotion rules between regions (regencies and provinces) with one nationally.
6) Schools and agencies should facilitate the resolution of obstacles faced by teachers in promotion

Based on the results of research and discussion, the following research findings were obtained:

1) Many physics teachers (44%) have Masters (S2) education
2) Most physics teachers teach in high school (62%) and in junior high school (30%)
3) Physics teachers still give a lot of "Unfavorable" assessments of the current promotion system accompanied by some supporting evidence.
4) The promotion of physics teacher was constrained mainly at the IIIC and IVB levels due to the application of scientific publication requirements and the increase in the number of scientific publication credits. This is in accordance with previous research conducted by Subarkah, 2020 [8], Nizar, 2020 [20], and Dihamri, 2018 [21].
5) The level of difficulty in fulfilling the credit score is the two highest assessment aspects, namely scientific publications (66%) and innovative works (23%)
6) Most of the physics teachers have done CAR, but only a few have compiled their CAR reports into scientific articles.
7) The two main obstacles, according to physics teachers in writing scientific article publications are not having enough time and not knowing the steps in publishing scientific articles.
8) In general, physics teachers' understanding of scientific publications is still very low, with an average score of 2.12 which is included in the "Knowing little" category. Kondisi ini sesuai dengan hasil penelitian Melati, 2020 [22].
9) Three scientific publication materials that physics teachers really expect are the writing of scientific journal manuscripts, important aspects of journal scientific publications, and best practice of journal scientific publications.

4. Conclusion

Based on the results and discussion above, it can be concluded that: (1) Physics teachers still give a lot of "Not Good" assessments of the current promotion system accompanied by some supporting evidence. (2) The promotion of physics teacher position level is mainly constrained at the IIIC and IVB levels, due to the application of scientific publication requirements and the increase in the number of scientific publication credits. (3) The level of difficulty in fulfilling the credit score is the two highest aspects of the assessment promotion, namely scientific publications and innovative works. (4) In general, physics teachers' understanding of scientific publication material is still very low which is included in the "Knowing little" category. (5) The advice given by the physics teacher includes two things, namely suggestions for improving the promotion system and suggestions for the main promotion obstacles experienced by teachers, namely the fulfillment of scientific publications credit scores. In fulfilling the credit score for scientific publications, physics teachers hope that the school and the education office will facilitate more by collaborating with universities and conducting regular training. The findings of this study are expected to contribute ideas in unraveling the problems of promotion of physics teachers, both from improving the promotion system and unraveling the main problems of physics teachers in fulfilling the requirements for credit scores for scientific publications.

Acknowledgments
This research was supported by Universitas Negeri Surabaya through the scheme PKM Kebijakan FMIPA with contract number 823/UN38/HK/PM/2021.

References

[1] Indonesia R Undang-Undang Republik Indonesia Nomor 20, Tahun 2003, tentang Sistem Pendidikan Nasional
[2] Nasional D P 2005 Undang-undang nomor 14 tahun 2005, tentang guru dan dosen *Jakarta: Depdiknas*
[3] Indonesia R 2010 Peraturan Menteri Negara Pendayagunaan Aparatur Negara dan Reformasi Birokrasi Nomor 16 Tahun 2009 tentang Jabatan Fungsional Guru dan Angka Kreditnya
[4] Widana I W, Suarta I M and Citrawan I W 2019 *JPM (J. Pemberdaya. Masy.)* **4** 365–75
[5] Saputra D H 2019 *J. Kependidikan Penelit. Inov. Pembelajaran* **3** 291–301
[6] Kellogg R T 2008 *J. Writ. Res.* **1**
[7] Kemdikbud 2016 *Pedoman Kegiatan Pengembangan Keprofesian Berkelanjutan Bagi Guru Pembelajar* (Jakarta: Kementerian Pendidikan dan Kebudayaan)
[8] Subarkah I 2020 *Ar-Rihlah J. Inov. Pengemb. Pendidik. Islam* **5** 89–98
[9] Wong J L N 2014 *Int. J. Educ. Manag.*
[10] Gillham B 2008 *Developing a questionnaire* (A&C Black)
[11] Rowley J 2014 *Manag. Res. Rev.*
[12] Hunter L 2012 *Nurse Res.* **20**
[13] Stokes D and Bergin R 2006 *Qual. Mark. Res. An Int. J.*
[14] Griffee D T 2005 *J. Dev. Educ.* **28** 36–7
[15] Gill P, Stewart K, Treasure E and Chadwick B 2008 *Br. Dent. J.* **204** 291–5
[16] Leonardo M C 2020 TA: Rancang Bangun Aplikasi Penilaian Kinerja Guru Berbasis Website pada SMA Gracia Surabaya Menggunakan Metode 360 Derajat (Doctoral dissertation, Universitas Dinamika).
[17] Sapri S 2018 *J. Media Infotama* **14**
[18] Agustina L 2018 Aplikasi Perhitungan Kredit Point untuk Pengajuan Kenaikan Pangkat Guru pada Dinas Pendidikan Kabupaten Tanah Datar Berbasis WEB
[19] Adisetiawan M N 2020 Aplikasi Pengelolaan Berkala Guru pada Dinas Pendidikan Kabupaten Tapin
[20] Nizar U K, Guspatni G and Gazali F 2020 *Pelita Eksakta* **3** 156–9
[21] Dihamri D, Haimah H and Srifitriani A 2018 *J. Pengabdi. Masy. Borneo* **2** 60–6
[22] Melati I S and Harnanik H 2 *Indones. J. Devot. Empower.* **2** 31–6

Analysis of the livewire software application on students' learning understanding and motivation on ohm's law concepts

Nurussaniah, I N Sari, M Rudi

Program Studi Pendidikan Fisika, IKIP PGRI Pontianak, Jalan Ampera Nomor 88 Pontianak Kalimantan Barat, Indonesia

nurussaniah@gmail.com

Abstract. The study aimed to analyze the conceptual understanding and motivation to learning physics in schools with the help of livewire software. An experimental method with a post-test-only nonequivalent control group design was used to examine Ohm's Law. The sample for the study consisted of a control class with 36 students and an experimental group comprising 37 learners drawn from a population of 145 through random sampling. Data collection tools used were tests and questionnaires to collect data, which was analyzed using the non-parametric statistics technique. The results showed that the livewire software made students' understanding of Ohm's Law concepts better compared to learning without the help of the software. Livewire software can motivate students to study Ohm's Law.

1. Introduction

Physics is an interesting Natural Science, which is further supported by the rapid technological advancement across the globe. Physics is also taught in schools to help students expand frontiers of knowledge about nature and help them in everyday life. However, in reality, many students think that this subject is difficult. Among the challenges students face while studying physics include connecting the concepts from the practicum with story problems and calculations [1].

The observations made in a senior high school in West Kalimantan Province found that students have difficulties understanding Ohm's Law material concepts. According to the study, many students could not determine the relationship between the quantities contained in Ohm's Law and the value of substitute resistance in an electrical circuit. Generally, abstract material has been pointed out as the cause why students have problems with the Ohm's Law concept. The daily tests given to students confirm that only a few students are able to score above the Minimum Completeness Criteria while the rest fall below the Minimum Completeness Criteria [2]. Therefore, there is a need to find lasting solutions to help students understand the concept. Being an abstract matter, Ohm's Law requires teachers to use visual tools to enable students to see the concepts taught [3]. For instance, the use of simulation media in learning has a positive impact on students' knowledge levels besides influencing the attitudes of tutors towards teaching the subject [4].

The efforts to improve students' understanding of electrical concepts on electrical materials including Ohm's Law have been conducted by a number of previous researchers. For instance, the use of augmented reality (AR) - based media [5][14], android - based media [7] and python script media [8]. The advantages of this technology is that it is able to display scientific phenomena and simulations in real life. Simulation - based multimedia that introduces electrical circuits suitable for elementary and middle school students [9]. This can be done by students together so as to increase collaborative activities in learning. Collaboration is very important to improve the understanding of the concept of electricity compared to cooperative and lecture methods [10].

Namun penggunaan media ini membutuhkan keterampilan teknologi, membutuhkan perangkat yang kompatibel dan akses internet sehingga kurang cocok untuk siswa yang tidak memiliki hal tersebut. However, the use of such media requires technological skills, compatible devices and internet access. Nonetheless, the media is less suitable for students who do not in possession of those things. Therefore, the media needed in this context is a media that can simulate the concept of electronics, but does not require the internet and high -tech skills in its operation. Livewire software can be installed on low - specification computer devices, can be used offline and is easy to operate.

Livewire software has many components such as switches, transistors, and diodes that can be connected to investigate the behavior of a circuit [11]. The simulation displayed in the livewire-professional edition makes it easier for students to do inquiry activities such as investigating, making hypotheses, formulating problems, experiments, and conclusions. Other livewire simulation software suitable for teaching electronics concepts include Phet simulation [12] and Physlab [13].

2. Method

The experimental method with a post-test-only nonequivalent control group design was used in the study. This method involved the experimental and the control classes whereby the experimental class was taught Ohm's Law using the livewire software as the visual aid, while the control class was only given a theoretical approach regarding the concept. After learning, both classes were subject to a post-test concerning what had been taught.

The sample population for the study involved 73 students who were randomly selected from a total population of 145 learners. The sampled population was put into two groups, including the experimental class with 37 students, while the remaining 36 were put into the control group.

The data collection techniques used included measurement and communication methods. Measurements were used to determine students' understanding of Ohm's Law material, while indirect communication determined students' learning motivation after applying livewire software in learning. The tool for data collection involved the concept of understanding test and a learning motivation questionnaire. The students' understanding of Ohm's Law concepts was obtained from the post-test results and analyzed by non-parametric statistics. The hypothesis for this study was that there were differences in students' understanding of Ohm's Law concepts after applying learning with livewire software. The results of the calculation of the percentage score are matched with categories according to [14], as presented in Table 1.

Table 1. Category value percentage

Interval	Category
0%-20%	Very low
21%-40%	Very low
41%-60%	Medium
61%-80%	High
81%-100%	Very high

3. Result and Discussion

Data on the acquisition of post-test scores in the experimental and control classes are presented in Table 2.

Table 2. Post-test scores for the experimental and control classes

Description	Experiment class	Control class
Total Students	37	36
The highest score	100.00	100.00
Lowest value	0.00	0.00
Average score	74.86	64.03
Standard deviation	20.09	23.42

Based on Table 2, the students in the experimental class performed better with an average of 74.86 compared to the control class, whose score was 64.03. additionally, the standard deviation of the experimental class was 20.09, while that of the control group was 23.42. This indicated that the average student in the experimental class scored marks almost similar to the average class score. In the control group, the average student had a greater diversity of data than the average in the experimental class. The analysis prerequisite test includes the normality test and homogeneity test of the data presented in Table 3.

Table 3. The results of the post-test value normality test

Class	χ2count	χ2table	conclusion
Experiment	94.12	7.81	Not Normal
Control	311.49	7.81	Not Normal

The results above show that the calculated chi-square value of the experimental and control classes is greater than the chi-squared table. This confirms that the student learning outcomes data from the experimental and control classes are not normally distributed hence, a homogeneity test is not carried out. The hypothesis testing results using the Mann-Whitney U test are presented in Table 4.

Table 4. Calculation Results of U Mann Whitney Test

Class	Total samples	Total ranks	Value Z_{count}	Value Z_{table}
Experiment	37	1563.50	-2.15	±1.54
Control	36	1137.50		

Based on the hypothesis test data results using the Mann Whitney U test, the value of Zcount $<$ Ztable (-2.15 $<$ -1.54) means Ho is rejected and Ha accepted. Thus, it can be concluded that students who learned using livewire software understood the Ohm's Law concept better. This is because livewire software attracts attention, and that is why students who usually make circuits with electronic components are challenged to use an application. Through simulations on the livewire software, students are able to analyze quantities in Ohm's Law and how they affect each other. Furthermore, livewire software helps students see the effects of the resistance given to the intensity of the lamp and a given voltage source on the resulting lamp. Additionally, students make connections between sources of voltage, current, and resistance in a circuit, which allows them to formulate Ohm's Law. Conclusively, developing interest in learning positively impacts the outcomes since students become attentive in class, which results in grasping the concept being taught [15,16].
The study also analyzed students' motivation after learning using livewire software, and the recapitulation of the results is presented in Table 5.

Table 5. Student learning motivation

Motivational aspect	Percentage	Category
Pleasure	76.69%	High
Attention	75.34%	High
Interested	77.70%	High
Curiosity	58.45%	medium
Enthusiasm	62.61%	medium
Average	70.16	medium

The above results indicate that students' learning motivation is in the medium category but ranks high in pleasure, attention, and interest. When students are engaged during learning, they become active and are able to ask areas questions regarding difficult concepts. Therefore, the use of livewire software is enough to motivate students to learn Ohm's Law concepts. The simulations displayed on livewire and freedom to express their ideas increase concentration during the process of learning. Furthermore, the use of livewire software educates students to have constructivist thinking patterns and visualize physics concepts. This is in line with several studies conducted, which affirm that the use of simulations and virtual labs in learning increases students' learning motivation [17–19] .

4. Conclusion

In conclusion, livewire software simulation enhances a better understanding of Ohm's Law concept compared to ordinary learning. Furthermore, the software motivates students to learn and achieve more in education, especially in physics.

References

[1] Wieman C and Holmes N G 2015 *Am. J. Phys.* **83** 972
[2] Azzarkasyi M, Halim A and Ilyas S 2015 *J. Pendidik. Sains Indones.* **3** 107–13
[3] Spodniaková Pfefferová M 2015 *Informatics Educ.* **14** 279–89
[4] Ulukök Ş and Sari U 2016 *Univers. J. Educ. Res.* **4** 465–74
[5] Ismail A, Festiana I, Hartini T I, Yusal Y and Malik A 2019 *J. Phys. Conf. Ser.* 32049
[6] Permana A H, Muliyati D, Bakri F, Dewi B P and Ambarwulan D 2019 *J. Phys. Conf. Ser.* 32027
[7] Mujiono M and Sarah S 2021 *Berk. Ilm. Pendidik. Fis.* **9** 115–25
[8] Cristiano K L, Estupiñán A and Triana D A 2019 *J. Phys. Conf. Ser.* 12044
[9] Debowska E, Girwidz R, Greczyło T, Kohnle A, Mason B, Mathelitsch L, Melder T, Michelini M, Ruddock I and Silva J 2013 *Eur. J. Phys, 34 L47–L54*
[10] Ponto H 2020 *Discuss. Collab. Int. J. Innov. Creat. Chang.* **12**
[11] Ibrahim A B, Yahya F H, Hanafi H F and Omar N H 2019 *Int. J. Acad. Res. Bus. Soc. Sci.* **9**
[12] Anon PhET: Free online physics, chemistry, biology, earth science and math simulations
[13] Anon PhysLab | Physics for a change
[14] Arikunto S 2012 *Dasar-dasar Evaluasi Pendidikan* (Jakarta: Bumi Aksara)
[15] Nurhasanah S and Sobandi A 2016 *J. Pendidik. Manaj. Perkantoran* **1** 128
[16] Ricardo R and Meilani R I 2017 *J. Pendidik. Manaj. Perkantoran* **2** 79
[17] Muthmainnah M and Rokhmat J 2017 *J. Pendidik. Fis. dan Teknol.* **3** 40–47
[18] Irawan E and Suryo T 2017 *Beta J. Tadris Mat.* **10** 33–50
[19] Khaerunnisak K 2018 *J. Penelit. Pendidik. IPA* **4** 7–12

PhET-assisted electronic student worksheets of physics (eSWoP) on heat for inquiry learning during covid

A S Budi[1], S W Sari[4], L A Sanjaya[1], F C Wibowo[1], I M Astra[1], R W Puspa[2], M Misbah[3], B K Prahani[4] and W A Pertiwi[5]

[1] Department of Physics Education, Universitas Negeri Jakarta, Jakarta 13220, Indonesia

[2] Kartika VIII-1 Senior High School, Jakarta, Indonesia

[3] Physics Education, Universitas Lambung Mangkurat, Indonesia

[4] Physics Education, Universitas Negeri Surabaya, Indonesia

[5] Islamic Education Manajemen Study Program, Faculty Islamic of Education and Teacher Training, Muhammad Azim Institute, Jambi, Indonesia

lari@unj.ac.id

Abstract. Now, this is referred to as the industrial era 4.0, which is present simultaneously as the era of disruption. Therefore, this study aims to produce Electronic Student Worksheets of Physics (eSWoP) with the help of Sway to use PhET Simulations on physics learning about temperature and heat. This research was conducted using ADDIE (Analyze, Design, Development, Implementation, and Evaluation) research methods. The sample for the application of eSWoP used in the study was 86 students aged 16-18 years from a province in Indonesia. The result of this study is a PhET-assisted eSWoP teaching material that can be used through various electronic devices, using PhET simulations that can help students understand physics subjects at temperature and heat. In the questionnaire given to students, information was obtained that eSWoP media is useful, easy to use and can increase students' learning motivation. It was suggested that eSWoP technology could be a powerful and effective tool to enable students from the physics lesson into the inquiry process. Moreover, the implications of using eSWoP for teaching physics and recommendations for further studies are also discussed in this study.

1. Introduction

In the era that has entered industry 4.0 today, everything cannot be separated from the influence of technology [1]. Industrial era 4.0 coincided with the age of disruption [2]. This technology exists because of the emergence of many innovations and scientific developments [3]. The advancement of this technology provides many conveniences and benefits for human life, especially in education. One of the benefits of this technological development in education is the emergence of learning media and teaching materials, making it easier for students and teachers in the learning process [4]. By using the existing ones, it is hoped that students can better understand and assist in following the learning process to improve the quality of education.

Physics is one of the branches of natural sciences, which studies various natural phenomena in everyday life [5]. To investigate this nature, students will face natural wonders, laws, and formulations for solving problems faced. Physics does not only rely on books, but experiments need to be carried out to support the theories, laws, and concepts of physics contained in books [6]. The results of observations made at 1st Public Islamic High School (MAN) Kerinci are known that 87% of students have difficulty studying physics. Students reasoned that physics is a subject that is difficult to reach and difficult. Other results show that 68% of students find it difficult to learn physics on temperature and heat materials so that more innovative teaching materials are needed [7]. Based on data from observations and interviews with students and physics teachers at 12th Public High School of Banda Aceh, information was obtained that the average value of students' daily tests, especially on the topic of temperature and heat, was still below the Minimum Completion Criteria (KKM). This is shown from the KKM set by the school, which is 75 only 70% of students who pass the KKM [8].

By taking advantage of technological developments to carry out experiments besides being carried out directly in the laboratory, these experiments can be conducted virtually. This virtual experiment uses a virtual laboratory. A virtual laboratory provides laboratory tools and materials through computer programs so that students can do experiments or practicum [9]. Virtual laboratories have several advantages, such as this laboratory can be used anytime and anywhere, does not require equipment and chemicals, and this virtual laboratory can see even small and abstract things [10]. One example of a virtual laboratory is Physic Education and Technology Simulation (PhET).

The research results conducted by a Rwandan secondary school stated that using PhET simulation in learning could improve students' understanding of studying physics [11]. PhET simulation in education can make learning more effective because, with PhET simulation, teachers can use PhET as a demonstration material, and students can play while learning by using PhET simulation [12]. In addition to research conducted at At-Taqwa Islamic Secondary School at Maumere, PhET simulation can improve students' understanding of physics concepts. This study showed that there were differences in the learning outcomes of students who were taught using PhET simulation media and those who were not taught using PhET simulation media. The result is learning using PhET simulation is better than control class learning [9]. Thus, it can be concluded that using PhET simulation media in learning can improve students' understanding of physics concepts.

In conducting practicals both virtual and directly in the laboratory, students need a Student Worksheet. Student Worksheet is one of the teaching materials that can help students to be able to develop thinking skills and encourage them to do a practicum to concrete the concept [13]. Based on the observations made at 95th Public High School of Jakarta, the Student Worksheet used did not increase the motivation of students to do a practicum. It is also supported by another research conducted at 8th Public High School of Bengkulu City in which 61.36% of students agree that the use of The Student Worksheet provided by the teacher still did not help students to understand the material in terms of students' concepts and process skills [14]. The implementation of practical activities requires Student Worksheet, which can make students design their investigations independently so that participants can understand and find concepts [15]. So, in this case, it is necessary to develop a Student Worksheet in accordance with the responses of students at 7th Public High School of Jambi City as much as 90% of students agree that the development of an electronic Student Worksheet of Physics (eSWoP) will be made [16]. One alternative to making eSWoP can be made using Microsoft Sway software.

Microsoft Sway is one of the online media that is practical and easy to use. Sway can be accessed from various devices such as smartphones, tablets, laptops, or computers when connected to the internet; thus, the use of Sway can be done without being limited by space and time [17]. Sway can also create and display interesting content [18]. Because of these advantages, Sway can be used to develop an e-Student Worksheet. Another advantage obtained is that an eSWoP with Sway will make it easier for students to learn in this era of disruption or online learning.

Regarding the explanation, it is necessary to develop teaching materials in the form of eSWoP assisted by Microsoft Sway to use PhET in physics learning about temperature and heat material. It is expected that this media can positively impact students and help students understand the learning material. Therefore, the author aims to develop an eSWoP assisted by Microsoft Sway for the use of PhET in physics learning on temperature and heat materials. The eSWoP is intended for students who are in the upper secondary education level. In addition, this eSWoP will be used as teaching materials that can be applied in the learning process by teachers.

2. Method

This research was conducted using the Analyze, Design, Development, Implementation, and Evaluation (ADDIE) research method. The ADDIE model was developed by Dick and Carry to design a learning system. The ADDIE development model consists of five stages of development, namely Analysis, Design, Development, Implementation, and Evaluation. Schematic of the steps in the ADDIE development model as shown in Figure 1 [19]. The sample for the application of eSWoP used in the study was 86 students aged 16-18 years from a province in Indonesia. In the questionnaire given to students, information was obtained that eSWoP media is functional, easy to use and can increase students' learning motivation.

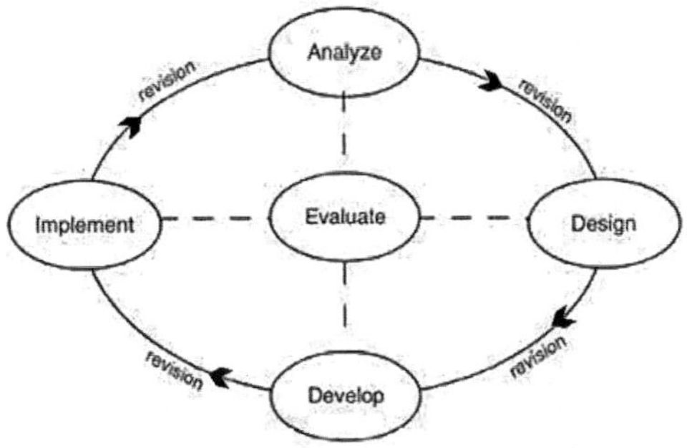

Figure 1. ADDIE Development Model Scheme

3. Result

The Student Worksheet resulting from this development is called Electonic Student Worksheet of Physics (eSWoP). This eSWoP discusses temperature and heat material for the use of PhET simulations made with the help of Microsoft Sway to make this teaching material more contemporary and make it easier for students to learn physics. This eSWoP will be published to students by sharing this eSWoP link to be accessed from all types of devices connected to the internet network. The display of this eSWoP is responsive, meaning that the resulting eSWoP display will adjust to the user's screen. This eSWoP can make it easier for students to do practicum because this eSWoP contains the title of the experiment, the purpose of the experiment, the theoretical basis, tools and materials, experimental procedures, tables of observational data, questions, and conclusions so that students can work independently. In addition, this Student Worksheet is equipped with a PhET simulation so that students no longer need to access the PhET simulation at a different website address. In addition, it is also equipped with an observation table that can be filled in directly by students via Microsoft Form. This will make it easier for students to work on eSWoP on Heat.

After accessing the shared link, students will see the eSWoP on Heat display in Figure 2.

Figure 2. eSWoP on Heat Cover Display

After that, students will find instructions using eSWoP on Heat, shown in Figure 3.

Figure 3. Display of eSWoP on Heat Instructions for Use

After students understand the instructions for using eSWoP, then students will see the Basic Competence display, shown in Figure 4.

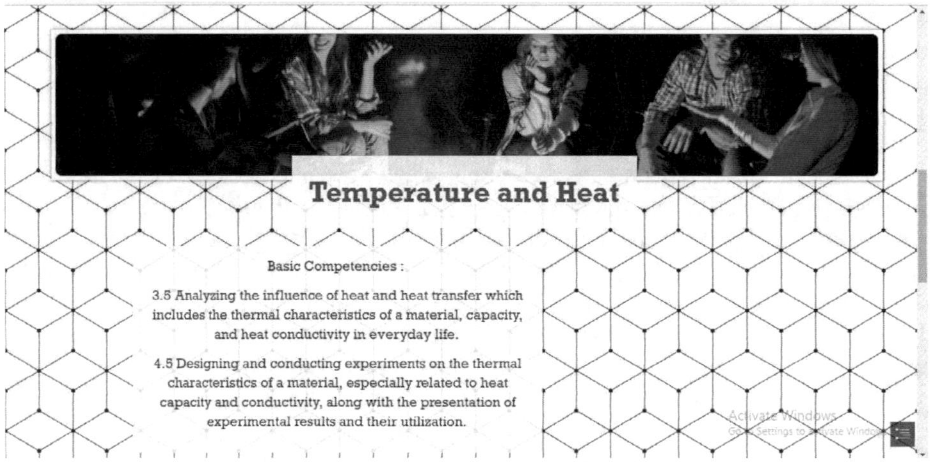

Figure 4. Basic Competency Display

Because this eSWoP has been integrated with the PhET simulation, students don't need to access the PhET simulation at a different tab browser or website address, as shown in Figure 5.

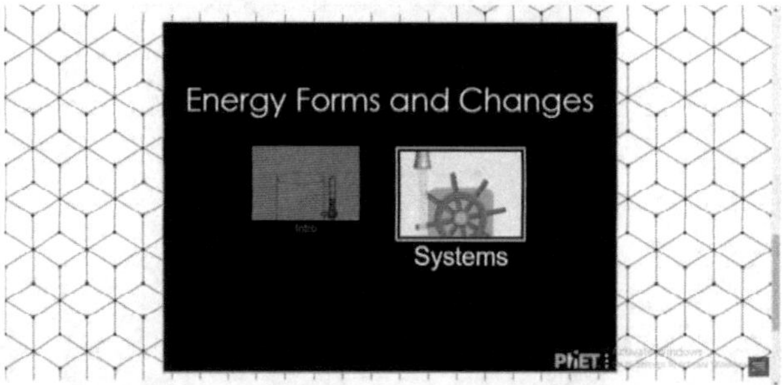

Figure 5. PhET Simulation Display on eSWoP on Heat

eSWoP on Heat Questionary to Students

The results showed that eSWoP on Heat contributed to students' concepts in physics subjects, with 86 students as respondents. In the questionnaire given to students, information was obtained that this eSWoP on Heat was useful and interesting for their learning. This is in line with the research results conducted in 2017, which showed that the application of LKPD can increase student motivation and learning outcomes at State Senior High School 4 Singaraja [20]. It has been suggested that eSWoP on Heat can be a powerful and effective teaching material for students studying physics subjects. The full results are shown in Figure 6.

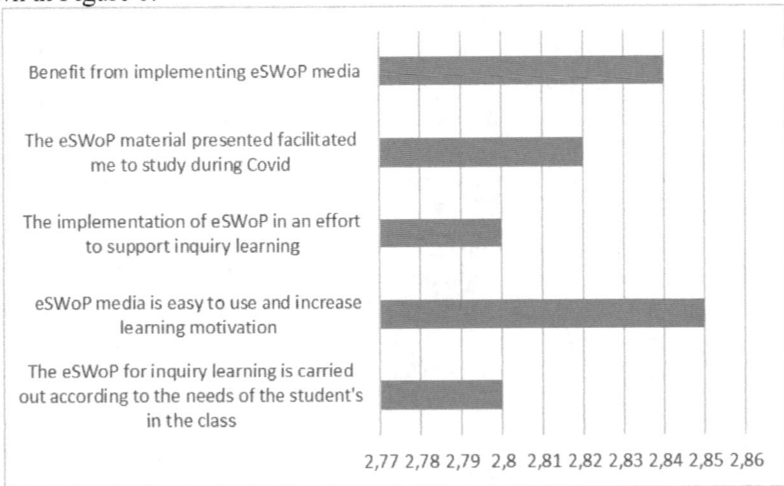

Figure 6. Result of eSWoP on Heat Questionary to Students'

Based on Figure 6, it can be seen that eSWoP on Heat is implemented according to the needs of students in grade 2.80 with good categories on a scale of 1-3. While eSWoP media is easy to use and can increase learning motivation by 2.85 with a good category on a scale of 1-3, and the application of eSWoP on Heat to support investigations is 2.80 with a good category on a scale of 1-3. The eSWoP on Heat material that was delivered was facilitated for learning during the COVID-19 (Corona Virus) pandemic of 2.82 with a good category on a scale of 1-3. Students feel the benefits of implementing eSWoP media are 2.84, with good categories on a scale of 1-3. Research in 2021 also results that electronic student worksheets can be used for distance learning which is helpful for the current COVID pandemic conditions [21].

4. Conclusion

eSWoP on Heat learning media with the help of Microsoft Sway for the use of PhET Simulation in physics learning about temperature and heat using the ADDIE research method. This eSWoP on Heat can be used by sharing a link to be accessed from all devices, mobile phones, laptops, and tablets connected to the internet network. This research still needs further research related to the validation of eSWoP on Heat assisted by Microsoft Sway for the use of PhET simulation in physics learning about temperature and heat until finally, the media is suitable for use by teachers in the learning process. Researchers' suggestions are to be followed up in further research, namely the need to develop eSWoP on other physics materials so that eSWoP can support physics learning as a whole at the senior high school level.

Acknowledgements

The research was funded by the Faculty of Mathematics and Natural Sciences, the State University of Jakarta, the Ministry of Education and Culture, through a Community Service grants 2021, with contract number: 60 /SPK PENELITIAN/5.FMIPA/2021.

References

[1] Adiansah W, Setiawan E, Kodaruddin W N and Wibowo H 2019 *Focus J. Pekerj. Sos.* **2** 47-60.

[2] Fitriani Y dan I A A 2019 *Senasbasa* **3**

[3] Effendi D and Wahidy A 2019 *In Prosiding Seminar Nasional Program Pascasarjana Universitas PGRI Palembang*

[4] Jamun Y M 2018 *J. Pendidik. dan Kebud. Missio* **10** 48-52

[5] Gani A A 2016 *J. Pembelajaran Fis.* **1** 754–67

[6] Yanti D E B, Subiki and Yushardi 2016 *J. Pembelajaran Fis* **5** 41-46.

[7] Purnawati W, Maison M and Haryanto H 2020 *Tarbawi: J. Ilmu Pendidik.* **16** 126–133

[8] Fithriani S, Halim A and Khaldun I 2016 *J. Pendidik. Sains Indones.* **4** 45-52

[9] Masita S I, Donuata P B, Ete A A and Rusdin M E 2020 *J. Penelit. Pendidik. Fis.* **5** 136-141

[10] Setiadi R and Muflika A A 2015 *J. Pengajaran Mat. dan Ilmu Pengetah. Alam* **17** 258-268

[11] Ndihokubwayo K, Uwamahoro J and Ndayambaje I 2020 *African J. Res. Math. Sci. Technol. Educ.* **24** 253-265

[12] Bryan J A and Slough S W 2009 *Phys. Educ.* **44** 264

[13] Zahro U L, Srevina V and Astra I M 2017 *J. Wahana Pendidik. Fis.* **2** 63–68

[14] Sari O B mila, Risdianto E and Sutarno S 2020 *PENDIPA J. Sci. Educ.* **4** 85-93

[15] Nuritasari A L, Wardani S and Supartono, S 2012 *J. Innov. Sci. Educ.* **5** 54-62

[16] Apriyanto C, Yusnelti Y and Asrial A 2019 *J. Indones. Soc. Integr. Chem.* **11** 38-42.

[17] Wihartanti L V and Wibawa R P 2017 *Din. Pendidik.* **12** 53-60

[18] Usman R K 2020 *J. English Lang. Teach. Cult. Stud.* **3** 82-88

[19] Branch R M 2009 *Approach, Instructional Design: The ADDIE* vol 53

[20] Muthoharoh M, Kirna I M and Indrawati G ayu 2017 *J. Pendidik. Kim. Indones.* **1** 13-22

[21] Fitriasari D N melati 2021 *Bioedu* **10** 510–22

Meta-analysis of the application *inquiry* learning models in physics classes

G S Budi

Physics Education Study Program, Faculty of Teacher Training and Education, Universitas Palangka Raya

gunarjosbudi2@gmail.com

Abstract. This research aims to analyze the *inquiry* learning model in the physics learning process. The samples of the study were 10 theses done by physics education students of Palangka Raya University. The theses were written by the students based on their research in applying the *inquiry* model in Palangka Raya Senior High School and in Junior High School. The theses were selected from among hundreds of reports based on the similarity topics started from the year 2015 until 2019. The method of this research is called meta-analysis, i.e., collecting the data from 10 selected thesis and analyzing them. To ensure the result, the researcher also carried out an interview with the students who wrote the thesis. The result of this meta-analysis research is that the *inquiry* learning model can improve student learning outcomes and also able to improve student skills such as Science Process Skills. However, the average score of the physics classes was still below the Minimal Achievement / Completeness Criteria, which is 75 %. It was found from the interview that the low achievement score of the physics classes was likely due to the lack of experience in teaching as well as lack of mastering the inquiry model.

1. Introduction

The 21st century has had a significant influence on the world of education in Indonesia. The 21st century requires everyone to have the skills to compete in facing the changing times. That is what makes the government in Indonesia to change the learning curriculum, namely from KTSP to Curriculum 2013 (K-13). This in line with research conducted by Andrian, which shows that the implementation of the 21st-century learning process is classified as quite good using K-13 [1].

K-13 aims to encourage students to be better at making observations, have questioning skills, have reasoning power, and be able to communicate what is obtained and known after receiving learning material at school. K-13 is used in the learning process in schools with the help of a scientific approach or contextual approach, which expects students to have a balanced competency between cognitive (knowledge), affective (attitude), and psychomotor (skills) [2] . These three competencies are demands that must be met by schools in accordance with the K-13 provisions in the learning process.

Learning is a system consisting of various components that are interconnected with one another. These components include objectives, materials, and evaluation [2] . The learning process at this time is expected not to make students just memorize facts and general concepts of subject matter delivered by the teacher, but students are expected to be able to observe and learn process skills, problem-solving skills, and thinking skills when learning a knowledge [3]. Important elements in the learning process include students who learn, teachers who teach, learning materials, and the relationship between teachers and students [4] . Teachers should pay attention to important elements in the learning process, especially in learning physics.

Physics is a part of Natural Sciences (IPA) which is developed through scientific steps to organize knowledge in predicting natural phenomena [5] . The important thing in learning activities in science learning, especially physics, is that students are active, so in this case the teacher must direct and encourage students to find their own knowledge. Teachers are expected to be able to master the teaching materials to be taught and also understand the situation of students so that they can teach according to the circumstances and students' development. In order for the knowledge to be given to be accepted by students well, teachers are expected to be able to package Physics learning well, and one way is by choosing the right learning model.

The Learning Model is a plan or pattern that can be used to form long-term learning plans, design learning materials, and guide learning in class or otherwise [6]. The learning model can be in the form of a conceptual framework that writes systematic steps in organizing learning experiences to achieve specific learning objectives and serves as a guide for instructional designers and teachers in designing and implementing the learning process [7]. This means that each learning model serves as a teacher's guide to carry out the teaching-learning process. Each model used by the teacher in the teaching and learning process also determines the devices used in the learning. In accordance with the previous statement that physics learning requires students to be active in the learning process so that teachers in choosing a learning model should choose a model that can also make students active. One learning model that can make students more active during learning is the inquiry learning model.

The learning model inquiry a learning model that gives free space for students to find a passion and a way of learning, respectively. Students are no longer forced to learn in a certain style or way. Students are developed to become creative and productive learners. The positive value, students will not only know (know), but also understand (understand) the essence and potential development of certain learning materials [8]. Students who understand the material being studied are expected to have high knowledge competency values.

The inquiry learning model affects understanding, knowledge development on a topic, scientific process skills, attitudes towards learning, learning motivation, and communication skills. The general syntax of the inquiry learning model in the implementation of learning is as follows: presenting questions or issues, making hypotheses, data collection, interpretation of data, developing conclusions, and repetition

Based on these descriptions, this study aims to analyze research related to the effectiveness of learning using the inquiry learning model on students' knowledge competencies. The analysis is viewed in terms of learning materials and research results. Like other meta-analysis research, this research is expected to be useful in the field of education, especially physics teachers in selecting appropriate learning materials and measuring student competencies if they apply the inquiry learning model in schools.

2. Method

The type of research used in this research is Meta-analysis. Meta-analysis is a research of several documents with similar research themes. Meta-analysis is also defined as research conducted by summarizing research data, reviewing and analyzing research data from previous research results[24]. Data collection in this research was carried out by tracing the thesis of students of the physics education study program at the University of Palangka Raya who applied the inquiry learning model. From the search, I obtained several theses and then randomly selected 10 theses with the category one of the variables seen is student learning outcomes.

Coding in meta-analysis is the most important requirement to be able to collect and analyze data [9]. Therefore, the instrument in this meta-analysis was carried out by presenting the code (coding category). It is intended that the variables used in research such as research name, educational level of research subjects, materials used in research, and research variables can be easily analyzed. After coding, data analysis was carried out. Data analysis in this research used quantitative data analysis with percentages and qualitative data analysis for the data from the narrative study of the studies encountered.

3. Results and Discussion

3.1. Research Results

Researches on the application of the *inquiry* learning model were collected as many as 10 research results from students of the physics education study program, Palangka Raya University. The analysis was carried out based on the learning material and the research results obtained.

3.1.1. Inquiry Learning Model Based on Learning Material.

The relationship between learning models and learning materials is very important. Not all learning models are suitable for use in all materials, so in this case the teacher is required to be able to choose a model that is appropriate to the material to be taught so that the results obtained can be maximized.

The inquiry learning model requires students to be creative, collaborative, and active in the learning process [10]. The teacher plays an important role in packaging learning so that it can make students more active. Therefore, it is necessary to know the materials that have been tested using the inquiry learning model.

The learning model is a systematic step that serves as a guide for teachers in carrying out the learning process [7]. One important factor to be able to achieve maximum learning objectives is the suitability of the learning model used with the material being taught. Therefore, it is necessary to analyze the materials that have been used in the inquiry learning model as in Table 1.

Table 1. Material Analysis Results

No.	Learning materials	Number of Thesis
1.	Temperature and Heat	3
2.	Expansion of Substance	1
3.	Simple Plane	1
4.	Light and Optical Equipment	1
5.	Pressure	2
6.	Characteristics of Substances	2

Based on Table 1, it can be explained that of the 10 theses that were the research samples, there were variations in learning materials that used the *inquiry* learning model. The learning materials used are temperature and heat, expansion of substances, simple planes, light and optical devices, pressure, and substance characteristics. The materials most often used in research on the application of the *inquiry* model are temperature and heat materials. It can be seen from Table 1 that there are 3 theses that use temperature and heat material while the others are only 1 thesis, and for pressure material, there are 2 theses.

3.1.2. Results with Learning Model Inquiry

The learning model greatly affects student learning outcomes. The learning model is one of the things that the teacher guides during the learning process. Therefore, if the learning model used as a guide is not correct in its implementation, it can cause student learning outcomes to be not good either.

Results of the study of the 10 students theses of physics education at the University of Palangka Raya can be seen from the aspect of knowledge in the form of individual completeness, classical, and Special Learning Objectives. The results can be seen in Table 2.

Table 2. Analysis of Research Results (Knowledge Aspects)

Essay	Completeness		
	Individual	**Classical**	**TPK**
1	64.10%	64.10%	76.67%
2	77.78%	77.78%	72.22%
3	65.38%	65.38%	65.55%
4	78.78%	78.78%	-
5	68%	68%	81.18%
6	70%	70%	77.14%
7	66.67%	66.67%	33.33%
8	16%	16%	25%
9	70%	70%	51%
10	76.67%	76.67%	-

Based on Table 2, it can be explained that the completeness of learning outcomes in the application of the inquiry learning model varies. The results of the existing research from 10 samples are in the form of completeness of individual, classical, and Special Learning Objectives learning outcomes. The highest completeness of individual and classical learning outcomes is 78.78% [11]. The highest completeness of specific learning objectives is 81.18% [12]. Meanwhile, the lowest completeness of individual and classical learning outcomes is 16%, and the lowest specific learning objectives as 25% [13].

Apart from being based on the knowledge aspect, 9 out of 10 research results were analyzed, measuring aspects of students' skills during learning. The skills measured include students' science process skills, psychomotor skills and student activities during learning. Research conducted by Juhji states that students' skills cab be improved with the inquiry learning model [14]. The results of research data measuring aspects of skills can be seen in Table 3.

Table 3. Analysis of Research Results (Skill Aspects)

Essay	Aspect of Skills	Research result
1	Science Process Skills	SB : 17.07% B : 43.90% TB : 24.39% STB : 7.32% TH : 7,32%
2	Science Process Skills	SB : 71.43% B : 25% CB : 3.57%
3	Psychomotor	SB : 19.23% B : 76.92% CB : 3.85%
4	Science Process Skills	SB : 72.9% B : 21.62% CB: 5.41%
5	Science Process Skills	81.65%
6	Science Process Skills	77%
7	Student Activities	- Conducting experiments and collecting data: 26.17% - Data analysis: 22.16% - Formulating problems: 13.33% - Present: 12.50% - Applying: 10.57% - Listening: 9.72% - Irrelevant behavior: 5.55%
8	Science Process Skills	- Observe: 82.60% - Collecting Data: 76.54% - Predicting: 65.45%

		- Summing up: 60.37% - Communicating: 62.94%
9	Science Process Skills	78.18%

Information:
SB : Very good
B : Fine
CB : Good enough
TB : Not good
STB : Not very good
TH : Not present

Based on Table 3, it can be seen that with the *inquiry* learning model many student skills can be measured. The results obtained from several analyzed studies also varied.

3.2. Discussion
The results of the meta-analysis in terms of the suitability of the material with the model show that some materials are suitable for using the *inquiry* learning model. One of them is the temperature and heat material, which in the learning process can be carried out a lot of experiments. Experiments that can be carried out on this material include experiments on thermometers, heat transfer, *black* principles, and changes in form. At the level of understanding, students are able to state problems in their own words and are able to provide examples [8]. This fits with the *inquiry* learning model , which helps students to be active, collaborative, and creative in learning.

In general, the inquiry learning model is very helpful for the practicum-based learning process. The steps in the inquiry learning model lead students to be able to find their own knowledge and be able to carry out investigations. The steps of the inquiry learning model according to Bronnstetter are problem orientation, formulating problems, proposing hypotheses, planning exoeriments, conducting experiments, analyzing data, communicating, and concluding [15]. Therefore, materials that are suitable for using the inquiry learning model are materials that have sub-topics for conducting experiments and will greatly help improve student skills.

The research sample is part of the number of characteristics possessed by the research [16]. The sample in this research was a student's thesis which was randomly selected related to the application of the inquiry learning model with the category one of variables seen being student learning outcomes. Educational research conducted in general tends to look at student learning outcomes or the level of achievement of learning competencies.

The application of the inquiry learning model can improve student learning outcomes. This is because the steps of the inquiry learning model guide students to be able to find and carry out investigations so that learning becomes more meaningful [10]. The average classical completeness of students increased after the application of the inquiry learning model was compared to before the learning model was applied [17–19]. Although classical completeness in the research analyzed, there are still many that have not reached the predetermined minimum classical completeness criteria, but when compared with classical completeness before this model is applied, the results obtained have increased. Many factors cause incomplete classical learning. These factors include that there are still students who are less active and less serious during the learning process, the lack of teachers in class mastery, and the inquiry learning model is a new learning model for students so that students still need adjustments [17–19].

The completeness of the Special Learning Objectives on the analyzed research results on average has a fairly good value, namely above 50%, there are only two research results whose ROR completeness value is below 50%. This is due to the fact that the evaluation questions given are quite good, although there are still some that still need to be improved. While the results of the research whose completeness score was below 50% were caused by the evaluation questions compiled by the researchers were still not optimal so that many students were still confused in working on these questions. This is in line with research conducted by Widianie, which stated that one of the factors that caused the completion of specific learning objectives was that the instruments used were still not optimal to measure student learning outcomes [13].

The results of the research analyzed in addition to measuring student achievement in the cognitive aspects, also measuring student achievement in the aspects of skills. Juhji stated that by using the inquiry learning model, students' skills can be improved [14]. The aspect of skills that is measured among others is Science Process Skills, there are 7 research results that measure, there is 1 research result measuring psychomotor skills, and student activity during the learning process there is 1 research result that measures.

The results of research that measure science process skills are 7 out of 10 research results. The average results obtained for science process skills are in the good and good enough categories. The results of Ramadany's research and Ul's research showed that with the inquiry learning model, the value of science process skills was 77% and 78.18% in both categories [20,21]. This is because during the learning process there are still quite a lot of students who are active in carrying out experiments. The results of the research that measured psychomotor skills were 1 in 10 of the research results, and the results obtained were categorized as good, namely with a percentage of 76.92%. The results of the research that measured student activity during the learning process took place are there were 1 in 10 research results, and the results obtained were the activities of conducting experiments and collecting data to obtain the highest percentage value, namely 26.17%.

4. Conclusion

Based on the results of the meta-analysis in this study, it can be concluded that various studies on the inquiry learning model improve student learning outcomes. The highest completeness of individual and classical learning outcomes is 78.78%. The highest completeness of specific learning objectives is 81.18%. Meanwhile, the lowest completeness of individual and classical learning outcomes is 16%, and the lowest specific learning objectives is 25%.

The material used in the inquiry learning model is material that requires experimentation in the process of finding concepts such as temperature and heat. In addition to improving student learning outcomes, the inquiry learning model can also be used to measure student skills such as Science Process Skills when conducting experiments. The results of Ramadany's research and Ul's research showed that with the inquiry learning model, the value of science process skills was 77% and 78.18% in both categories.

Research using the meta-analysis method will be more valid the results obtained if more journals or research results are reviewed. Therefore, suggestions for the future are to be able to add research results from both national and international journals related to the inquiry learning model so that the results obtained can be maximized.

References

[1] Andrian Y and Rusman R 2019 *J. Penelit. Ilmu Pendidik.* **12** 14–23
[2] Hosnan M 2014 Pendekatan saintifik dan kontekstual dalam pembelajaran abad 21: Kunci sukses implementasi kurikulum 2013 (Jakarta: Ghalia Indonesia)
[3] Jufri W 2013 *Belajar dan pembelajaran sains* (Bandung: Pustaka Reka Cipta)
[4] Suparno S 2007 *Filsafat penelitian kuantitatif kualitatif dan R&D* (Bandung: Alfabeta)
[5] Permendikbud 2014 Peraturan Menteri Pendidikan dan Kebdayaan Nomor 59 Tahun 2014 Bagian C PMP FIS-Minat SMA
[6] Rusman 2012 *Model-model pembelajaran* (Bandung: Raja Grafindo Persada)
[7] Trianto T 2015 *Model Pembelajaran terpadu* (Jakarta: PT. Bumi Aksara)

[8] Anam Khoirul M 2016 *Pembelajaran Berbasis Inkuiri* (Yogyakarta: Pustaka Pelajar)
[9] Sari A R and Hardini A T A 2020 *J. Ilm. Pendidik. Profesi Guru* **3** 1–8
[10] Bayram Z, Oskay Ö Ö, Erdem E, Özgür S D and Şen Ş 2013 *Procedia-Social Behav. Sci.* **106** 988–996
[11] Fazrin M 2013 *Penerapan motode pembelajaran inkuiri terbimbing untuk melatih keterampilan proses sains pada siswa kelas xi ipa semester ii sma negeri 4 palangka raya tahun ajaran 2012/2013* (Thesis (University of Palangka Raya)
[12] Artika I H 2019 *Penerapan model inkuiri terbimbing pada materi pesawat sederhana di kelas viii smp negeri 1 palangka raya tahun ajaran 2018/2019* (Thesis (University of Palangka Raya)
[13] Widianie P 2017 *Penerapan model inkuiri terbimbing pada materi tekanan kelas viii semester genap smpn 1 palangka raya tahun ajaran 2016/2017* (Thesis (University of Palangka Raya)
[14] Juhji J 2016 *J. Penelit. dan Pembelajaran IPA* **2** 58–70
[15] Basri Z, Nursalam N and Suharti S 2018 *Auladuna J. Pendidik. Dasar Islam* **5** 94–104
[16] Sugiyono S 2015 *Metode penelitian pendidikan (pendekatan kuantitatif, kualitatif, dan R&D* (Bandung: Alfabeta)
[17] Hidayati N 2020 *Penerapan model pembelajaran inkuiri terbimbing pada materi tekanan di kelas viii semester ii smp negeri 8 palagka raya tahun ajaran 2019/2020.Thesis* (University of Palangka Raya)
[18] Gea M 2019 *Penerapan model pembelajaran inkuiri terbimbing pada materi cahaya dan alat optik di kelas viii semester genap smp kristen palangka raya tahun ajaran 2018/2019.Thesis* (University of Palangka Raya)
[19] Widiastuti F 2020 *Implementasi pendekatan inkuiri terbimbing pada materi suhu dan kalor di kelas vii semester i smp santa maria palangka raya tahun ajaran 2019/2020* (Thesis (University of Palangka Raya)
[20] Rahmadany A 2019 *Penerapan model pembelajaran inquiry terbimbing pada materi suhu dan kalor di kelas vii semester i smp negeri 4 palangka raya tahun ajaran 2018/2019* (Thesis (University of Palangka Raya)
[21] Ul' N U 2014 *Pembelajaran ipa menggunakan metode inkuiri terbimbing pada materi pokok pemuaian zat di kelas vii semester ii smp negeri 9 palangka raya tahun ajaran 2013/2014* (Thesis (University of Palangka Raya)

AUTHOR INDEX